ON SEWAGE TREATMENT AND DISPOSAL

D1882737

Owing to the antiquity and rarity of this book, certain pages are regrettably absent, and readability is compromised due to the blurring of the original text. If you possess any insights regarding the availability of these missing sections, kindly inform us. Your assistance is crucial in ensuring the completion of this significant work for the benefit of future generations.

ON SEWAGE TREATMENT AND DISPOSAL

FOR CITIES, TOWNS, VILLAGES, PRIVATE DWELLINGS, AND PUBLIC INSTITUTIONS

THOMAS WARDLE

GYAN

Published by

Gyan Publishing House
5, Ansari Road
Daryaganj, New Delhi-110002
Phone: 011-47034999, 9811692060
E-mail: books@gyanbooks.com

Distribution Network
gyanbooks.com
India, USA, Canada, UK, Australia

© Publisher
All rights reserved. No part of this work may be reproduced, stored, adapted, or transmitted in any form or by any means, electronic, mechanical, photocopying, micro-filming, recording or otherwise, or translated in any language, without the prior written permission of the copyright owner and the publisher.

ISBN: 978-93-7447-620-8 (PB)

First Published, 1893

2nd Impression 2025

Printed at: Gyan Press, Delhi.

The book is sold subject to the condition that it shall not, by way of trade or otherwise, be lent, resold, hired out, or otherwise circulated without the prior publisher's written consent.

ON SEWAGE TREATMENT AND DISPOSAL
Author: THOMAS WARDLE

ON

SEWAGE TREATMENT AND DISPOSAL:

FOR

CITIES, TOWNS, VILLAGES, PRIVATE DWELLINGS, AND PUBLIC INSTITUTIONS.

BY

THOMAS WARDLE,

Fellow of the Chemical Society; Fellow of the Geological Society; Fellow of the Imperial Institute: Member of Council of the Palæontographical Society; Chevalier de la Legion d'Honneur of France; Officier d'Academie of France; Membre du Jury de l'Industrie de la Soie aux Expositions Universelles, at Paris, 1878 and 1889; Honorary Superintendent of the Indian Silk Culture Court of the Colonial and Indian Exhibition, London, 1886; Chairman of the Silk Section of the Royal Jubilee Exhibition, Manchester, 1887; one of the Examiners to the City and Guilds of London Institute; President of the Silk Association of Great Britain and Ireland; President of the North Staffordshire Field Naturalists' Club and Archæological Society; Chairman of the Standard Sewage and Water Purification Company, Limited.

WITH TABLES, PLANS OF SEWAGE PRECIPITATION WORKS, AND MANY ILLUSTRATIONS.

JOHN HEYWOOD,
DEANSGATE AND RIDGEFIELD, MANCHESTER.
2, AMEN CORNER, LONDON, E.C.
33, BRIDGE STREET, BRISTOL.

DEDICATED,

WITH SINCERE ESTEEM,

TO MY OLD FRIEND AND SCHOOLFELLOW,

CHARLES LOWE, ESQ., F.C.S.,

SUMMERFIELD HOUSE, REDDISH, NEAR STOCKPORT.

PREFACE.

Most of the chapters in this book originally appeared in the columns of the *Leek Post* as weekly papers, contributed whilst I was engaged in conducting a series of sewage investigations, and extending over a period of about two years.

They form a record of studies and of inquiries into the subject, which commenced when I was a member of the Sanitary Committee of the Leek Board of Improvement Commissioners, and also a member of the Sewage Inquiry Sub-committee of that Board. After a limited series of visits to sewage works, it was found inconvenient further to pursue joint investigation. I then embarked upon a more extended inquiry on my own account, partly from a growing interest in the subject, and partly from a feeling that the subject was too important and too complex to be dealt with except in a more exhaustive manner. As each week passed, I recorded my observations in the *Leek Post*, hoping thereby to interest and inform the local public in a matter which, in consequence of County Council pressure, threatened deeply to affect their pockets and sanitary arrangements. During the issue of these weekly notices I frequently received requests from various friends and correspondents in and out of the county, who were in several ways interested in the question of sewage treatment and disposal, to republish them in book form. I have yielded to their desire, and hope my book may be found useful. I cannot pretend that it will prove superior to the many valuable works on the subject which have preceded it, but it may supply a want in several ways. I believe it is the only book which gives the methods of sewage treatment and disposal, in some detail, of a large number of towns. The bacterial aspect of the question has more importance assigned to it than in any other sewage works with which I am acquainted, whilst with regard to new ideas of sewage treatment, it may be said to be brought quite up to date in recording the latest thoughts on the subject. Further, a new method of my own, which has already met with immediate promise of useful application, will relieve me from any charge that I have not contributed anything new to a subject which is probably without parallel in the number and importance of contemporary inquirers.

Some of the chapters have been necessarily re-written, others remain almost as they were originally issued, and if these appear at times to have too colloquial a form, it may be explained that they were only written for local reading. It may also be objected, that too frequent mention is made of Leek requirements, but I think the localising of a consideration of this kind is not without its advantage, as it shows that the study has been conducted for practical application rather than from theoretical reasons. The discussion of the defects, difficulties, and peculiarities of the sewage sanitation of any one town must have, besides, a general application beyond its purely local bearing. Those parts, however, which are but purely local, I have placed in the appendix; to have omitted them would have made my book of less interest to the town of Leek and the neighbouring towns of the Staffordshire Potteries. But apart from such localisation, my interest in this large subject dates back to a time considerably anterior to the Act which caused the establishment of County Council government, as the following report of Mr. John Elliot, engineer, so far back as 1872, will show. He was requested by the Tunstall Local Board of Health to report on the various methods of sewage disposal practised at that time. He consulted me upon the chemical states of the sewage and of the effluents after treatment by the principal processes at that time practised. I extract the following from his report :—

"REPORT OF THE TUNSTALL SEWAGE QUESTION.

"*To the Chairman and Members of the Local Board of Health, Tunstall.*

"October 8th, 1872.

"In treating the question of the defecation or deodorisation of sewage, there are at the present time two acknowledged systems, one by irrigation, and the other by chemical precipitation. By either of these means, if properly carried out, the sewage of towns may be rendered innocuous for the purpose of turning the effluent water into an ordinary running stream.

"In reference to the town of Tunstall, to which this report particularly refers, the Board can have the option, under certain circumstances, of the two systems, as Mr. Sneyd, I believe, will entertain a proposition to lease to the Board 300 acres of Bradwell farm, at an annual rent of 45s. per acre for 21 years, provided the Board will farm it themselves; this, of course, would involve the pumping of the sewage to Bradwell. The greatest agricultural chemists of the day, such as Dr. Letheby, Dr. Frankland, Dr. Odling, &c., are very much against the system of irrigation, but it is being successfully carried out at Barking, Romford, and other places.

"Should the Board not think well to entertain this proposition, the other course is to adopt the best known system by chemical precipitation, and for this purpose I lay before you the result of my inquiries.

"There are five different processes of defecating sewage by chemical precipitation in operation at the present time: the A B C process in operation at Bolton, the ingredients of which are composed of alum, blood, clay, and charcoal; the phosphate of alumina and lime process, in operation at Tottenham; the sulphate of alumina and lime, known as Dr. Anderson's process, in operation at Nuneaton; lime and clay, known as General Scott's process, in operation at Ealing; and the lime, or Mr. Higg's process, in operation at Luton. In order to ascertain which of the different processes is most effectual in deodorising the sewage, Mr. Wardle, of Leek, has analysed samples taken by myself of the sewage and effluent water of the different systems, the result of which is as follows:—

		Total residue per gallon. Grains.	Organic matter. Grains.
A B C Process, alum, blood, clay, and charcoal	Sewage	84	28
Ditto ditto	Effluent water	49	28
Phosphate of alumina	Sewage	70	28
Ditto ditto	Effluent water	70	21
Anderson's process, sulphate of alumina and lime	Sewage	98	28
Ditto ditto	Effluent water	70	14
General Scott's	Sewage	63	28
Ditto ditto	Effluent water	49	7
Mr. Higg's	Sewage	75	35
Ditto	Effluent water	28	7."

These statements must be taken as recording Mr. Elliot's individual opinion of the methods described at that time. I have quoted them only to show my interest in the subject twenty years ago.

My reason for treating at so much length the subject of Agriculture is the importance of continued study and experimental work in fertilisation, and the necessity which is forced upon us by seeing that at present we are not equal to the proper utilisation of either the suspended matter or the dissolved organic matter in sewage.

The principle expressed in the old rhyme—

> Let nothing wasted be,
> No product useless lie,

is of peculiar applicability in this case, and I feel sure that when the functions and requirements of healthy plant-life are better understood, we shall not be hampered with the nuisances of either over-charged soil and its over-fed crops, or of the reeking and deleterious accumulations of sewage sludge. It is with this end in view that I have endeavoured to lay down the outlines of scientific agriculture, and to emphasise the necessity of increasing our knowledge in this respect, to aid us in utilising these products up to their fertilising powers, or to add to them such fertilising media as will cause them to be economically useful and sought for, possibilities which I feel sure are within the reach of skilled inquiry and experiment.

With regard to that part of my book which treats of zymotic disease so largely, surely no apology is needed for emphasising the importance of a more profound study and determined action in this branch of hygiene and sanitation.

These diseases are almost all, if not all, preventible, and ought not to be allowed any entry. I am well supported in this view by a statement to this effect on scarlatina, made by Mr. Ernest Hart, in his excellent article on cholera, in the last October number of the *Nineteenth Century*, in which he remarks "our customary and preventible autumnal epidemic of scarlatina."

I believe that not only do populations suffer largely by imperfect sewage treatment and disposal directly, but also indirectly, by being fed upon plant and animal food which have been grown and fattened upon crude sewage-fed land.

Upon this latter point I have adduced some remarkable evidence; and whilst admitting that pathological science is not sufficiently advanced to warrant very much more than inference, largely owing to the scanty provision in this country for facilities and means of research, there is authoritative evidence which shows very close connection between impure cattle-feeding and disease, and its transmission to man through the milk from cattle so diseased, as well as by eating flesh containing pathogenic germs.

My grateful acknowledgments and thanks are due to my friend, J. Carter-Bell, Esq., F.C.S., F.I.C., County Analyst of Cheshire, for several important chemical suggestions and corrections; to Edgar M. Crookshank, Esq., M.B., Professor of Comparative Pathology and Bacteriology in, and Fellow of, King's College, London, Lecturer on Bacteriology, Royal Veterinary College, London, for permission to reproduce Figs. 1, 2, 3, 4, 5, 6, and 7, Plate 1; Figs. 10 and 12, Plate 3; Fig. 18, Plate 5; Fig. 20, Plate 6, from his "Manual of Bacteriology." To E. Klein, Esq., M.D., F.R.S., Lecturer on General Anatomy and Physiology in the Medical School of St. Bartholomew's Hospital, London, for permission to reproduce Figs. 27, 29, 30, 31, 32, 33, and 44, from his work on "Micro-Organisms and Disease." To C. Flügge, Esq., O.O. Professor and Director of the Hygienic Institute at Göttingen, for permission to reproduce Fig. 8, Plate 1; Figs. 9, 11, 13, and 14, Plate 2; Figs. 15 and 16, Plate 3; Fig. 17, Plate 4; Figs. 19 and 21, Plate 5; Fig. 22, Plate 6; and Fig. 34, from his work on "Micro-Organisms: Etiology of the Infective Diseases." To Percy Frankland, Esq., Ph.D., F.R.S., for permission to reproduce Figs. 21, 22, 24, 25, 28, 29, 35, and 41A, from his Cantor Lectures on "Recent Contributions to the Chemistry and Bacteriology of the Fermentation Industries," from the "Journal of the Society of Arts." To W. Santo Crimp, Esq., and H. Robinson, Esq., C.E., F.G.S., F.R.G.S., for permission to copy some interesting and valuable tables from their respective works. To Kineton Parkes, Esq., editor of the "Library Review," in assisting me in the reading of proofs.

LEEK, STAFFORDSHIRE, *June 30th, 1893.*

ERRATA.

The following plans were omitted from the list of illustrations: Kettering, page 121, and Richmond, page 124.

On page 381 the following were omitted from the list of Appendices:—

H. London Sewage Treatment and Disposal.

I. On the Action of Infusoria in Purifying and Deodorising Sewage.

Page 110, second line from bottom, for "1837" read "1887."

Page 217, answer 28, for "4 and 6" diameter read "4 inches and 6 inches."

Page 408, line 9, for "genus" read "genera."

CONTENTS.

CHAPTER XXI.

CHAPTER XXII.

CHAPTER XXIII.

CHAPTER XXIV.

CHAPTER XXV.

CHAPTER XXVI.

APPENDICES.

LIST OF ILLUSTRATIONS.

PLATE 1.

PLATE 2.

PLATE 3.

PLATE 4.

PLATE 5.

PLATE 6.

PLATE 7.

PLATE 8.

PLATE 9.

PLATE 10.

Since this book was printed, I have found it necessary to withdraw my Patented Precipitation Process and my connection from the Standard Sewage Company, mentioned on page 160; consequently it will be necessary that any communications or consultations which any of my readers may require of me should be made to me direct, addressed Thomas Wardle, Leek.

THE

TREATMENT AND DISPOSAL

OF

SEWAGE.

CHAPTER I.

INTRODUCTORY.

I PROPOSE in the following chapters to lay before the public all the information I have been able to obtain on this difficult and important subject—information gained from four sources:

1. That derived by actual conversation and correspondence with some eminent authorities.

2. That derived from personal investigation in the laboratory, and by actual inspection of typical and other sewage works and sewage disposals in various parts of the country.

3. That derived from the perusal and study of some of the best books on the subject; and

4. The results of my own researches, now extending over several years, not alone in the laboratory experimentally, but on a practical scale at the Salford Corporation Sewage Works, where I have been chemically purifying 100,000 gallons of sewage per day for three months of the latter part of the past year, the results of which will be described in my last chapter. I am also purifying the whole of the sewage at Nuneaton at the present time.

If the outcome may be that more light is thrown upon the subject, and that rural and urban readers may derive some sound and accurate information, my end will be served. Besides, there can be no harm in helping to contribute to accurate information on such an important subject as this, and an explanatory description of the different systems may serve two purposes: one to help us which to avoid, the other that which may be best to adopt. On such a very complex and difficult question, there is no doubt that many people are in many minds, some prejudiced for this or that system; some thinking of it with imperfect acquaintance; others, like myself, desirous of treating it in a student-like way, and endeavouring to learn all that has hitherto been set before us either theoretically, practically or economically. I think such an unprejudiced examination and method as this, will, at any rate, help us to free ourselves from advocating the very costly systems which have been adopted in so many towns, many of which have proved to be at most very ineffective remedies, and which, under pressure from County Council authorities, will certainly have to be very much modified and improved at still further expense. It will be easily seen that the task I have set myself is that of an impartial critic, wedded to no particular scheme or system—for all are more or less defective, or wanting in absolute completeness—nor recommending any, at least until after a full examination has been made of those in practical operation, or being tried on a practical scale by urban authorities. Not only do I propose to examine and report on the various systems, but I desire to collect such information, and to make such tests and examinations, as will most likely enable us to form a sound judgment on a matter which so seriously affects the ratepayers of the country, both in pocket and in successful application of the best sanitary and hygienic principles in this most important respect. To this end I purpose to give to the subject the most thoughtful consideration in my power, actuated only by the principle, "*Salus populi suprema est lex.*" One thing I can assure those readers who care to be at the trouble to read my chapters is, that much of the information will be of a most interesting character, coming as it does from the study and observations of most thoughtful and competent men, whose studies of the laws which concern the health of populations living closely together are worthy of close attention; studies which, in consequence of the breakdown of many systems of modern sewage disposal, are more scientific, more accurate, and vastly more important than any of those which have preceded them.

Works on Sewage.

Amongst the many interesting works I have been able to find on this subject are:—

Corfield's Treatment and Utilisation of Sewage. 3rd edition, 1887. Revised and enlarged by the Author and Dr. Parkes. Macmillan and Co.

Sewage Disposal. By Henry Robinson, C.E., F.G.S., F.R.G.S. 2nd edition, 1882. E. and F. N. Spon. An excellent little book.

Intermittent Filtration. By J. Bailey Denton, 35 years Principal Engineer to the General Land Drainage and Improvement Company; 1885. E. and F. N. Spon.

The Municipal and Sanitary Engineer's Handbook. By H. Percy Boulnois, M.Inst.C.E., Borough Engineer of Portsmouth; 1883. E. and F. N. Spon.

Sewage Disposal Works. By W. Santo Crimp, Assistant Engineer of the London County Council; 1891. C. Griffin and Co. The latest work on the subject. It is an excellent book.

See also Appendix B for the Bibliography of this subject.

CHAPTER II.

Estimations and Standards of Purity.

It is now conceded by all scientific authorities that the best test results to be regarded are those which give the indications of the exact amount of organic matter in the effluent water which passes away after the sewage has been fully treated in a chemical or agricultural way, or by both, separately or combined. After such treatment, no matter what system is adopted, there remain certain impurities in this effluent water, the most important of which, so far as health is concerned, are organic nitrogen and organic carbon.

These compounds I shall fully explain further on, and I need now only to remark that in the analysing of sewages and effluents they are estimated for free ammonia and albuminoid ammonia, *ad est*, that the quantitative estimation of the organic nitrogen is obtained by subjecting it to a chemical process which converts it into ammonia, which is divided into free and albuminoid ammonia, in which state they are most convenient for accurate measurement and estimation.

In order to ascertain correctly the degree of purification of an effluent water which is the result of any method of sewage treatment, it is of absolute necessity to know the amount of these two bodies, namely,

Free Ammonia and Albuminoid Ammonia.

Without this knowledge no reliable opinion of the value of any process or method of sewage treatment can be formed, and it is quite impossible for any person, be he ever so experienced in municipal or agricultural matters, to give an opinion of any real value, who is not able to state from actual and most accurate analyses what proportions and quantities of these two bodies, free ammonia and organic nitrogen (which is represented by albuminoid ammonia), are contained in effluent waters; and I wish this to be borne in mind all along, as I shall make it the keystone of the arch of my argument throughout. This method of analysis is the one most generally relied upon. All reports of sewage disposal which do not give the quantities of free ammonia and albuminoid ammonia must be taken *cum grano salis*, and be considered wanting in the points which it is of the greatest importance to know; and not only that, it is not the analyses of effluent waters of newly-applied systems which are to be relied upon, but the analyses after such systems have been for some time at work.

It will help us to form a sounder opinion on the measure of sewage purification if we carefully study the limits proposed by the Rivers Pollution Commission of 1868. These limits are their standards, from conclusions arrived at after their inquiry as to the best means of preventing the pollution of rivers. It is true these standards, only adopted for a time, were ultimately abandoned or postponed by the Government, no doubt as being impracticable to insist upon in the then state of knowledge of sewage disposal, and from the difficulties attendant upon a compulsory treatment of sewage, and purification of rivers, in important manufacturing districts.

I will also give the standards issued by the Conservators of the River Thames, adding to both tables, for the sake of easy comparison, the conversion of their figures into parts per million in decimals.

Rivers Pollution Commissioners' Standards.

(*a*) Any liquid containing, in suspension, more than three parts by weight of dry mineral matter, or one part by weight of dry organic matter in 100,000 parts by weight of the liquid—*i.e.*, 30 parts of dry mineral matter per million and 10 parts of dry organic matter per million.

(*b*) Any liquid containing, in solution, more than 2 parts by weight of organic carbon, or 0·3 part by weight of organic nitrogen in 100,000 parts by weight—*i.e.*, 20 parts of organic carbon per million and 3 parts of organic nitrogen per million.

(*c*) Any liquid which shall exhibit by daylight a distinct colour when a stream of it 1in. deep is placed in a white porcelain or earthenware vessel.

(*d*) Any liquid which contains in solution in 100,000 parts by weight more than two parts by weight of any other material except calcium, magnesium, potassium, and sodium—*i.e.*, 20 parts per million.

(*e*) Any liquid which in 100,000 parts by weight contains, whether in solution or suspension, in chemical combination or otherwise, more than 0·05 part by weight of metallic arsenic—*i.e.*, 0·50 part per million.

(*f*) Any liquid which, after acidification with sulphuric acid, contains in 100,000 parts by weight more than one part by weight of free chlorine—*i.e.*, 10 parts per million.

(*g*) Any liquid which contains, in 100,000 parts by weight more than one part by weight of sulphur in the condition either of sulphuretted hydrogen or of a soluble sulphuret—*i.e.*, 10 parts per million.

(*h*) Any liquid possessing an acidity greater than that which is produced by adding two parts by weight of real muriatic acid to 1,000 parts by weight of distilled water—*i.e.*, 2,000 parts per million.

(*i*) Any liquid possessing an alkalinity greater than that produced by adding one part by weight of dry caustic soda to 1,000 parts by weight of distilled water—*i.e.*, 1,000 parts per million.

(*k*) Any liquid exhibiting a film of petroleum or hydro-carbon oil upon its surface, or containing in suspension in 100,000 parts more than 0·05 of such oil—*i.e.*, 0·50 part per million.

They add the following remark with reference to storm water: "We are of opinion that, however undesirable, it will be necessary to permit storm water to flow directly into rivers without preliminary cleansing. Unfortunately, chemical analyses show that storm water, so far at least as its earlier portions are concerned, is more polluting than dry-weather sewage, owing to all deposits in the sewers being then swept to the outfall; and it will be important to guard against any unnecessary use of this exceptional permission."

An explanation of some of the terms just given seems desirable to enable the non-scientific reader better to understand their argument.

(*a*) "Containing in suspension"—*i.e.*, not dissolved or in solution, but what could be removed by filtration.

(*b*) "Organic carbon": Carbon in the form of organic or non-mineral matter. "Organic nitrogen": Nitrogen in the form of organic matter, whether animal or vegetable, occurring in sewage. Mr. Santo Crimp on this states: "Ammonia, and the salts of nitrous and nitric acids, do not in themselves contribute to the pollution of water, but their presence in considerable proportion denotes with certainty its anterior pollution by animal matters." And speaking of the estimation of total combined nitrogen, he says: "This determination sums up the whole of the analytical evidence against the water as regards both past and present organic contamination."

THE THAMES CONSERVATORS' STANDARDS.

The standards recognised by the Conservators of the River Thames as applicable to districts below the intakes of the London Water Companies are as follow:—

1. It should be free from an offensive odour.
2. It should be free from suspended matters, or, in other words, be perfectly clear.
3. It should not be alkaline to turmeric paper, nor acid to litmus paper.
4. It should not contain per gallon more than 60 grains of solid matter, dried at 260 deg. Fahr.

5. It should not contain more than three-quarters of a grain of organic and ammoniacal nitrogen per gallon, or 10·710 parts per million.

6. It should not contain more than two grains of organic carbon per gallon, or 28·571 parts per million.

7. It should contain not less than one cubic inch of free oxygen in a gallon.

Of this Dr. Frankland, one of the Rivers Pollution Commissioners, says: "Whilst thus far agreeing with my colleagues, I wish it to be distinctly understood that, in my opinion, such (effluent) fluid can only be safely admissible into the Thames on condition that the water is not afterwards used for domestic purposes."

CHAPTER III.

Bacteria.

"The pestilence that walketh in darkness; the destruction that wasteth at noon-day." Ps. 91, v. 6.

The consideration of the part played by bacteria in sewage decomposition is of the highest importance, and I propose to give it the first place in this discussion on sewage treatment and disposal. The recent history of bacteriology shows it to have been lifted out of the domain of speculation such as that of spontaneous generation, to that of a profound biological science, in which it is proved that bacteria proceed from germs, each kind from its own species. The studies of recent years have relegated bacteria to a most important place in the vegetable kingdom, and have shown that most diseases are inevitably accompanied, if not caused, by specific forms and accumulations of bacteria. The functions of bacteria are to change the nature and constitution of the useless or waste products of the organic world; their work is effectuated in the grave of organic matter, they being the causation of decomposition and decay. They also feed and multiply in living tissues, giving rise to pathogenic disturbances, and are often the sources of the greatest danger. They are at once the chief causes of infection, fermentation, and putrefaction. There are also non-pathogenic, or harmless species, unassociated with disease.

Pathogenesis—from πάθος, disease; γένεσις, generation—is a term used to denote a branch of pathology which deals with the scientific study of the generation and development of disease. It is therefore evident that the pathogenic biology of bacteria must form a necessary part of the inquiry into the dangers of bacteria in sewage, and the prevention of diseases caused or promoted by their action.

Dr. Fraenkel describes pathogenic bacteria as those which, by their vital action, produce excretions injurious to the bodies of men and animals.*

The nitrification of the soil, so necessary for healthy plant growth, is a kind of putrefaction, which entirely depends on the action of bacteria. It is, in fact, purely the decomposition of organic and mineral matter by bacterial influence. Some species of bacteria have the power of developing gases from decaying organic matter, and are the cause of the occurrence of smells, very offensive and infectious when from putrefactive matter; hence the danger from crude sewage deposits, and the necessity for this branch of natural history to take important standing in the consideration of sewage treatment and disposal.

Bacteria require carbon for their structure and growth, but it is proved they cannot secrete it from carbonic acid; they therefore have to take it directly or indirectly from organic matter.

In order to obtain a knowledge of the results of modern bacterial research, it would be necessary to study the works of the specialists who are labouring in the various branches, pathogenic and non-pathogenic, of this great subject. Of foreign labourers, amongst many others, I would mention Baumgarten, Cohn, De Bary, Flügge, Fraenkel, Hueppe, Koch, Kopf, and Pasteur; and in our own country, Crookshank, Dr. Percy Frankland, Dr. Klein, and Dr. Watson Cheyne. The bibliography of bacteriology comprises not less than 1,200 publications, chiefly papers, a list of which may be found in Dr. Flügge's "Micro-Organisms, with Special Reference to Infective Diseases," published by the Sydenham Society, in 1890. There is also a comprehensive list classified under the headings of the various diseases in which micro-organisms are an important factor in Crookshank's "Manual of Bacteriology." For further information on the morphology and physiology of bacteria, I refer the reader to these publications, most of them being research papers on special branches of the science; also the excellent books referred

* Fraenkel, "Text Book of Bacteriology."

to in this chapter, to the authors of which I desire here to express my indebtedness.

In giving greater prominence to this important branch of biological study in relation to sewage contamination than any former writer on sewage has done, it will not be possible, even if it were desirable, to give in detail my own microscopic observations of the last two years. I propose to describe the more salient points of bacteriology, as far as possible in their relation to sewage treatment and sanitation generally, and to show the important rôle they play in the organic world in health and disease.

Under alkaline or neutral food conditions and a warm temperature bacteria multiply enormously. They cannot thrive in acid media like the mould fungi, which is food for them. They possess great reducing powers. Some species can educe acid, others alkaline products; others, again, as will be shown further on, possess the power of producing great chemical changes in their surroundings; and as is well said by Carl Fraenkel, Professor of Hygiene at the University of Könisberg, "Bacteria are the cause of a whole series of phenomena, which are of deep and wide-spread importance in the economy of nature." We strike the keynote of their operations in their relation to the subject of sewage disposal on learning that it is chiefly the anaërobic (or existing without oxygen) bacteria which are the agents in causing the putrefaction of organic matter, some species of aërobic (or requiring oxygen for their growth and existence) bacteria having also the same powers, and it is to their decomposition of the albumen in such matter that much of the offensive odours and sewage gases are due, such as ammonia, sulphuretted hydrogen, and many uncomely stenches and gaseous results which are the products of the reducing processes of bacterial agency. Amongst the powers possessed by some species of bacteria is the secretion in themselves of colour; the term chromogenic is applied to them. They are of various bright hues, red, blue, and even black, as well as a variety of other colours. Some species are also strongly phosphorescent; but this chapter must confine itself to discussing the subject from an etiological point of view, and with this object I hope to enlist the interest and attention of medical officers of health and sanitary committees, both rural and urban, in this, a branch of my subject which I think has not hitherto received the attention its great importance demands. Bacteria may be thus provisionally summarised, in the absence of a completely-accepted classification, the basis of which we do not at present possess: they are comprised under the term Schisomycetes or fission fungi, the lowest members of the vegetable kingdom, and closely related to the lower algæ, and belong to the fungoid class, of which they are the lowest members. There are an immense number of well-known and defined species in the forms of globular bacteria, or micrococci, as in figs. 1 to 8, plate 1; rod-shaped bacteria, or bacilli, figs. 9 to 22, plates 2, 3, 4, 5, and 6; and screw-shaped, as spirilla, figs. 12 to 14, plate 2. They are all cells, and grow as such. They divide by fission, forming (with exceptions) spores; their central mass is albumen, or protoplasm, enclosed in a membrane or cell wall, which in due time bursts, and sets the young bacilli free, generally as spores.

It may, however, help the student of bacteriology if I describe three of the most recent attempts at classification, those of Hueppe, Baumgarten, and of Kopf, all of which are at present but provisional.

Hueppe's Classification.

Cocci Forms (Figs. 1 to 8, Plate 1).

Arranged in chains—in fours, in eights, or in irregular heaps, and in Zooglœa. These comprisethe genera Streptococcus, Artho-streptococcus, Leuconostoc, Merista, Sarcina, Micrococcus, and Ascococcus.

Rod Forms (Figs. 9 to 22, Plate 2).

Smaller or longer threads—wavy or straight; with and without endogenous spores; with and without change of cells; with and without presence of sulphur.

Threads with and without divisions—branched and not branched. These comprise the genera Artho-bacterium, Spirulina, Bacillus, Clostridium, Leptothrix, Beggiatoa, Crenothrix, Cladothrix.

Plate 1.

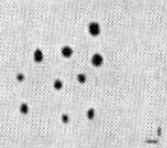

FIG. 1. COCCI, singly and varying in size.

FIG. 2. COCCI in chains of rosaries (streptococcus).

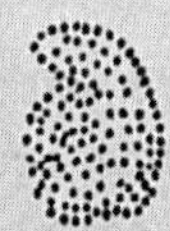

FIG 3. COCCI in a mass or swarm (zooglœa).

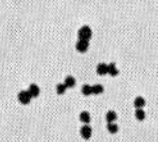

FIG. 4. COCCI in pairs (diplococcus).

FIG. 5. COCCI in groups of four (merismopedia).

FIG. 6. COCCI in packets (sarcina).

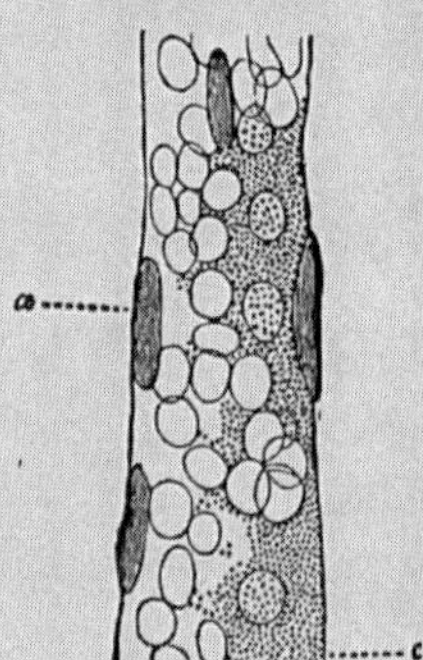

FIG. 7. MICROCOCCUS OF PYÆMIA IN RABBITS; VESSEL FROM THE CORTEX OF THE KIDNEY, × 700. (a) Nuclei of the vascular wall; (c) Masses of micrococci adherent to the wall and enclosing blood-corpuscles.

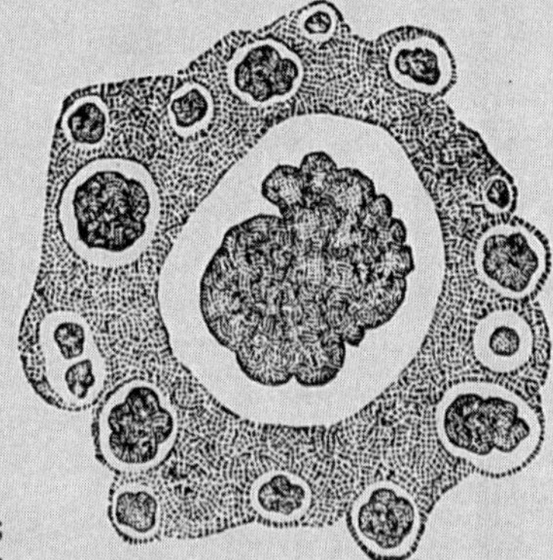

FIG. 8. ASCOCOCCUS BILLROTHII (after Cohn). Immense numbers embedded in tissue.

Screw Rods.

Screw-like threads—flexible or rigid; with or without change of form; comprising the genera Spirochæta, Vibrio, Spirillum. (Figs. 12, 13A, 14B, Plate 2 and 17, Plate 3.

Baumgarten's Classification.

Group I.—Monomorphic.

Genera Coccus (Figs. 1 to 8, Plate 1), Bacillus (Figs. 9, 10, 11, 15, 16, Plate 2), Spirillum (Figs. 2, 12, 13A, 13B,

Group II.—Pleomorphic.

Genera Spirulina Leptothrix (Fig. 21, Plate 2), Cladothrix (Fig. 22, Plate 2).

Kopf divides them into four groups, and Crookshank, in his "Manual of Bacteriology," adopts Kopf's classification, which I abridge from his valuable work.

Group I. Coccaceæ, with 5 genera—Streptococcus, Merismopedia, Sarcina, Micrococcus, Ascococcus.

II. Bacteriaceæ, with 6 genera—Bacterium, Spirillum, Leuconostoc, Bacillus, Vibrio, Clostridium.

III. Leptotricheæ, 5 genera—Crenothrix, Beggiatoa, Phragmidiothrix, Leptothrix.

IV. Cladotricheæ, 1 genus—Cladothrix.

On plates 2, 3, and 4 will be found figured type specimens of one or more of the above-named genera, a description of which will be found below.

Description of Plates.

Bacteria, Schisomycetes, or Fission Fungi.

Plate 1.

Group I.—Coccaceæ.

Fig. 1. Cocci singly, and varying in size.
„ 2. Cocci in chains or rosaries, Streptococcus.
„ 3. Cocci in a mass or swarm, Zooglea.
„ 4. Cocci in pairs, Diplococcus.
„ 5. Cocci in groups of four, Merismopedia.
„ 6. Cocci in packets, Sarcina.
„ 7. Cocci of pyæmia in rabbits, Micrococcus parvus ovatus.
„ 8. Ascococcus billrothii.

PLATE 2.

Group II.—BACTERIACEÆ.

Fig. 9. Bacterium termo, a collective name for a number of known species causing putrefactive smells, × (*i.e.*, magnified) 650 diameters.
„ 10. Bacterium termo, showing flagella, × by 4,000 (after Dallinger and Drysdale).
„ 11. Bacterium lineola, × by 650 (after Cohn), found in wells, and forming slimy masses on potatoes.
„ 12. Spirillum undula (after Cohn).
„ 13A. Spirillum volutans, showing flagellum at each end, × by 650 (after Cohn).
„ 14B. Spirillum sanguineum, showing flagellum at each end, × by 600 (after Cohn).

PLATE 3.

„ 15. Leuconostoc mesenteroides.
„ 16. Bacillus subtilis (after Prazmowski), × by 1020.

PLATE 4.

„ 17. Vibrio rugula (after Prazmowski), × by 1020.
„ 18. Clostridium butyricum (after Prazmowski).

Group III.—LEPTOTRICHEÆ.

PLATE 5.

Fig. 19. Crenothrix kuhniana, × by 600 (after Zopf).
„ 21. Leptothrix buccalis, × by 1000.

PLATE 6.

Fig. 20. Beggiatoa alba, × by 560 (after Zopf).
„ 22. Cladothrix dichotoma (after Cohn).

The Coccaceæ occur in one or other of their species in abscesses, erysipelas, gangrene, diphtheria, puerperal fever, scarlet fever, cattle plague, strangles, yellow fever, measles, whooping cough, typhoid fever, pneumonia, necrosis, anthrax, meningitis, the flacherie of silkworms, and in other diseased tissue; also in soil, the air, and in putrefactive matter. One of the most characteristic forms is Streptococcus pyogenes.

The Bacteriaceæ occur in diphtheria, fowl-cholera, croup, pneumonia, pebrine, relapsing fever, asiatic cholera, leprosy, syphilis, typhoid fever, gangrene, tuberculosis, glanders, anthrax, ulcers, tetanus, swine erysipelas, malaria, choleraic diarrhœa, woolsorters' disease, &c.

Of the Leptotrichieæ there are three genera unassociated with disease, and one pathogenic, Leptothrix parasitica, found in the teeth-slime in man and some of the lower animals. The non-pathogenic genera are found in water. Sulphur granules are found in the Beggiatoa, which occur in stagnant and impure

Plate 2.

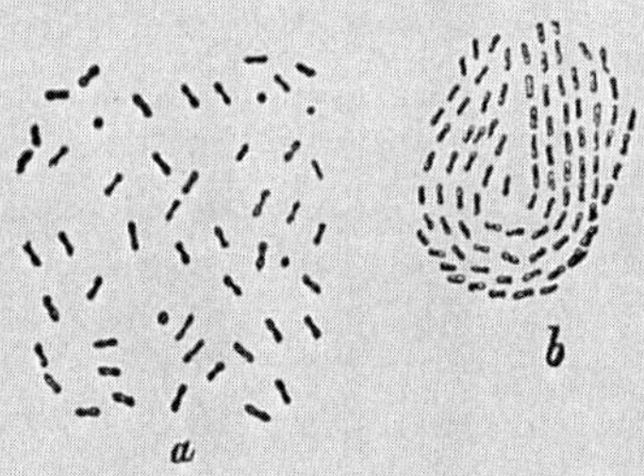

FIG. 9. BACTERIUM TERMO × 650.
(*a*) Isolated Bacteria; (*b*) Group of Bacteria.

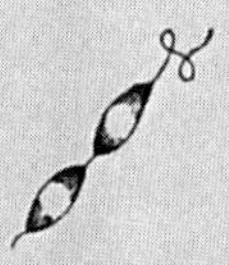

FIG. 10. BACTERIUM TERMO × 4000. (after Dallinger and Drysdale.)

FIG. 11. BACTERIUM LINEOLA (after Cohn) × 650.

FIG. 12. SPIRILLUM UNDULA (after Cohn.)

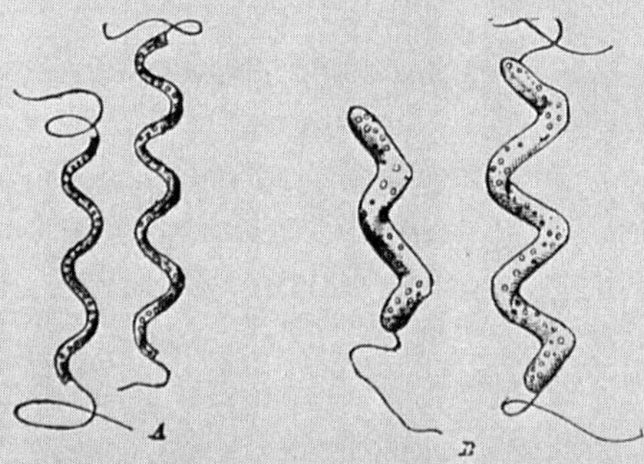

FIGS. 13, 14. A. SPIRILLUM VOLUTANS × 150. (After Cohn.)
B Spirillum sanguineum (Ophidomonas sang.) × 600.

Plate 3.

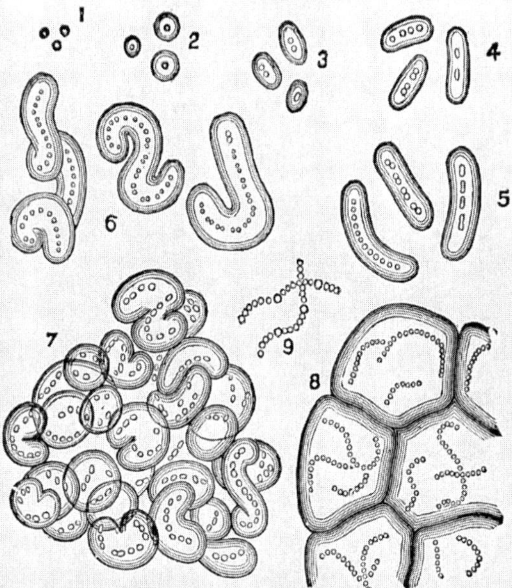

FIG. 15. LEUCONOSTOC MESENTEROIDES.

1. Spores. 2 Spores after germination, showing gelatinous envelope. 3, 4, 5, 6. Increase by division. 7. Glomerular form of zooglœa. 8. Section of an old mass of zooglœa. 9. Cocci chains with arthrospores [after Tieghem and Cienkowski].

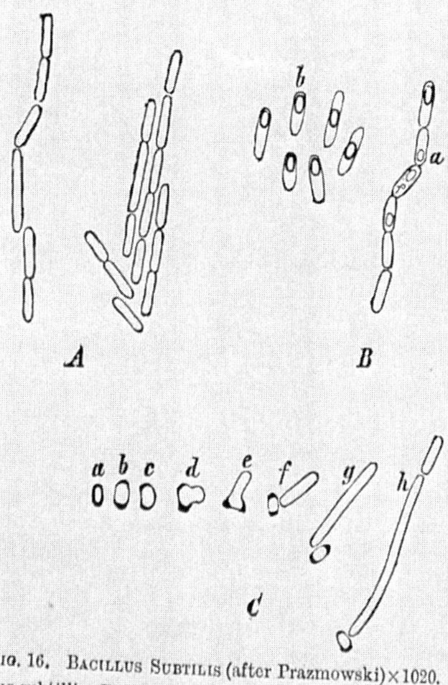

FIG. 16. BACILLUS SUBTILIS (after Prazmowski)×1020.

A. Colonies of Bacillus subtillis. B. Spore formation ; (*a*) in a pseudo-thread ; (*b*) in individual rods ; (*c*) germination of the spores ; (*a*—*h*) successive stages.

Plate 4.

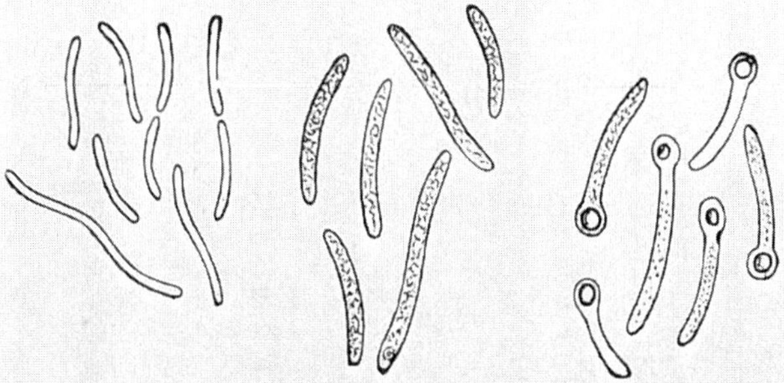

FIG. 17. VIBRIO RUGULA (after Prazmowski) × 1020.
(*a*) Young rods ; (*b*) thicker rods ; (*c*) spore-bearing rods.

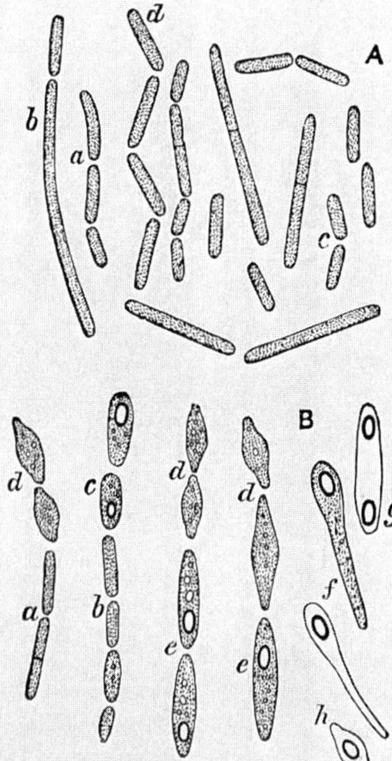

FIG. 18. CLOSTRIDIUM BUTYRICUM.

A. Active stage. (*a*, *b*) Bent rods (vibrio-form) and threads. (*c*) Short rods. (*d*) Long rods.
B. Spore-formation. C. Spore-germination. (After Prazmowski).

Plate 5.

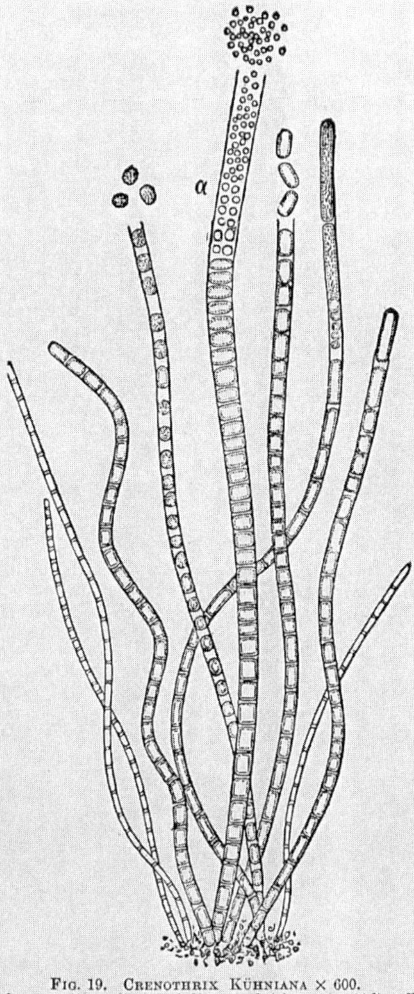

FIG. 19. CRENOTHRIX KÜHNIANA × 600.
Thread forms with cylindrical discs and (a) cocci. (After Zopf.)

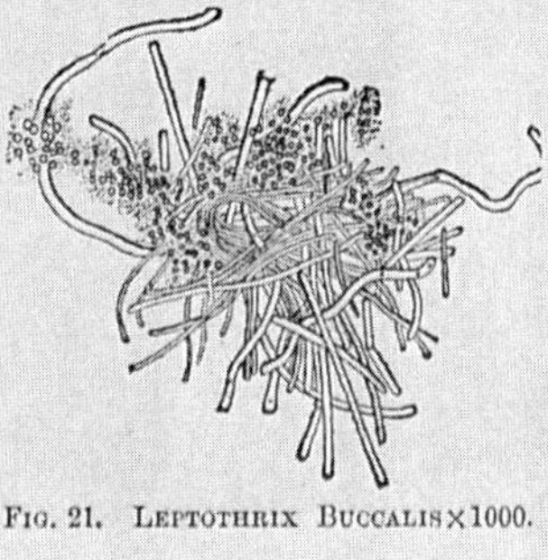

FIG. 21. LEPTOTHRIX BUCCALIS × 1000.

water, developing sulphuretted hydrogen, and destroying fish (fig. 20, plate 6). Crenothrix is found in tufts in drain pipes, and renders drinking water foul (fig. 19).

Cladotricheæ. Only one genus is described, two species of which are saprophytic (feeding on dead organic matter) and one pathogenic, Cladothrix dichotoma, Cohn (fig. 22), is found in straight or spiral threads of considerable length. Crookshank says they are the commonest of all bacteria, and are found in running water in which organic matter is present.

Of the genus Bacterium at least 22 species are known to be either pathogenic or saprophytic, and at least 27 species unassociated with disease.

Of the genus Spirillum five species are known to be associated with disease, some being pathogenic in man and saprophytic in animals, or *vice versa*. To this genus belongs the Comma bacillus or Spirillum Choleræ asiaticæ. Nine species of this genus are known to be unassociated with disease.

Of the genus Leuconostoc only one species is mentioned by Crookshank, under the name mesenteroides. It is found in molasses, and grows rapidly, converting them into a gelatinous mass in a few hours.

Of the genus Bacillus 18 species are recorded as being associated with disease in man and animals, and 27 unassociated. Amongst these are the microbes of leprosy, which occur in great numbers in tuberculous leprosy in most parts of the body except the blood.

This list epitomises the bacterial world; but besides these organisms, there are found in sewage a large number of species of yeasts and moulds. In Dr. W. T. Sedgewick's "Report of the Biological Work of the Lawrence Experiment Station, Massachusetts, U.S.A.," it is stated that yeast cells are the most constant micro-organisms in the Lawrence sewage, and are derived from the breweries of the city, giving no doubt a formidable body of albuminoid matter to be dealt with. Dr. Sedgewick states that he found from 1,000 to 300,000 cells of

Saccharomyces cervisia, Meyer (the yeast plant) per cubic centimetre always present.

The microscopic examinations of Dr. Sedgewick of the Lawrence sewage are so interesting and instructive as to tempt me to mention them here. Besides bacilli and yeast cells and other fungi, he found infusoria—the ciliata and flagellata; paramœcium is conspicuous, and of the flagellata monads and euglema. No doubt similar organisms exist in British sewage generally, but much depends whether it is of a domestic kind, or largely mixed with chemical and manufacturing and dye refuse. I found the Salford and Pendleton sewage, which is largely manufacturing and chemical, contained a very representative quantity of the sub-kingdom infusoria, both in quantity and number of species, especially in the vicinity of sewage fungus, Krenothrix Kuhinana (Fig. 19), in and about which they formed an *entourage.* I observed many monads: Urella glaucoma, in circling clusters; Kerona polyporum; Chilodon cucullus; Paramæcium aurelia; and one species of a small rotifer.

The average number of bacteria in the Lawrence sewage was 708,000 per cubic centimetre; the extremes were 102,400 in the cold weather of February, and 3,963,000 in April. A cubic centimetre is less than the $\frac{1}{6}$th of a cubic inch. They were found to multiply enormously in the upper parts of the sewers and in drains, depriving the sewage of all free oxygen.

Mr. E. O. Jordan, Chief Assistant Biologist to the State Board of Health, Massachusetts, U.S.A., who examined the Lawrence sewage for bacteria, has published a list of the species, which he found both in the sewage and in some experimental effluents.

He describes twelve species. They are all bacilli. He did not find micrococci present. The two most prevalent forms were Bacterium coli commune and Bacterium cloacæ.

B. coli commune is a species which inhabits the human alimentary canal, and in the fæces of infants when fed exclusively on mother's milk. Inoculation into the

Plate 6.

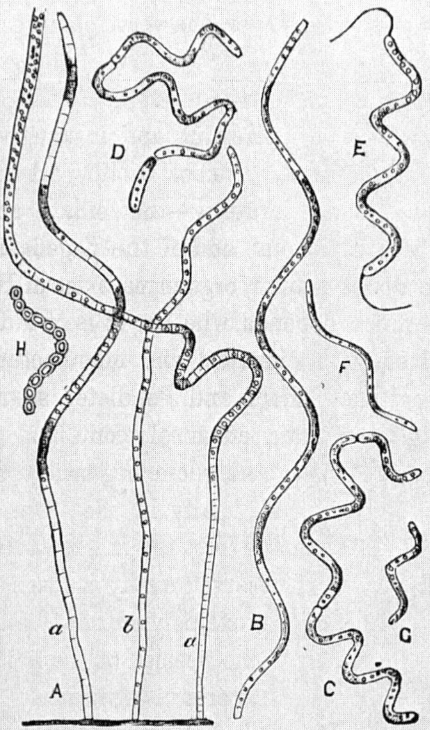

FIG. 20. BEGGIATOA ALBA.

A Threads at base distinctly linked, partly spiral. B. A thread, spiral in its whole length. C, D. Fragments detached from threads; immotile. E. Active spirillum-forms, with a flagellum at either end. F, G. Thin and short spiral forms. H. A spiral showing the individual links. × 540. (After Zopf.)

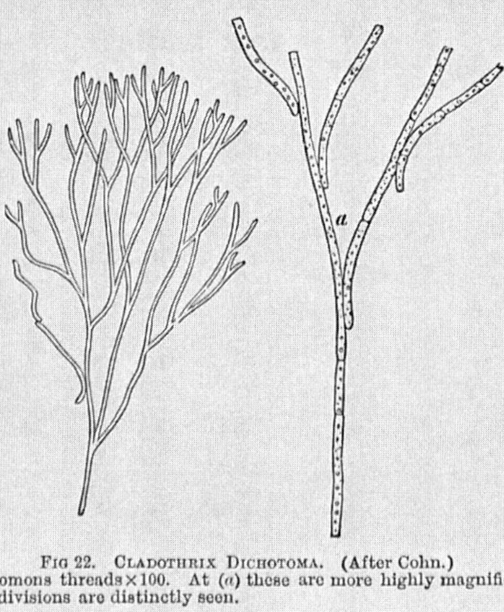

FIG 22. CLADOTHRIX DICHOTOMA. (After Cohn.)

Apparent Dichotomous threads × 100. At (a) these are more highly magnified (× 600), and the false Dichotomous divisions are distinctly seen.

veins of rabbits or guinea pigs is fatal. It is difficult to distinguish from the typhoid bacillus.

B. cloacæ appears to be a new form. It has very lively movements, and a strong acid reaction; it coagulates and acidifies milk in four days.

The other species he enumerates are—

Bacterium delicatulus, lively movement, always acid reaction, coagulates milk.

Bacterium ubiquitus, non-mobile, acid reaction, milk coagulated in eighteen hours. Preparation from a two days' old agar culture. Fig. 23, × 1,000.

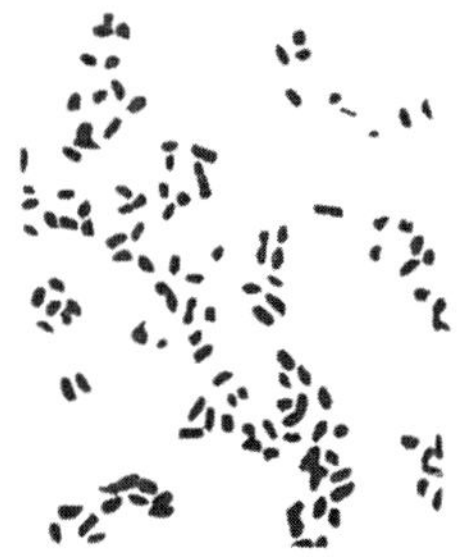

FIG. 23. BACTERIUM UBIQUITUS × 1,000.

Bacterium cyanogenus, not very mobile, alkaline reaction, milk slowly becomes chocolate coloured. Fig. 24, × 1,000.

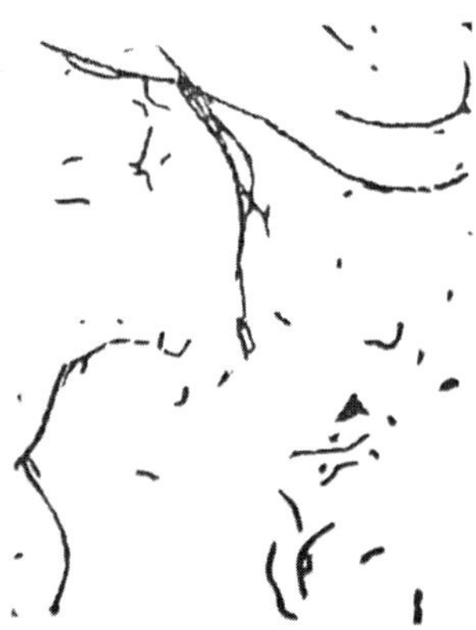

FIG. 24. BACILLUS CYANOGENUS × 1,000.

Bacterium hyalinus, lively movements, acid reaction, milk coagulated in seven days. Fig. 25, × 1,000.

Fig. 25. Bacillus hyalinus × 1,000.

Bacterium superficialis, independent movement, slightly acid reaction.

Bacterium reticularis, slow sinuous motion, acid reaction, milk slowly coagulated.

Beggiatoa roseo-persicina (B. rubescens) or peach-coloured bacterium, slow, sluggish movement, they give to milk a pinkish tinge, and have an alkaline reaction.

Bacillus violaceus laurentius, a new species, found in great numbers in tank effluent. It has very lively movements. Mr. Jordan remarks of this species, that milk seems to be an especially favourable medium for this bacillus. There is a rapid and luxurious growth, colouring the milk a deep blue violet, and soon causing it to coagulate and to acidify.

Proteus Zenkeri (Hansen) has movement in some stages of development, in others not. It produces no change in milk in thirty days.

Bacillus janthinus (Zoft) was found in effluent, but not in sewage; has movement, grows vigorously in milk, colouring it a deep violet, but not coagulating it.

Nine of these species were found to convert nitrates into nitrites, some with great completeness and rapidity. The species which failed to reduce nitrates were Bac. superficialis, Bac. rubescens, and P. Zenkeri.

"Wherever dead organic materials, excreta of man and animals, dead bodies, dead plants, household refuse, &c., accumulate on the surface of the ground, in stagnant or running water, or within dwelling-houses, in the presence of sufficient moisture and temperature, bacterial growth occurs, and ultimately occasions the complete destruction of these materials, the place of which is taken by an enormous number of newly-formed individuals."*

"The refuse of houses and of cattle stalls forms also a favourable place for the accumulation of bacteria. The intestinal excreta contain at times infective agents mixed with the large numbers of saprophytes; for example, typhoid bacilli, cholera bacilli, tubercle bacilli, anthrax bacilli, the bacilli of swine fever, of chicken cholera, &c.; further, probably the infective agents of dysentery and of the epidemic diarrhœa of children."†

Turning now from American investigations, interesting as they prove to be, I will enumerate those bacteria which chiefly play such a harmful part in the economy of animal life, both in man and in the lower animals, namely, the pathogenic, or those associated with disease, and it will be instructive, and I hope helpful, if I illustrate this part of the chapter as fully as the present state of this branch of natural history permits. I will also include a few forms which are saprophytic but unassociated with disease, but which play a great part as the burners and buriers of dead organic matter, and I hope the addition of one or two additional plates on this subject will emphasise the importance which I believe ought to attach to it in connection with sewage treatment and disposal, and disarm a criticism

---

* Flügge's "Micro-organisms," p. 689.

† *Ibid.*, p. 720.

which might charge me with too great a treatment of bacteriology. I do so only because I think the importance of the study of the connection between bacteria and sewage cannot be overrated.

Pathogenic Bacteria.

One of the most fatal and destructive diseases to which cattle are subjected is—

Anthrax (*Bacillus anthraces*), the bacillus of which is the sole

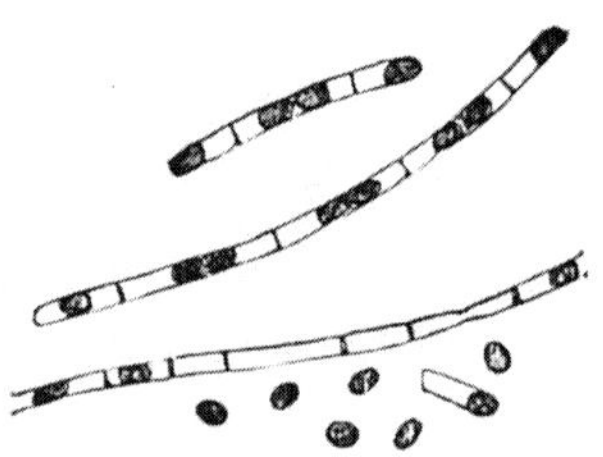

Fig. 26.—Spores of Bacillus anthracis, × 1200.

Stained with gentian-violet, after passing the cover-glass twelve times through the flame.

cause of this, is seen in fig. 26. Its spores remain in the soil for a long time after the cattle have died, and constitute a serious source of danger. Concerning the Anthrax bacilli, it may be stated that when present in the soi they form an abundant crop of spores, which contaminate grass and spread this direful disease, which speedily infects the blood, kidneys, liver, spleen, skin, lungs, mucous membranes, and capillaries of cattle, causing speedy death. So quickly do they multiply in such nutrient media, that sheep die in a few days, mice and guinea pigs in 48 hours, after subcutaneous injection of these spores.

Œdema, or dropsical swelling (*Bacillus œdematis maligni*). Death occurs in animals after inoculation in 24 to 48 hours. Fig. 27.

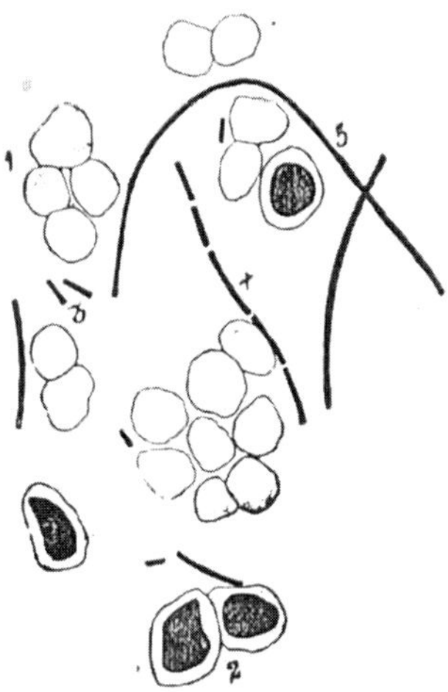

FIG. 27. BLOOD OF GUINEA-PIG DEAD OF KOCH'S MALIGNANT ŒDEMA.

(1) Red blood discs; (2) white corpuscles; (3) single bacilli; (4) chain of long bacilli: (5 leptothrix. Magnifying power 700 (stained with gentian violet).

Tuberculosis (*Bacillus tuberculosis*). Fig. 28.

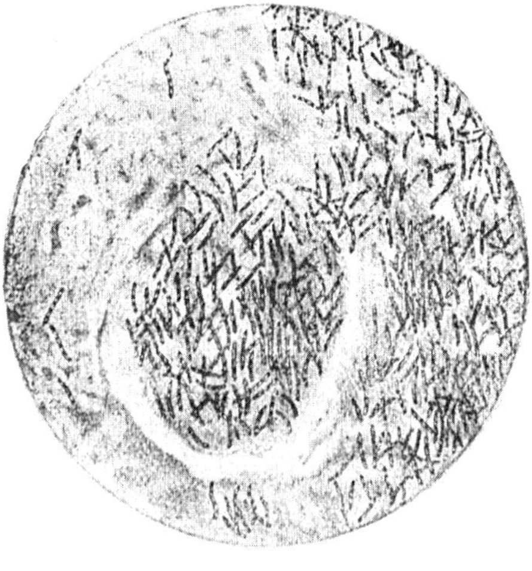

FIG. 28. Bacillus tuberculosis, artificial cultivation in guinea pig, cat, &c, × 1500.

Leprosy (*Bacillus lepræ*). Fine rods 0·004 millimetres long, and less than 0·001 thick, pointed at the ends. Figs. 29 and 30.

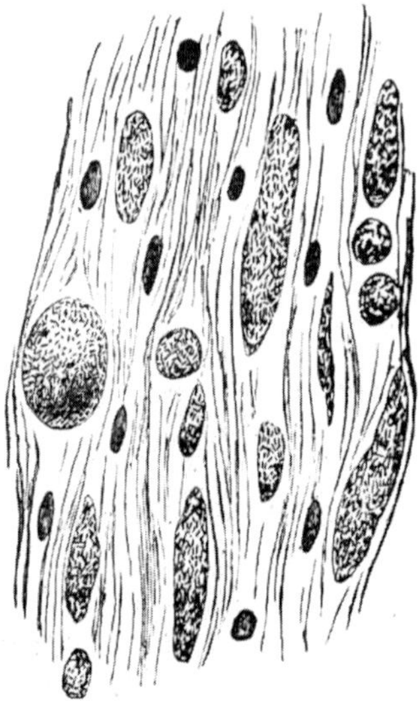

FIG. 29. FROM A SECTION THROUGH THE LARYNX OF A PATIENT DEAD OF LEPROSY.

Huge cells in fibrous connective tissue; the cells are filled with the leprosy bacilli. Magnifyin power 600 (stained with magenta and vesuvin.)

FIG. 30. FROM AN ARTIFICIAL CULTURE OF BACILLUS OF LEPROSY. (After Neisser.)

Glanders (*Bacillus mallei*). Fig. 31.

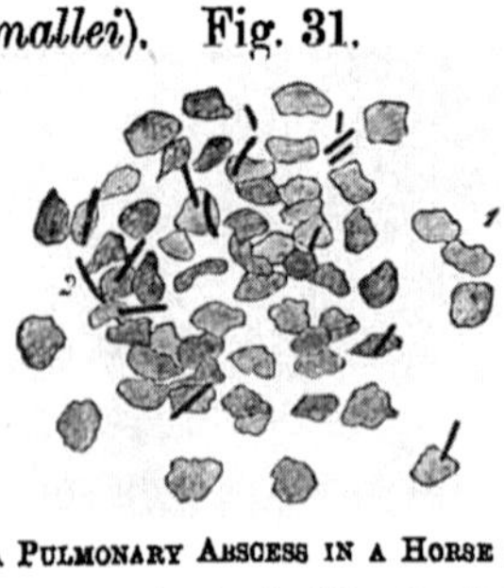

FIG. 31. PUS OF A PULMONARY ABSCESS IN A HORSE DEAD OF GLANDERS.

1. The nuclei of puss cells. 2. The glanders-bacilli. Magnifying power 700. (The preparation had been stained with methylene-blue).

Swine plague. Figs. 32 and 33.

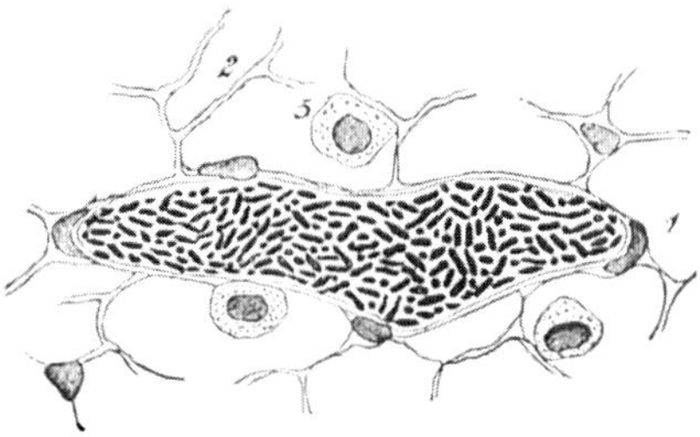

FIG. 32. FROM A SECTION THROUGH THE INFLAMED INGUINAL LYMPH-GLAND OF A PIG DEAD OF SWINE PLAGUE.

1. A capillary blood-vessel filled with bacilli. 2. Reticulum of adenoid tissue. 3. A lymph cell. Magnifying power 700. (Stained with Spiller's purple).

FIG. 33. BACILLI OF SWINE PLAGUE, FROM AN ARTIFICIAL CULTURE, DURING SIXTH DAY OF INCUBATION.

1 and 2. Bacilli. 3. Bacilli in which spores have been formed. Magnifying power 700. (Fresh specimen).

Cholera (*Spirillum cholera Asiaticæ*), or comma bacillus. Fig. 34.

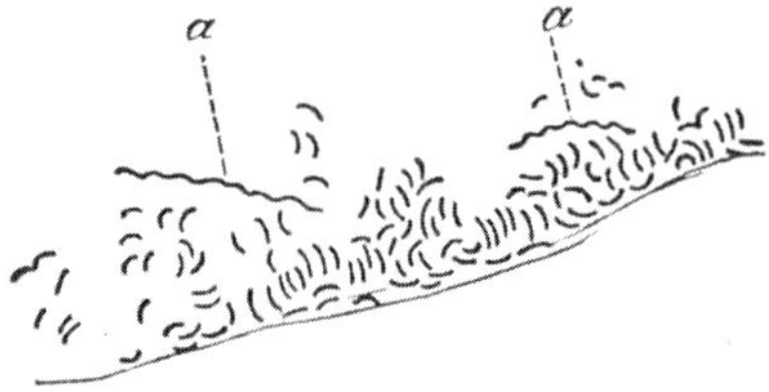

FIG. 34. COVER-GLASS PREPARATION OF THE EDGE OF A DROP OF MEAT INFUSION, containing pure cultivation of comma bacilli, with (*a*) spirilliform threads, × 600 (after Koch).

Curved rods actively motile. The commas are about half the length of the tubercle bacillus; they develop into spirilla, as shown at *a* in fig. 34. In an excellent paper in the "Nineteenth Century," of October last, Mr. Ernest Hart graphically writes on cholera and our protection against it. He says: "Cholera is a filth disease, carried by dirty people to dirty places," and that it and other kindred diseases are caused by polluted water supplies, pollution of the soil, and waterlogging, and that in India, the home of cholera, it is generally propagated by means of specifically polluted water. He quotes Dr. Furnell and Dr. Simpson to effective purpose, and shows 250,000 people in the suburbs of Calcutta have no water supply except wells and tanks, no drainage, nor building regulations. The tanks are defiled by the excretions of the body by the daily baths. This I can verify from personal observation at Calcutta and many other parts of India in 1885 and 1886.

Typhoid fever (*Bacillus typhosus*). Rods with rounded ends. Migula has observed that they have flagella, or organs of locomotion, as seen in fig. 35. The one on the right

FIG 35. TYPHOID BACILLI, showing the Flagella or organs of Locomotion (after Migula).

shows the fission of a spore or young bacillus. In typhoid fever the bowels also contain immense quantities of micrococci; also in the glands and spleen.

Relapsing fever (*Spirillum obermeieri*). Rapidly moving threads in the blood. Fig. 36.

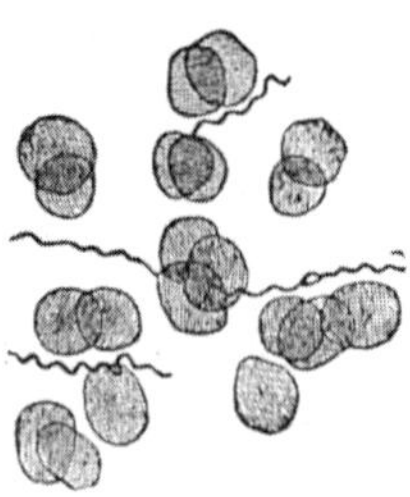

FIG. 36. SPIRILLA OF RELAPSING FEVER IN BLOOD × 500

Malaria or intermittent fever (*Bacillus malaria*). Rods growing into twisted threads. First found in the soil of the

Roman campagna. In cases of malaria other micro-organisms are also found in the blood. Fig. 37.

FIG. 37.

Pneumonia (*Bacterium pneumoniæ crouposa*). Figs. 38 and 39.

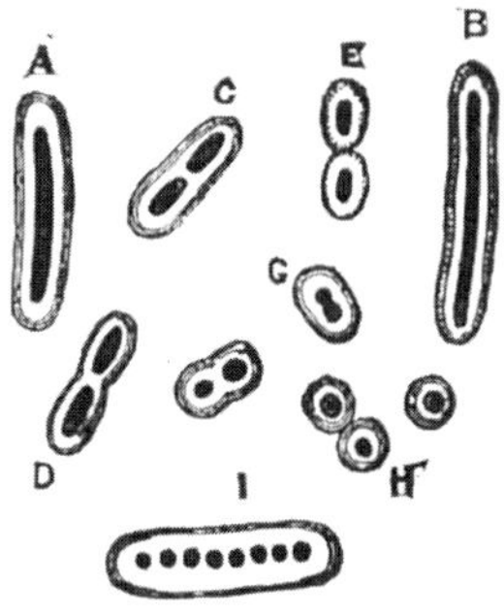

FIG. 38. BACTERIUM PNEUMONIÆ CROUPOSÆ, FROM PLEURAL CAVITY OF A MOUSE, × 1,500.

A, B. Thread-forms. C, D, E. Short rod-forms. G. Diplococci. H. Cocci. I. Strepto-cocci. (After Zopf.)

FIG. 39. BACTERIUM NEAPOLITANUM, × 700.

(*a*) From intestinal contents in a case of cholera; (*b*) from peritoneal fluid of an inoculated guinea-pig (after Emmerich).

Crookshank states that thirty-two mice, having been injected with this microbe, all died. Others, which had been subjected to inhalation with its spores, in mouse cages, died in the proportion of three out of ten. Fraenkel states that the microbe of this terrible disease is to be found in healthy persons, and that it is inert until certain preparatory but yet unknown factors enable it to make its attacks.

Diphtheria (*Micrococcus diphtheritica*). Found in large numbers in the organs of the throat. Bacteria are also found associated with these micrococci. Slightly bent rods, but of variable form. Fig. 40.

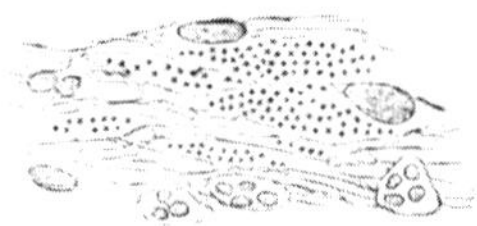

FIG. 40. PORTION OF A DIPHTHERITIC MEMBRANE.

Numerous micrococci present.

Tetanus. Long, slender, motile rods, with terminal spores and rounded ends. Found in the pus of a wound which

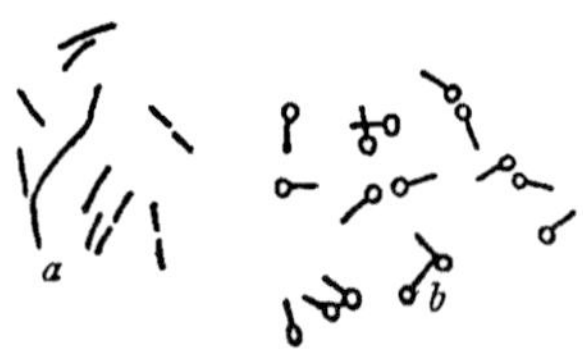

FIG. 41a. TETANUS BACILLI (KITASATO).

(*a*) Sporeless bacilli, (*b*) bacilli, each bearing a spore at its extremity, giving rise to club-shaped forms.

gave rise to tetanus. Death shortly follows the inoculation in mice, guinea pigs, and rabbits, the symptoms of the disease appearing in one and a half to two and a half days after inoculation. The bacilli form in threads and irregular groups, and produce spores. They are slightly thicker than the bacilli in mouse septicæmia. These bacilli are found in garden mould, in ruined walls, in putrefying fluids, and in manure. They are anaërobic.

Cholera in fowls—chicken cholera (*Bacterium choleræ gallinarum*). An aërobic micrococci and short bacilli, absorbing oxygen from the blood, and producing asphixia. Figs. 41 and 42.

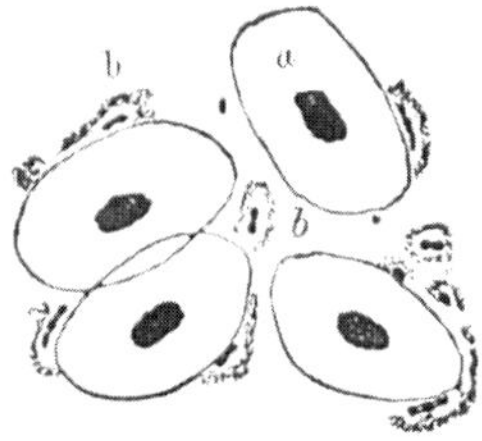

FIG. 41. BACTERIUM OF CHICKEN CHOLERA, BLOOD OF INOCULATED HEN, × 1,200.

(*a*) blood-corpuscles, (*b b*) Bacteria.

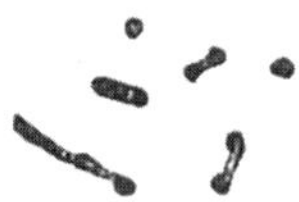

FIG. 42. BACTERIUM OF CHICKEN CHOLERA: FROM MUSCLE JUICE OF INFECTED HEN, × 2,500 [from a photograph]

Pyœmia, pyogenic or pus-forming bacteria, both micrococci and bacilli, as Staphylococcus pyogenes aureus, Staphylococcus pyogenes albus, Staphylococcus pyogenes citreus, Streptococcus pyogenes, Gonococcus bacillus pyocyaneus.

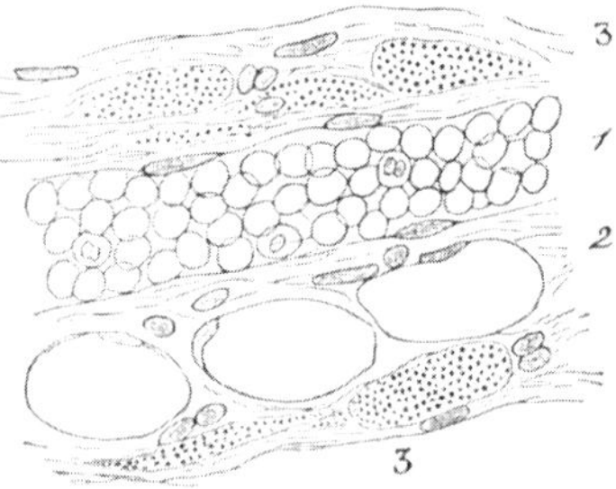

FIG. 43.—FROM A SECTION THROUGH THE TAIL OF A MOUSE INOCULATED INTO THE SUBCUTANEOUS TISSUE OF THE TAIL WITH ARTIFICIALLY CULTIVATED MICROCOCCUS.

The part here illustrated is a good distance from the ulceration.

1. A capillary blood-vessel filled with blood-corpuscles.
2. Fat cells.
3. Groups of micrococci filling the lymph-spaces of the connective tissue.

Septicæmia, or contamination of the blood by micro-organisms.

FIG. 44.—FROM A SECTION THROUGH A LYMPHATIC GLAND OF MAN DEAD OF SEPTICÆMIA.

1. A blood vessel which at one place is distended by and filled with minute bacilli.
2. Lymph corpuscles.
3. Degenerated lymph-corpuscles.

Magnifying power 700.

In health the blood of man and animals is bacteria-resisting. Healthy blood and tissues do not contain them. In white rats the blood is so alkaline as to destroy the bacilli of anthrax, and the disease cannot be communicated to them by simple inoculation.

Besides the foregoing pathogenic bacteria, biologists have found bacteria intimately connected with actinomycosis, atrophy, cattle plague, meningitis, dental caries, endocarditis, gonorrhœa, hydrophobia, measles, opthalmic diseases, pleuro pneumonia, puerperal fever, pyæmia, scarlatina, whooping cough, syphillis, yellow fever, and others.

Bacteria are found in the skin, under the finger-nails, in the mouth, the respiratory mucous membrane, in the stomach; they occur in the intestines, but chiefly anaërobic species, the aërobic forms not finding sufficient oxygen to favour their existence or growth.

Sericiculture has received a new and undying impulse in the discovery, by Pasteur, that the maladies of silkworms were owing to parasitic bacteria, the prevalence of which, twenty-five years ago, almost destroyed the valuable sericicultural industries of Italy and the South of France. He found a special bacteria preying upon the silkworm, its chrysalis and moth, and even the silkworm eggs, caused by filthy and unhealthy surroundings in the close and dirty village magnanaries of the silk-producing districts. A very large percentage of the worms died, and the industry was threatened with the extinction analagous to that occasioned in some wine-growing districts of France by the phylloxera. Fortunately, Pasteur's aid was invoked, and now, by a microscopical selection of those silkworm eggs which are free from bacteria, the percentage of loss of silkworms by disease has been reduced to less than 5 per cent. It is now estimated to be not more than 2½ per cent. In 1886 I was asked by the Government of India to go out to inspect the silk districts of Bengal, which were seriously suffering from a diminution of output from causes then obscure. I found that 60 per cent of the silkworms died of disease, chiefly pebrine. This enormous loss was entirely caused by the uncleanly surroundings and want of pure air in the village silkworm-rearing districts, of which Murshidabad, the ancient capital of Bengal, may be considered the centre. At my recommendation, an educated native, Mr. Nitya Gopal Mukerji, was sent over to study under Pasteur, and to ascertain the best sericicultural methods of France and Italy, by a prolonged stay during the silkworm-rearing periods. He has since been established in a sericicultural laboratory in Berhampur, and is successfully teaching the natives to rear healthy silkworms. In some of the silkworm moths I microscopically examined I found their bodies an entire mass of the characteristic bacteria of pebrine, Panhistophyton ovatum of

Lebert; called also Nosema bombycis, Micrococcus ovatus, Corpuscles du ver à soie. The microscopic appearance of this micro-organism is shown in Fig. 45.[a] There is also another

FIG. 45.[a] MICROCOCCUS BOMBYCIS (after Cohn)×600.

FIG. 46.[a] NOSEMA BOMBYCIS×500. (a) Nosema cells; (b) Urates which are usually present in the preparation. (After Duclaux.)

silkworm microbe, which causes the disease called flacherie, or maladie de morts blancs, which is at times very destructive. It is a micrococcus. Fig. 46.[a]

It is when in the form of spores that the bacteria are the most indestructible organisms of all life-world products. They can exist safely in the extremest cold, and many species can bear even boiling heat without being destroyed. Long-continued boiling, however, destroys them, and in dry air they are annihilated at a temperature of 280 deg. Fah., long continued. It is in this comparatively indestructible property of the spores of pathogenic bacteria that zymotic danger exists; and as we know the air is laden with them, and that dust, water, and soil are also more or less impregnated with them, the necessity of my attaching such high importance to this part of my subject is therefore apparent, and the absolute necessity of providing in sewage treatment that the atmosphere should not be supplied with or contaminated by them from any sewage sources, whether of effluents or sewage sludge, is scarcely necessary to be insisted upon.

The following changes of organic constitution are amongst the many wrought by bacteria on chemical compounds and living tissue:—

Urea into carbonate of ammonia by Bacterium ureœ.

Sugar into alcohol by Saccharomyces cerevisiæ.

Alcohol into vinegar by Bacterium aceti.

Liquefaction or peptonising of gelatine by Micrococcus pyogenes alba, and many others.

Albumen into peptone by Micrococcus pyogenes aureus.

Acid fluids into alkaline by Ascococcus billrothii.

Milk turned acid by Baccilus acidi lactici, and others.

Lactic acid into butyric acid by Colstridium butyricum.

Nitrates into nitrites by Bacillus ramosus, pestifer, and other species.

Nitric acid into nitrous acid by Bacillus violaceus, plicatus, and others.

Health into splenic fever or anthrax in cattle by spores of Bacillus anthracis.

Health into lung tubercle in man by Baccillus tuberculosis.

Health into typhoid fever by Bacillus typhosus. Regarding typhoid fever, how frequently has it not been traced to milk infected with Bacillus typhosus, such milk not seldom from sewage-fed cows. Marvellous results, effected by invisible action.

Nitric acid into ammonia.

A modern method of dyeing with indigo is its reduction and solution in the indigo vat by bacteria.

Bacterium amylozymicus ferments starch with the production of amyl alcohol.

Flügge gives the following table of yet other changes produced by various micro-organisms*:—

Species of bacterium.	Putrefactive products as yet demonstrated.
Bacillus putrificus coli............	Peptone, ammonia, fatty acids, tyrosine, phenol, indol, scatol, &c.
Bacillus saprogenes, 1, 2, 3......	Foul smelling gases.

* Flügge, "Micro-organisms," p. 611.

Bacillus coprogenus fœtidus ...	Foul smelling gases.
Proteus vulgaris, mirabilis, zenkeri	Peptone, foul smelling gases.
Bacillus pyogenes fœtidus	Foul smelling gases.
Micrococcus fœtidus..............	Foul smelling gases.
Miller's bacillus.....................	Sulphuretted hydrogen, ammonia.
Bacillus fluorescens liquefaciens (Bacterium termo)..............	Peptone, volatile fatty acids, green colouring matter.
Bacillus butyricus (Hueppe) ...	Peptone, leucin, tyrosine, ammonia, substances with a bitter taste.
Bacillus unrae	Trimethylamine.
Bacillus prodigiosus	Trimethylamine.
Bacillus pyocyaneus...............	Peptone, ammonia.
Bacillus fluorescens putidus ...	Trimethylamine.
Bacillus janthinus	Peptone, ammonia.
Various anaërobes	Foul smelling gases.
Bacteria from mud or the intestinal contents of ruminants (Tappeiner)	Carbonic acid, marsh gas, sulphuretted hydrogen, &c.

In putrefacive processes are formed—Alkaloids, sulphuretted hydrogen, oxalic acid, phosphuretted hydrogen, formic acid, carbonic acid, nitrogen, sulphide of ammonia, amine bases, and many other bodies by their vital action, the removal of oxygen from albuminous matter causing putrefaction; the decidedly acid reaction of Anthrax bacilli is the sign of virulent attacks. Emmeniti's Neapolitan bacillus, found in cholera cases associated with the common bacillus, produces from alkaline solutions of gelatine an acid reaction. Bacillus typhosus is also an acid producer.

The sources of many zymotic diseases are living germs floating in the air, which in populous places is highly charged with them; animal and vegetable bodies are immediately attacked by them; the putrefaction and decomposition which follow the attack are absolutely and entirely due to them. Going still further, lifeless bodies or parts of bodies would not undergo decay, decomposition, or putrefaction were there no spores or seeds of bacteria and of fungoid growth present in the air, and this is sufficiently proved by the non-decomposition of frozen meat, which is kept at too low a temperature for bacteria to propagate in, and also in the case of tinned meat, which can be kept perfectly fresh and good for many years if it is hermetically sealed from the air. It is these micro-organisms in the air, and not the air itself, which are the cause of decomposition and decay. These facts ought to lead the governing bodies of all populations to consider how far sewage sludge is allowable to aggregate. Few, if any, substances putrefy more rapidly than sewage. In summer time the quantity of bacteria increases in three days in a cubic centimetre of imperfectly treated effluent from 2,000 to 5,000,000. Crude sewage ordinarily contains about 786,000 per cubic centimetre. To what an enormous extent, then, does the danger of infection grow when the wind, drying this accumulated filth in summer time, distributes the germs, spores, or seeds of the worst forms of bacteria, and even the bacteria themselves. What of cholera germs in the East, always present in the air? The sewage and garbage disposal arrangements of Calcutta, which I carefully examined on the spot, are simply horrible. They consist solely in running them into a huge canal, where numbers of animals of various kinds lie decomposing. whilst the garbage and dead animals brought down by the garbage train every day are so laden with decomposing and putrefactive matter, that it is one of the most wonderful of sights to see the cloud of vultures and other huge scavenger birds which gather round and upon them. On the arrival of the train the wagons quickly become covered with these birds. The stench at this outfall is such as to be almost unbearable; the air is simply charged with putrefactive germs and odours. It was several days before I recovered from my visit to this place, and I can no longer wonder at the periodical occurrences and decimative power of cholera, smallpox, measles, scarlatina, typhus, diphtheria, whooping-cough, erysipelas, influenza, dysentery, diarrhœa remittent and other fevers, and the other infectious zymotic diseases. Ought, then, sewage sludge to be allowed to accumulate? I think not, for I feel sure that the health of the adjacent populations is greatly impaired and lowered, and the death-rate made higher in consequence.

If any argument were needed to prove the absolute necessity of my emphasising this part of my subject, one comes lately from Manchester with terrible force, showing the insidious power of Bacillus typhosus and the dangers of sewage farming.

The following extract is from the *Manchester Guardian* of October 11th of the past year:—

"THE EMPLOYMENT OF BOYS ON THE MANCHESTER SEWAGE FARM: COMPLAINT BY A MEDICAL OFFICER.—At the fortnightly meeting of the Altrincham Union of the Rural Sanitary Authority, on Wednesday, Dr. Fox, medical officer, said a month ago he called the attention of the Authority to the case of a boy who had been seized with typhoid fever, and who worked for the Manchester Corporation on their sewage farm at Carrington. He mentioned this case specially, because boys of the age of 16 or thereabouts were susceptible to such infection. He called the attention of Dr. Tatham, medical officer of health for Manchester, to the matter, and he quite agreed with him that a boy should not be allowed to work under such conditions, and promised it should not occur again. It was with profound regret he had to report the death of another boy of 16 from typhoid. He was a strong, healthy boy, and they would have to enter his death under the head of preventible disease. If he could not get some assurance that these boys should not be employed, he should take action. It was decided to call the serious attention of the Manchester Corporation to the matter."

A short time ago an unexpected occurrence in my laboratory took place. I had several stoneware gallon bottles of sewage and effluents waiting analysis. One bottle of sewage, three months old, exploded violently with a loud report. The laboratory was immediately filled with a most intolerable stench. The adjoining offices had to be vacated, all the windows opened, and it was some time before the stench was got rid of. This was no doubt the result of bacterial growth, probably increasing by millions a day, causing putrefactive gases in such pent-up volume that at last the resistance of the stoneware bottle had to yield to their expansive force.

If this could occur with a single gallon, what must be the danger of bacterial putrefaction in the slowly-flowing or imperfectly flushed sewers of towns not having much fall, or imperfect flushing in any town ? No wonder at the rapid spread of zymotic disease germs and of the deterioration of health.

To give an idea of the occurrence of bacteria and their enormous fecund development under suitable conditions, there were in the River Irwell water above the Salford Sewage intake at Weaste, after one week's incubation, above 100,000 bacteria per cubic centimetre (less than $\frac{1}{6}$in. cube), and in the Pendleton Sewage, before treatment and after a week's incubation, more than 10,000,000 per cubic centimetre.

The illustrations in this chapter are inserted by kind permission of the following authors from their works: Dr. Crookshank, Dr. Klein, Dr. Flügge, and Dr. Percy Frankland, fuller acknowledgments of which will be found in the preface.

CHAPTER IV.

ON METHODS OF SEWAGE DISPOSAL.

THE considerations in the last chapter pave the way for discussing first that mode of sewage disposal which is called broad irrigation, the examination of which, with certain and rare exceptions, has deepened my conviction that crude sewage ought not to be allowed to flow on grazing land or upon green crops in its crude state, and without some previous method of treatment.

In this chapter I propose also to describe the other most important systems of sewage treatment, the several methods of which (other than those which do not concern inland towns—for example, the discharge of crude sewage into the sea or tidal waters) are :—

BROAD IRRIGATION.

1. Broad or surface irrigation, which means passing the crude sewage upon land, for fertilising purposes, by gravitation or by pumping, or by both combined, as at Longton, Fenton, Croydon, Banbury, Burton-on-Trent, Burslem, and some other towns, the land being drained or not, according to circumstances.

Broad irrigation has its advocates, but their number is decreasing in ratio of the failure of the prolonged efficiency of sewage farms.

One of the conclusions as to crude sewage irrigation arrived at by Mr. Forbes, F.R.S., who discovered the phosphate process, although it has not my full concurrence, is so well expressed that I quote it in full: "There can be no question whatever but that when the local circumstances of climate and soil are favourable to irrigation, and the conditions essential to its successful application properly observed, that sewage irrigation is the most natural and effective system for the utilisation of sewage, since it is only by this means that we can render available the whole of the ammoniacal salts, upon which so very much of the fertilising value of sewage depends."

This opinion is fully shared in and lately endorsed by Dr. Carpenter in an able lecture, in which he says that with referenec to the sewage farm at Beddington, which deals with the sewage of Croydon, "we are able to raise 15,000 tons of produce." I have extracted the following particulars from his papers: "This sewage has been distributed on the farm for 30 years; the subsoil, at a depth of 3ft. in the cultivated fields, shows no signs of sewage pollution; and the soil is doing its work as well as it did at first. The quantity grown is according to the sewage applied. Forty tons per acre of green food is obtained, and the quantity of sewage applied is 5,000 tons, by applications of about two days per fortnight. There is no nuisance. The death-rate at Beddington is under 11, and the birth-rate over 30. The sewage is purified, and the effluent produced far above the requirements of the Rivers Pollution Commission. The imperial advantages are—(1) five times the amount of produce from each acre of land, as compared with ordinary agricultural land; (2) five times the amount of labour for the agricultural classes at full wages, doing in another way what the allotment promoters are asking for; (3) the production of meat and milk in a corresponding ratio; (4) there is a higher rateable value on the property, and so, in addition to increased wealth from the produce, there is increased wealth for the nation; (5) this is effected by an actual increase in the public health, cheaply purchased at a 2d. rate."

In his address at the annual dinner of the Association of Public Sanitary Inspectors, held at Liverpool in January of last year, Dr. Carpenter alluded to the yearly increase of diphtheria.

He attributed this to the preservation and growth of disease germs, to cesspools and other midden storages, and to sewers becoming seriously infected; and said that it only needed a hot dry summer, following upon a severe winter, to develop this disease to an extent never before experienced. He strongly recommended the hurrying away of all excreta to the land, and said that sewers which are sewers of deposit must be relaid on safe principles; house drains, too, must be similarly treated, and must be flushed and ventilated in a simple and scientific manner. He stated that in his district of Croydon, by the steady application of these precautions, the death-rate has been reduced within four years from 16·8 per 1,000 to 10·8, whilst the zymotic deaths had during the same period declined from 17 per annum to 2. He stated that disease germs can no more increase and multiply on a properly-managed sewage farm than gunpowder can explode if kept wet or mixed with moist sand; and that the presence of organic life in the humus in the soil, and the embrace of the minute radicles of a growing crop, set up a series of changes which are absolutely fatal to disease germs. He advocated the application of sewage to rye-grass for feeding milch cows, and said that for every 10,000 of the population of a water-closet town there should be 35 acres of land provided. There would be from 5,000 to 6,000 tons of sewage annually for each acre of land, and each acre would grow from 42 to 50 tons of produce. There would then be a complete purification of the water; the land would be free from excess of nitrogenous matter in its surface earth, for it would not find its way into the deep subsoil if properly laid out; and cultivated by a proper rotation of rye-grass, three years roots, one year cereals, one year the latter without sewage, there would be complete health to those who live on the farm and in the neighbourhood. His arguments are based on the principle, which he strongly insists on, that it has been ordained that plant life shall be nourished by means of animal exuviæ and animal decay, and that animals shall be nourished by the aid of the vegetable world. He estimates the annual loss to this country, since two centuries ago, by its improper treatment of sewage, to be sixteen thousand millions of pounds sterling. Such opinions from a man so distinguished in sanitary science, and living in the immediate vicinity of arrangements successfully carried out in conformity with his teaching, deserve serious attention before the decision is arrived at for the adoption of any system.

There are numbers of proved instances of the milk of cows being poisoned by the disease germs of typhus and other diseases in cases where the animals have fed upon lands to which sewage in its crude state had been applied. This part of my subject will be more fully investigated in a later chapter. At best sewage farming is but a filthy condition of things. I have, therefore, come to the conclusion that only effluent waters freed from suspended matter should be allowed to flow upon land, and that the bulk of the organic matter in solution should be separated by precipitation, and then burnt, or used as manure after having been denuded of its dangers. I am sure that the time will come when it will be the bounden duty of county councils to insist upon some method of sewage-purifying treatment before land is fertilised with it; but so long as it is allowed to run on land in its crude state, I say it should be done so sparingly and intermittently that land is not clogged with it, nor roots too highly stimulated by such forcing food.

Opinions vary as to the quantity of land necessary to dispose of crude sewage by irrigation. An acre of land would be required for every 50 persons; some authorities fix the limit as 25 persons per acre. For a town of 15,000 persons, 300 acres would be required at 50 persons per acre; and even then, to quote the words of Mr. Hutchinson, "It has become from experience a well-known fact that the purifying effect of an area of land continuously used for purposes of irrigation is not lasting, and that ultimately the soil becomes so saturated that a serious nuisance, both disagreeable to the senses and injurious to health, is the result." On this head the experience of Mr. Hawksley may be quoted: "Irrigation carried on in warm weather is exceedingly unhealthy. I can speak positively of it from repeated observation in different places, that the odour, particularly at night, and particularly upon damp evenings in autumn, is very sickly indeed, and that in all these cases a great deal of disease prevails. The sewage forms a deposit on the surface of the ground; that deposit forms a cake of organic matter, and that organic matter, when it is in a damp state, as it usually is, gives off in warm weather a most odious stench."

Dr. Carpenter, however, considers that an acre to every 250 persons is, with proper precautions, enough. Each acre produces annually 40 to 50 tons of produce, but it must be a condition that the storm water be separated from the sewage. Also Dr. Carpenter insists that the application of sewage in this way must be capable of rapid removal from such areas; in fact, that it must be rapidly intermittent in its application. He says also that the sewage should be applied to the land within twelve hours of its discharge. These seem to me to be conditions attended with great risk and cost, but in the hands of such an able and enthusiastic sanitarian, they are no doubt attended with success.

We must consider that this opinion is the result of successful sewage farming at Croydon, over which Dr. Carpenter has full superintendence, and the management of which is as strict as it is unquestionably exact and successful.

In a report to the Town Council of Glasgow, in 1878, I read: "Great hopes were a few years ago entertained that in irrigation had been found the grand solution of the sewage question . . . All that is changed now. Sewage farms on which immense sums of money have been expended have been reluctantly abandoned, and irrigation is no longer regarded as anything more than a means of obtaining a good effluent at a moderate outlay."

Such is the discordance of these two extremes of opinion. The sewage farms I have seen I cannot consider successful. Even from the model farm at Wimbledon may be seen examples which to my mind settle the question in favour of precipitation, and after irrigation by effluent only. On a part of that farm sewage and offal refuse are applied without treatment; the stench, when I was there, was simply horrible, even at a distance; but on the part on which the Amines process was applied there was no offensive odour, not even at the sewage works.

It may be said that the application of this crude decaying matter was imperfect, but the farm manager assured me the conditions on which the farm had to receive the untreated matter were uncontrollable, and they did the best they could.

No doubt the application of crude sewage to land is more fertilising than that of effluents, and not much objection could be made to its use if it could be always applied laterally, and without much actual contact with the lower parts of crops, and an entire freedom of contact with the upper parts, especially such as are used for human food. As far as I have yet been able to examine into the various systems of sewage disposal and the effluent results, the fertilising powers of their effluents are by no means destroyed. I do not mean to assert that effluents possess the same amount of fertilising power as raw sewage with its suspended matters. The state of the case was well put some years ago by Dr. Corfield, in speaking of the Stoke Workhouse plan of filtration through charcoal, when in the following sentence he defined the effluent as a dilute sewage: "For although," he said, "the suspended matters were removed, and the ammonia and organic nitrogen much reduced in quantity, no oxidation took place, as no nitrates were found in the effluent water, which was to all intents and purposes a dilute sewage." On the other hand, he states "it is possible, when the filtering medium is not renewed with sufficient frequency, for the effluent water to pass away with even more valuable fertilising elements than the raw sewage itself possessed."

I have noticed a plan in Lancashire which is most successful in raising celery: Two closely adjoining furrows are dug; into one is put human excreta, into the other the celery plants. There is no contact; it is broken by the ridge between the two furrows. Both furrows are soiled over, and as the celery grows nitrate of soda is added from time to time, from a tub in which it has been dissolved, in the proportion of about 4lb. or 5lb. to each 100lb. of water.

The finest celery is grown in this way, realising as much as 2s. 6d. per dozen roots. Could sewage be administered in the same manner, no objection save the storage and accumulation of bacilli could be urged against crude sewage fertilisation, but until this can be done there can be no doubt that a chemical treatment by precipitation is the safest and cleanliest method, especially if such treatment effects microbic sterilisation.

Dr. Carpenter meets this difficulty by contending that the pathogenic or disease germs are annihilated on a sewage farm. He endeavours to prove this thesis by adducing the absence of enthetic diseases in the population which is fed from the root-crops grown on the sewage-fed farm of Beddington, Croydon. I do not consider this as proof; the very opposite would probably occur on carelessly managed sewage farms. Disease germs certainly have a limit placed to their rapid multiplication by being immediately brought into contact with soil, and by being carried by rain to the subsoil; but as to the total destruction of them I should like to see more evidence in proof, for it is well known that their vitality is so great as to outlast burial for indefinite periods. My experience of sewage farms is that they are more or less of a nuisance, with rare exceptions, owing to exceptionally careful management. Sewage farms get clogged, and sewage sick. In 1888, at the Warwickshire Assizes, an architect and member of the Town Council of Stratford claimed and obtained £1,000 damages for a sewage farm nuisance of six months' standing. This led to the discontinuance of sewage farming, and the adoption of precipitation.

It has been proved that soils have the power of separating and absorbing ammonia from both its solutions and salts, but it is a mistake to suppose that land is made more fertile by applying larger quantities of sewage than the roots of plants can assimilate. Any surplus either passes away by the subsoil drainage, or remains with an injurious effect upon the crops; besides which it is necessary to give soils intermittent rest' so as to regain aeration and power of nitrification. An intermittent application of sewage of alternate weeks, with alternating weekly rests, is considered a good method of fertilisation.

Broad irrigation, if carefully administered, is doubtless a fertilising mode of using the sewage, if sufficient land is available, and with the precautions I have stated.

In France fæcal matter is highly prized for agricultural purposes, especially in the Maritime Alps. In the Riviera I have noticed the care taken to collect and treat it for manure. Its value is about three shillings per cubic yard. It is applied directly to the roots of orange and lemon trees, in small pits. Another form of it is in what is locally termed taffo, which is a mixture of excrement and clayey soil, made into bricks and dried. Mr. Corfield relates that "the peasants exhibit an astonishing zeal in collecting excreta, and they will even walk a league 'pour ne pas perdre leur dejection.'" The fæcal matter is not used until fermentation has set in. In that state it is applied to crops generally, and at Grasse and Nice it is used as the manure for the flowers which are so extensively used in the manufacture of perfumery; notably, says Mr. Corfield, for violets and roses.

In a future chapter, when treating of pathogenic dangers from sewage, I shall describe the objections to broad irrigation with crude sewage more fully.

Downward Filtration.

2. Downward filtration, or intermittent downward filtration, the surface of the land having to be constantly aërated by drying and ploughing. By this method the crude sewage is put upon land previously drained to a depth dependent upon the greater or less porosity of the soil or subsoil, generally about three feet to six feet. This is applied irrespective of any effect upon the crops which may be growing upon the land; or, as Mr. Bailey Denton, the advocate of this system, says, "Making the produce of secondary importance." Chemical changes take place. According to Mr. Crookes, F.R.S., "The fæcal matters coat the earthy particles they come in contact with, and in this state are attacked by the oxygen of the air. Their organic carbon becomes carbonic acid, their nitrogen, nitrous and nitric acids, which unite with the lime, magnesia, and other basic matters in the land, leaving the effluent in such a fair state of purity as to be quite inoffensive to the senses, though not fit for domestic purposes." Clay soils are not suitable for this method of treatment.

At Dewsbury, where Mr. Bailey Denton's system has been adopted, the sewage is pumped to two levels to a sewage farm of 70 acres. The sewage is not allowed to come into contact with the leaves of crops, but it is absorbed laterally through specially cut channels or furrows through sandy land suitable for such filtration; the surface water is excluded from the sewage, the trade wash regulated, the solids are not separated.

The effluent is stated by Mr. Crimp to be so good that the mill hands come to bathe at the outlet into the Calder, where a few years ago the River Commissioners found the river so black that they used it as ink in writing their memoranda.

Hitchen, Abingdon, Malvern, Barnsley, Kendal, Merthyr, and Forfar are examples of Mr. Bailey Denton's downward intermittent filtration, but, I am informed, are not of equal success.

Mr. Corfield considers that intermittent downward filtration is a valuable method for the purification of sewage, but not for the utilisation of it, as the area of ground is too small, and the quantities of sewage turned on to it too large. The difficulty of applying this system in hard frosts must be self-evident.

Sewage Settling Tanks.

3. The settling tank system, by which the crude sewage is run into tanks and allowed to settle mechanically, the liquid part or effluent water being passed over land or into a river.

Sewage Filtration

4. Ordinary filtration of sewage without chemical treatment, the solid part, as in the last, being dealt with separately, and the effluent run upon land or into a river. Such filtration is through sand, gravel, coke, charcoal, peat, and other substances, but neither this method nor the preceding one is considered very satisfactory.

Chemical Precipitation.

5. Precipitating the sewage by lime or other chemical processes, which will be hereafter described. The solid part or precipitate, or, as it is more generally termed, the sludge, is removed from the precipitating tanks and treated by itself in various ways, and the effluent water, containing a reduced quantity of organic matter, is either run upon land drained to receive it, as at Birmingham, or upon land not drained, or run into a river or water course.

Where sewage, whether crude or effluent, is put upon land in large quantities, there can be no doubt it is better that the land should be well drained, but very light, sandy land, which could absorb the whole sewage over large areas intermittently, would not need drainage. That, however, depends upon the population and the quantity of sewage to be disposed of. Mr. Henry Robinson, C.E., in an able work entitled "Sewage Disposal," shows in the cases of nineteen towns the number of inhabitants, the quantity of sewage, and other particulars relating to this point, statistics of such interest that I reproduce them in their entirety.

Name of Town.	Population.	Daily flow of Sewage, in gallons.	Annual Rateable Value.	Area of Land Irrigated, in acres.	No. of Inhabitants to each acre irrigated.
			£		
Abingdon	5662	150000	16361	30	189
Banbury	12127	400000	49526	138	87
Bedford	19552	900000	74500	139	140
Blackburn	104012	2800000	307410	500	208
Cheltenham	43972	1250000	264809	{ 131 330 }	95
Crewe	24371	900000	69978	257	95
Croydon	78000	3000000	430000	525	149
Doncaster	23098	700000	96316	262	88
Kendal	13697	1800000	49563	16	856
Leamington	26000	800000	123274	469	55
Merthyr Tydvil	48500	1200000	135000	{ 380 20 }	121
Northampton	51847	1000000	149625	327	158
Norwich	88000	6000000	260000	500	176
Rugby	10000	400000	35629	78	128
Tunbridge Wells	24309	650000	177212	285	85
West Derby	26000	780000	188456	207	125
Wigan	48196	2000000	139000	273	176
Wolverhampton	70000	2500000	216081	330	212
Wrexham	10978	300000	42000	84	131

From this important table the following averages are deduced:

1. 137 people to each acre of land irrigation.

2. 195 acres to each million gallons of sewage, or 5,128 gallons per acre.

3. 38 gallons of sewage per head of population per day. 100 tons of sewage will cover an acre one inch deep, and its value as manure is said to be ½d. to 2d. per ton.

CHAPTER V.

ON THE VARIOUS METHODS OR SYSTEMS OF SEWAGE TREATMENT.

THERE have been many plans, systems, and patents for the chemical treatment of sewage since it was first tried in Paris in 1740. Mr. Boulnois states that more than 400 patents were taken out between 1865 and 1875, and but few of them had much practical value. Amongst the best known methods are the following: The lime process, or Higg's; the Northampton, or modification of the lime process; Sellars', or the A B C process; Goodall's, or the M. and C. process; Forbes' and Price's process; Hansen's process; Hille's process; Holden's process; General Scott's process; the Anderson process; Weare's charcoal filtration process; Whitehead's process; the Hertford system; Dibdin's process; Spence's process; the Amines process; the Electrical process; the Ferozone and Polarite process. I propose to give some particulars of the above methods, dealing more especially with the three last-named from their greater scientific interest and from their being of later date. The last two years have been prolific in new systems or new applications of older ideas. Amongst the more recent processes are those of C. E. Bell, Lockwood, Grimshaw, Spencer's Carbide of Iron, Barry, Mather and Platt, Kaye-Parry, and lastly, my own, now in successful operation at Salford and at Nuneaton. Besides the list I have given there are other older ones, known as Bird's, Blythe's, Campbell's, Collins', Fulda's, Lenk's, Lundy's, Manning's, Smith's, Stothert's, Suvern's, and Wickstead's, but they do not need description here.

THE LIME PROCESS (said to be Mr. Higg s process).

This is simply running into the sewage, as at Salford, a certain proportion, about 15 or more grains per gallon, of milk of lime, which throws down a copious precipitate. The Sewage Commission report upon the method as follows: "Though very simple, and the least costly of any, it could not be profitable in an agricultural sense, and did not purify the sewage." The Rivers Pollution Commission condemn it, and say the effluent is not admissible into a river, and further remark: "In all these places the plan has been a conspicuous failure, whether as regards the manufacture of valuable manure or the purification of the inoffensive liquid." Secondary decomposition and putrefaction invariably follow precipitation by lime alone.

NORTHAMPTON.

Here lime and chloride of iron are used in the proportion of 12 bushels of lime to 6 gallons chloride of iron per million gallons of sewage. Putrefaction is deferred by the iron salt, but it ultimately takes place and fouls the river. Professor Way reports it as unsuccessful, besides being expensive.

GOODALL'S PROCESS, OR M. AND C.

Lime, refuse carbon, house ashes, soda, and perchloride of iron. The Committee of the Local Government Board on Modes of Treating Town Sewage, in 1875, after examining the process in use at Bolton-le-Moors, reported as follows: "The M. and C. process, like its twin the A B C process, merely removes the grosser parts of the suspended matters in sewage, but fails to remove the putrescible organic matters in solution, and therefore the clarified sewage cannot be admitted into rivers without causing pollution."

FORBES' AND PRICE'S PROCESS.

(Experimented upon at Tottenham.)

The precipitant is phosphate of alumina dissolved in sulphuric or hydrochloric acid, "a powerful antiseptic and disinfectant, completely arresting further putrefaction, and depriving the most fœtid sewage of its offensive smell." Dr. Voelcker says it possesses valuable fertilising properties, chiefly due to the phosphoric acid put into it. The effluent is not likely to come up to the required standard.

HANSEN'S PROCESS.

By old black-ash waste, a by-product or waste in the manufacture of soda. In use at Aldershot with the lime process. This process is not one of precipitation, but of oxidation. I cannot hear of any opinion on the process. Black-ash waste is a very abundant and very cheap waste product, but it is extremely doubtful if it has the power to oxidise the organic matter, particularly that in suspension.

HILLE'S PROCESS.

Was treated at Wimbledon, and consisted of lime, tar, chloride of magnesium, and another substance as a precipitant. It is a deodorising process, destructive of fertilising properties. The process is in use at Tottenham.

HOLDEN'S PROCESS.

A French method, described in the Rivers Pollution Commission first report, consisting of sulphate of iron and coal dust as a precipitant. The Commissioners report of it as follows: "It separates the whole of the suspended matters; it not only fails to remove the putrescible organic matters in solution, but actually (as measured by the organic nitrogen contained in these organic matters) increases their quantity. This it does by causing some of the putrescible organic matter in suspension in the original sewage to pass into solution. The effluent water could not therefore be admitted into rivers without causing pollution."

GENERAL SCOTT'S SEWAGE CEMENT PROCESS.

The principle of General Scott's process is to arrest the flow of the sewage by tanks, the suspended matters being precipitated by means of lime and clay, which are added to the sewage in the sewer previous to its arrival at the tanks, the proportion of lime so added being about 10 cwt., and of clay 5 cwt., to 400,000 gallons of sewage. After the sludge has sufficiently accumulated in the tanks it is drawn off, placed in a kiln, and burnt by intense heat, and then ground into cement. It is used at Ealing. The Committee's report is that the effluent is too impure to be sent into a river, and too valuable to be wasted. It was tried at Birmingham, and abandoned.

The Anderson Process.

Patented in 1872 by the General Sewage and Manure Co., who contracted to purify the sewage at Coventry—two million gallons daily. Process: Pass through extractors to remove suspended matters into circular mixing tanks, and agitate with a saturated solution of crude sulphate of alumina heated to boiling point, It then is treated in another series of tanks with milk of lime. and flows into subsidence tanks, and then on to four and a half acres of a loamy soil, to filter it into drains five feet deep, from which it passed into an outlet into the river. The effluent was clear and inodorous. The sludge is dried by artificial heat by mud-drying machines. The sludge was difficult of sale at 3s. per ton, and it cost £4 10s. per ton. The loss was £1,500 per annum for a million of gallons of sewage per day treated.

At Hertford and Leyton the sewage is also precipitated by lime and sulphate of alumina. Major Flower reported that this process is not capable of giving a pure effluent, but gives rise to secondary decomposition, becoming offensive after discharge into the river Lea, and bringing about a very filthy state of things.

Whitehead's Patent Process.

The precipitant is a mixture of dicalcic and monocalcic phosphates. The British Association Committee experimented upon sewage with this mixture. They report rapid precipitation, the organic matters in solution and suspension being almost entirely removed.

Professor Corfield reported upon it in the following words: "The process may be relied upon for the purification of the sewage of Luton, so as to render the effluent water of sufficient purity to comply with the standard of the Rivers Pollution Commission, and it is the only precipitating process that I am acquainted with of which I could say this. As to the value of the manure, it is impossible to estimate it with any approach to accuracy."

The Hertford System.

Lime and chloride of lime; 56lb. of the latter per million gallons of sewage. Dr. Tidy speaks favourably of this process, as also of the lime and sulphate of alumina process.

Dibdin's Process.

Lime and proto-sulphate of iron, or green copperas, in the proportion of 3·7 grains of lime in solution and one grain of copperas. This has been thoroughly tried upon Metropolitan sewage, and reported upon by the Metropolitan Board of Works as follows: "Precipitation of suspended matters satisfactorily effected." Four distinguished chemists were asked to report also, and they said the method of precipitation was a good one, but it left a sufficiently unpleasant smell to prohibit the effluent water being discharged into the river at all states of weather and tide. They thought there would not be immunity from secondary fermentation and development of offensive gases in hot weather, and required a further treatment with manganate of soda and sulphuric acid to render it fit to be discharged into the river. It is used at Leicester, but with lime only as a precipitant.

The Native Guano Co., or A B C Process for the Utilisation of Sewage.

This process is in use at Kingston-on-Thames, which includes Surbiton, a joint population of 38,068, also at Aylesbury and Wellington College.

It consists, shortly, of a screening chamber and grating to intercept solids likely to injure pumps. The sewage is received into a well, where it receives a portion of the substances necessary for its treatment, alum, charcoal, or carbonaceous refuse from prussiate of potash works, which is an impure carbon (the waste product used at Newcastle-under-Lyme), and clay. In the original specification of the patent, blood and other substances were named, but which do not now appear to be useful. At Kingston, the sewage treated is raised 12ft. by three centrifugal pumps, capable of 1,650 gallons per minute, and runs into settling tanks, receiving on its way the other portions, which act as precipitate agents.

The precipitate forms in tanks, and the purified sewage passes, it is said, clear, bright, and odourless into the Thames.

The proportion of the ingredients used are from 28 grains per gallon to 224 grains.

The quantity of precipitate or sludge is very large, but it is dried, ground, and sold under the name of Native Guano at £3 10s., and is said to form an excellent manure.

This company publishes several analyses, from which I select one of the crude sewage and four of the after effluent, from which it will be seen that variable results occur. The quantities are unfortunately expressed in grains per gallon instead of in parts per million, but the standard, 21 grains per gallon of albuminoid ammonia, is approximately equivalent to 260 parts per million.

A B C Process, Aylesbury.

Parts per million.		
	Free Amm.	Alb. Amm.
Crude sewage..................	40·00	06·60
Effluent	10·20	00·70

ANALYSES OF EFFLUENTS, BY PROFESSOR WANKLYN.

Date of collection of sample.	Parts per million.	
	Free Amm.	Alb. Amm.
4th October, 1880	22·00	0·76
5th "	12·10	0·60
6th "	5·30	0·36
7th "	4·30	0·32
8th "	4·10	0·20
9th "	5·00	0·40

Any one of these effluents may be (according to Prof. Wanklyn) with safety discharged into a river.

ANALYSIS.—GRAINS PER GALLON.

Suspended matters.	Sewage.	Effluent water.				
	Dr. Wallace.	Dr. Tidy.	Dr. Wallace.	Prof. Wanklyn	Dr. Keates	Standard of Rivers Pollution Committee for effluent.
Organic	24·9	·04	traces	...	...	·70
Inorganic..................	12·9	·18	"	...	...	2·10
Total......	37·5	·22	...	...	...	2·80
Free ammonia	5·18	·60	2·73	·61	1·75	no standard
Albuminoid ammonia.	·58	...	·12	·03	·18	·21

CHAPTER VI.

The Amines Process.

In November, 1891, I called on the inventor of this process, Mr. Hugo Wollheim, Analytical Chemist, at 101, Leadenhall Street, London, and discussed his system, and examined his method and results thoroughly, during the two hours I spent with him in his laboratory. I satisfied myself by microscopic observations that sewage matter abounds with micro-organisms which more or less generally may be classified under the name of bacteria and fungi, which rapidly increase and cause putrefaction both of sewage and sludge. The object of the Amines process is to kill these organisms, and to make the media in which they live and propagate impossible for them to feed upon; in fact, to extirpate and destroy them entirely. The re-agent is produced by the action of lime on certain organic bases belonging to the Amine group, notably trimethylamine. When these organic bases are acted upon by lime, a very soluble gas is evolved, which spreads rapidly through every part of the liquid, and is held in solution therein, with great tenacity. The gaseous re-agent has been found to be antagonistic to the existence and propagation of every species of bacteria occurring in sewage and other similar or impure waters, for it is claimed that it thoroughly extirpates them in a remarkably short space of time. The effluent from such waters, after treatment by this process, is actually sterilised; it shows no living micro-organisms whatsoever, even under the most powerful microscope, and its sterility is further confirmed by that latest and most severe test known to modern science, viz., inoculation and growth on gelatine broth and other nutrient media, and by "plate" cultivation. As a natural consequence of the sterilising action of this re-agent, decomposition of organic matter, whether incipient or far progressed, and the objectionable phenomenon of putrefaction, attendant on such decomposition, is completely arrested, and even new infection cannot beget fresh putrefaction, as long as there is a sufficiency of the gas remaining in solution. Absolute sterility and immunity from new infection are claimed for this process, and this has been repeatedly proved by the Government Analysts at Somerset House, upon Metropolitan outfall sewage. When trimethylamine is added to milk of lime, this volatile gaseous product is formed, to which the name of aminol has been given. Its composition is not yet ascertained. When the milk of lime to which trimethylamine has been added, has been well stirred, it is added to the sewage, which immediately dissolves and absorbs this gaseous product aminol. The sewage remains suffused by it. I will state in detail the process and its results further on. Trimethylamine is an expensive liquid to prepare, but fortunately it is found present in sufficient quantity in the waste brine in which herrings have been cured, its presence being easily detected by its very characteristic odour. Herring brine containing the trimethylamine is said to cost not more than 10s. for each million gallons of sewage, whilst the quantity of lime is not much larger than is necessary to ensure the maximum of clarification. Both are expressed in the following quantities:—

One-fifth of a ton of herring brine	per million parts.
2½ tons of lime	

Valuable antiseptic properties are recorded of the Amines process by Mr. Cooper, Engineer and Surveyor to the Wimbledon Local Board, who reports that a layer of the Amines sludge 9in. deep, in drying, lost 75 per cent of its bulk while exposed in a pit for three weeks in September, 1891; there was no smell from it, neither was there any from the sludge tanks or press liquor. The effluent, after having been kept for months and exposed to the most trying tests, showed no signs of decomposition or bad smell; in fact, the micro-organisms which cause these putrefactive changes were completely killed, and the sewage matters sterilised. Not only is the effluent sludge and press liquor, he says, free from microbes, but they remain free from all such organisms. The precipitation of the organic matter by this process Mr. Cooper states to be so remarkable, that in almost every instance within 25 minutes after the sewage is admitted into the tanks the supernatant liquid from which the sludge has precipitated is transparent to a depth of nearly 6ft. This antiseptic treatment has been proved on the Wimbledon farm to have no injurious effects on the fertilising properties of the sewage.

The Amines treatment does not remove the whole of the organic matter, but the effluent is fit for application to land for fertilisation, and also fit to turn into a river, because it is sterilised and incapable of undergoing putrefaction. To my mind, a thoroughly sterilised effluent going on land, especially grass land where cows are fed for milk supply to towns, is vastly preferable to sewage or effluent water being applied to such land directly, whether by simple or drained broad irrigation. It is conclusively proved that disease germs are conveyed to the blood and milk of cows by feeding on such polluted vegetation—that is, vegetation in contact with sewage above ground. Freedom from these germs is secured by the Amines process if it will sterilise the sewage to the extent claimed for it; that is, effectually and completely destroy and extirpate the bacteria, and by so doing prevent the possibility of any fermentation or putrefaction arising.

Chemistry of Trimethylamine.

It may be interesting in passing, if I give some account of the chemistry of this remarkable body, which, whilst being in itself probably a product of decomposition from the incipient or nascent decay of fish in the process of curing, possesses the remarkable power, after conversion into aminol, of preventing germ or microbe growth, an instance analogous in some degree to "like curing like," and of the compensating forces of nature. It will be necessary to dissect this compound word, the second syllable of which (methyl) is one of three well-known alcohols, and the last, amine, belongs to the compound ammonia group. Ethyl alcohol is the ordinary spirits of wine, of the composition indicated by the formula C_2H_5OH. It boils at 78 deg. C., becomes viscid at 100 deg., but does not solidify. Methyl alcohol, CH_3OH, or wood spirit, so called from its being a product of the destructive distillation of wood, a colourless mobile liquid, burns with a pale bluish flame, and, like ethyl alcohol, mixes readily with water. It boils at 66 deg. C. Amyl alcohol is the third alcohol, and is chiefly derived from corn and potato spirit; hence its name, from amylum, starch. It is a liquid with a penetrating and unpleasant smell. It boils at 132 deg., and solidifies at 20 deg. into a crystalline mass. The amines are compound ammonias, volatile bodies having an alkaline reaction and ammoniacal smell. They combine with methyl, and form the nitrogen bases of that body. Ordinary ammonia has the formula of NH_3, or

$$\frac{\frac{H}{NH}}{H} \quad \text{Methylamine has the formula} \quad \frac{\frac{NCH_2}{H}}{H}$$

It is a colourless inflammable gas, and is more soluble in water than ordinary ammonia. It occurs sparingly in the brine in which herrings have been cured. From this we come to dimethylamine, also a colourless gas, with the formula

$$N\left\{\begin{array}{c} CH_3 \\ \hline CH_3 \\ \hline H \end{array}\right.$$

i.e., two molecules of the methyl radical replacing two atoms of hydrogen in the ammonia. It has an ammoniacal smell, and condenses at 48 deg. C. into a mobile liquid. We now come to the product from which is obtained the steriliser of the Amines process, trimethylamine. Its composition is expressed by the formula,

$$N\begin{cases} CH_3 \\ CH_3 \\ CH_3 \end{cases}$$

or 3 molecules of CH_3 to one atom of nitrogen.

It is a mobile alkaline liquid, having a strongly ammoniacal smell, and when not pure has the odour of stale fish. It boils at 9·3 deg. C. It can be obtained from the flowers of the hawthorn and pear tree, and from many plants. It is also a product of the natural decomposition of nitrogenous animal and vegetable substances, and this is why herring brine contains it in such large quantities, thus forming a cheap and, so to speak, natural source of supply.

Dr. Klein, the Professor of Bacteriology at the College of State Medicine, London, and the author of an excellent treatise on this subject, has made several investigations on the sterilising properties of the Amines process, which are so full of interest that a short summary of them seems desirable. He made four examinations, one on sewage, one on press liquor, one on sludge, and one on the effluent. He took one cubic centimetre (about ½in. cube) of each of the four samples, and put them into test tubes. From No. 1, ordinary Wimbledon sewage, he took out 20 cubic millimetres, and mixed this very small quantity with nutrient gelatine spread on a large plate, food the microbes love and thrive upon. On the fourth day the number of colonies of bacteria was far too great to be counted. He then took a further small quantity of five cubic millimetres of original sewage, and diluted it 2,000 times with pure water. In six days he counted 24 colonies of bacteria, and estimated the number per cubic centimetre at 2,400,000 bacteria. He subjected No. 2, press liquor, to the same operations, and found a cubic centimetre to contain only 650 organisms. No. 3, after the same treatment, contained 400 organisms, whilst in No. 4, the effluent, none could be discovered. The number found in Nos. 2 and 3 is far below that found in ordinary London drinking water, for Dr. Percy Frankland found in a cubic centimetre of it 1,525 organisms. The number of living microbes in crude sewage he found averaged 768,000 per cubic centimetre. In another examination of Wimbledon sewage, operated upon by the Amines process some months later, he found the effluent to be quite free, and consequently sterile. He then examined sewage precipitated with lime only, and demonstrated that such treatment is not capable of producing the same effect as with lime and trimethylamine. Another most interesting examination of Dr. Klein's was to try the effect of the Amines process on several of the well-known forms of species of pathogenic microbes of more or less degrees of resistance. The bacillæ of typhoid fever, of diphtheria, of grouse disease, of anthrax, of scurf, and the micrococcus of ordinary suppuration, were totally destroyed. Of the bacillus of pneumonia, occurring by hundreds of colonies, only one colony was left after treatment. The streptococci of scarlatina and of erysipelas, by the same treatment, were affected to a very high degree—seven colonies out of 180 of the former and only one of the latter remained.

Mr. Cassal, public Analyst for the Parish of St. George, Hanover Square, Battersea and Kensington, reported in March last the results of his investigations at Wimbledon on the Amines treatment, to be as follow: 1. Clarification, rapid subsidence, and removal of offensive odour. 2. Removal of part of the organic matter. 3. The production of an effluent free from micro-organisms. 4. The production of a non-putrescible sludge. He states that this treatment applied to putrid pond water entirely destroys its offensive odour. Mr. Evelyn Pocklington, Officer of Health to the Wimbledon Board, states in his report that the process renders all the sewage products inodorous, and that the crops irrigated with the effluent water are quite as luxurious as those treated by other methods, and that the cost of the Amines treatment is not much in excess, if any, of that of lime and alumina. The men employed in the tanks, pumps, and wells find great benefit since the process has been adopted. Mr. Cooper, the Surveyor of the Wimbledon Local Board, replying to my questions on the 22nd December, informed me that the Amines process had been discontinued working there on account of the expense (a statement which I have since found is capable of another interpretation, and which I will explain further on). The system was resumed by order of the Board more than a year ago, and, I believe, has been since continued. In the Wimbledon printed report for the year ended March 25th, 1890, it is stated that during the months of July and August, 1889, permission was given to the Amines syndicate to conduct a series of experiments at the farm at their own expense, to demonstrate their system of sewage purification. Special demonstrations of the process were given before large companies of scientists, engineers, and representatives from Corporations and Local Boards in different parts of England. Owing to the successful experiments carried out by the Amines Syndicate with their process during the months of July and August, that process has since been applied to all the sewage chemically treated. The average cost of chemical treatment during the year amounted to 38s. 6d. per million gallons. With regard to Mr. Cooper's statement that the Amines process had been discontinued at Wimbledon on account of expense, I am informed by Mr. Hugo Wollheim that the following is an accurate explanation of the facts of the case. During the summer it was found that the sewage and sludge were freed from decomposition and smell by the Amines process, and the surveyor, thinking the result might be due to the lime only, and that the cost of the herring brine might be saved, obtained permission to try his own ideas during the winter; but by Christmas last it was found that the pipes conveying the sewage became clogged with fungoid growth, and the offensive smells of decomposing sewage returned even in the cold weather. The board then ordered the Amines process to be resumed. Mr.

Wollheim showed me a copy of the *Surrey Comet* newspaper, from which I copied the following confirmatory passage: "On the motion of Mr. Holland, seconded by Mr. Hellier, it was decided to re-adopt the Amines process at the Sewage Farm. The chairman said that, having tried the surveyor's scheme, the committee came to the conclusion that it would be better to fall back upon the Amines system pure and simple." This discovery deserves the serious attention of all men engaged in sanitary improvements. Independently of its aspects from municipal economists' points of view, it aims at the destruction of those causes of disease which are now so demonstrable, and which are known under the term of "the germ theory." If the process will accomplish but half what it claims to do, then zymotic and other infectious diseases can be kept at bay, and indescribable danger to populations in summer from the poisonous and putrefying masses of sewage sludge, to be seen in close proximity to such populations as Birmingham, Stoke, &c., so reduced as to cause no alarm or danger.

Laboratory Examination of the Amines Process.

On the 30th January I again, by arrangement, visited Mr. Hugo Wollheim, and spent the day in his laboratory. I was met by him and his colleague and chemist (Mr. Leo Taylor, F.C.S.), with whom I thoroughly investigated the efficiency of the Amines process as far as a chemical and microscopic investigation could take us. The first thing we did was to pour a small portion of Leek sewage into a beaker, from which we took a very small drop and placed it under the microscope, using a one-sixth objective, which revealed the presence of various kinds of bacilli and bacteria, as well as numerous spores or germs. There was also a number of protococci, but we did not observe any spirillæ. We then examined them with a one-twelfth objective, and identified several well-known species. Most of them were mobile, and apparently upon fæcal or other decaying particles. Their number in such a minute quantity of fluid was simply astounding. Some were motionless, but most of them were moving about freely, some slowly, others quickly. When we consider the rapid multiplication of these minute organisms, so rapid and prolific that in a few days, under favourable circumstances, it is impossible to compute their rate of increase and quantity, by such millions do they multiply, a serious reflection is forced upon us, namely, that this phase of this difficult subject is one that should have prior and most thoughtful consideration, and that there need be no wonder in finding serious health deterioration and disease wherever animal or vegetable refuse is permitted to accumulate and dry. After our microscopic examination of the raw sewage, we then poured out the gallon of sewage, which was considered a fair average sample. We then took about 50 grammes of freshly-burnt chalk lime, allowing about 10 per cent for earthy and silicious impurities in the lime. No more water was added to the lime at first than was necessary to dry-slake it. When the lime had fallen to powder, water was added so as to make 10oz. of milk of lime. To this we added 2 drams of herring brine, or at the rate of 3 grains per gallon of sewage. This mixture was well stirred, and gradually the new product was evolved, to which the name "Aminol" has been given. It became very perceptible by its odour, which differs considerably from that of herring brine or of trimethylamine. The gallon of sewage having been poured into a large vessel, we added to it 1oz. of the mixture of lime and herring brine. This proportion is greater than is necessary for towns having soft water. The mixture was then stirred. A dense flocculent precipitate began immediately to form, which in half an hour entirely fell to the bottom of the glass vessel, leaving the supernatant liquid quite clear but with a slightly yellowish tinge, which Mr. Taylor thought might be due to dye wash in the sewage. We then examined the clear liquor of effluent water for microbes, but not a trace of bacteria, bacilli, spores, or spirillæ could be detected, All the living germs, organisms, and suspended matter had fallen with the precipitate. This effluent was apparently pure enough to be turned into a stream or river or run on land. The use of lime to sewage is for clarification; that of herring brine for sterilisation. When lime alone is used, there must always be secondary precipitation, and as the microbes and germs are not destroyed by lime, they set up subsequent putrefaction in the effluent or in the watercourses or rivers by their rapid multiplication.

I make no apology for this lengthy review of the Amines process, because if such refuse matters are really and effectually sterilised and decomposed by it, the most important part of the problem of sewage treatment is solved, or at any rate it is shown that it is possible to disinfect sewage and prevent its being a danger to life and health. The greatest sanitary benefits are thus conferred on populations living in crowded areas, as well as by preventing cattle contagion and milk and flesh contamination.

So important are these considerations to the community at large, that no county council would be free from responsibility and blame if they forced any town to adopt any scheme hurriedly or without reassurable time being given for this and the other new discoveries being practically tested, especially when on examination of existing systems we find unmistakable evidence of imperfections, for which a remedy must be found, or costly failures, which must be avoided.

Notwithstanding the efficacy claimed for this system, it does not seem to be received with popular favour.

It formed one of the Salford trials, and was tested for some time on 100 thousand gallons daily; the results will be published in the second report of the sewage experiments conducted at the Salford Borough Sewage Works.

CHAPTER VII.

The Electrical Treatment of Sewage, or the Webster Electrolysis Process.

Of all the many ideas for the purification and disposal of sewage, that of the Electrical Purification Association, 22, Great George Street, Westminster, is certainly one of the most interesting. It has at any rate the merit of novelty, and is upon a strictly scientific basis. A year ago I went to London to inquire fully into the system, and have thoroughly investigated its action and manipulation upon a gallon of sewage, sent up to me there by the Sanitary Inspector of Leek in December, 1890, in the laboratory of the Company. The system has received the attention of eminent scientific men. Professor Armstrong, F.R.S., first informed me of it. He said, without pronouncing upon its merits, he was of opinion that it deserved serious attention. Sir Henry Roscoe has thoroughly investigated its powers, and further on I will give his opinion, and results. It is the discovery of Mr. Webster, F.C.S., contractor and former pupil of Dr. Letheby. Being of opinion that chemical precipitation was imperfect and inadequate, it occurred to him that the electric current, properly applied, would split up the component parts of sewage in solution, and effect an economical purification. After a long series of laboratory experiments, he found that oxidisable metallic plates produced the desired result. To avoid poisonous effects on land or in rivers, he selected iron as the best, as it is the cheapest, metal. Iron in the state of oxide has many valuable qualities, one of the chief being that sulphuretted hydrogen cannot exist when ferrous or ferric oxides are present. As an example, its action is well known and practised in freeing coal gas from sulphuretted hydrogen. The success of the experiments led Mr. Webster, and a number of friends whom he had convinced of the efficacy of his application, to set up plant at their own expense at Crossness, near the southern outfall of the Metropolitan sewage into the Thames. The operations there have been conducted for several months on a large scale. They find that cast-iron plates of the commonest quality employed as electrodes give the best results. The operations are most simple—in fact, just like the electrolysis of water—*i e.*, the decomposition of water into its elements, oxygen and hydrogen, by a galvanic current. The sewage flows through channels in which are placed these iron plates, set longitudinally, with the usual battery connections with the positive and negative terminals of a dynamo. Mr. Webster states that what occurs is as follows: The sewage, or other impure liquid, in its passage through these channels becomes entirely split up by the electric action. The matters in suspension in the sewage and part of the organic matter are not only removed by precipitation, but the soluble organic matter is oxidised and burnt up by the nascent oxygen and chlorine oxides evolved, and this oxidisation may be carried to any extent, according to the amount of purification required. The chemical changes that take place in sewage when it is electrolysed depend chiefly upon the well-known fact that sodium, magnesium, and other chlorides (which are always present in sewage) are split up into their constituent parts. At the positive pole the chlorine and oxygen given off combine with the iron to form a salt, which Mr. Webster believes is a hypochlorite of the metal. It is stated by the patentees that during the chemical action carbonate of iron exists in solution, and its formation is due to the presence of carbonates in the sewage, chiefly carbonate of ammonia.

In samples which are absolutely free from dissolved oxygen the ferrous oxide in the white form is precipitated, and on shaking up with air it changes to the usual pale green colour. The carbonate of iron at the same time being oxidised, the ultimate precipitate is red oxide, known as ferric oxide (Fe_2O_3), and it is noticed that sometimes this changes after a time back to the ferrous state (Fe O), thus showing that it has acted as a carrier of oxygen to the organic matter present. The exact chemical re-actions which take place are, however, impossible to define.

It is calculated that thirty horse power is, as a rule, sufficient for the full treatment of one million of gallons per 24 hours of sewage, or 25 ampère hours per gallon. Crossness sewage, treated with 25 ampère hours per gallon, showed a reduction of organic matter in solution which cannot be obtained by chemical precipitants except at a prohibitive cost, besides entailing a large addition to the bulk of sludge and organic matters in solution, which inevitably produce secondary putrefaction. A point of immense importance is that by this process the bulk of sludge is reduced to a minimum. Where suitable land is available for irrigation or filtration, a smaller reduction of organic matter in solution is sufficient; settling tanks are unnecessary, and the expense of the electrical treatment can be much reduced. Where suitable land is not available for this purpose, settling tanks are required. After the separated solid matter has been allowed to precipitate, the effluent so obtained is fit to discharge into any river. This effluent contains about three grains of suspended matter per gallon, which, as it consists almost entirely of iron oxide, is quite innocuous. It will be gathered from the foregoing description that this process aims more at sewage disposal than sewage utilisation. In the first place, the sludge has to be got rid of, but as there is less sludge left behind by this treatment, owing to no precipitating ingredients being added, as in other processes, the getting-rid-of necessity is consequently less than in the other, a fact of importance when it must be conceded that ordinary sewage sludge is not of very fertilising utility, and, as a rule, has but a very small marketable value, if any, in large quantities. The sludge of this process would have to be carted away to be filter-pressed, and perhaps dried to recover the oxide of iron, which could be afterwards used for gas or sewage purification. Precipitation is said to take place more quickly by this method than by any other, but I have found it is not so quick and thorough in my laboratory trials as by the Amines process. But the effluent soon separates, and can be run off with ease, and used for irrigation or turned into a river. In either case it is said to show no signs of after putrefaction. This process is found to destroy microphytic life, but it does not claim a sterilising power. Mr. Webster found, when studying under M. Pasteur, that the number of organisms was reduced from five millions per cubic centimetre to six hundred, a mighty massacre, it is true, but, not being complete, the numbers would very soon be made up, even to the amount of several millions, by the next day, such being his estimate of their rapid microbe growth.

The necessary plant consists of channels for the iron plates, copper conductors and measuring instruments, dynamos, engines, and boilers. For a town of 30,000 inhabitants, 450 tons of iron would be required, and thirty horse power for treating one million gallons of sewage per 24 hours. The quantity of iron is estimated as a ten years' supply, the consumption of iron being about 45 tons per million gallons per annum. The cost of plant for treating one million gallons per 24 hours of sewage, similar to that of London, not including iron, is about £2,000, providing for three engines and dynamos and two boilers, but any two engines and dynamos and one boiler will do the work. The coal consumption will be about 2lb. per indicated horse power. The cost of maintenance includes only coal and iron consumed, and labour. Two shifts of two men each would suffice for a million gallons.

This new company has not been afraid of scientific investigation. They have called in Sir Henry Roscoe, who has made an elaborate examination of the system and its results, from which I extract the following particulars. He has made a series of careful experiments of their treatment at the Crossness sewage. He operated upon 20,000 gallons of sewage altogether, collected from the flow every five minutes at the entrance into the shoot in which the sewage was treated electrically—that is, before it was treated; they were then mixed up together. The effluent he examined was drawn from the settling tank, 12 inches from the bottom. He states that the reduction of organic matter in solution is the crucial test of the value of a purifying agent, for unless the organic matter is reduced, the effluent will putrify and rapidly become offensive. To ascertain the extent of this reduction, it will be necessary to compare the analysis with the following result: Percentage reduction of organic matter, as measured by oxygen, absorbed from permanganate of potash after three hours.

Oxygen absorbed. Per Million Parts.	Albuminoid Ammonia. Per Million Parts.
70·3.	64·6.

The highest of these analyses gave 77·2 of oxygen, absorbed with 75·0 per cent of albuminoid ammonia, giving a mean of 70·3 oxygen absorbed, and 64·6 of albuminoid ammonia, as I have just stated. He further states: "I have not observed in any of the unfiltered effluents from this process which I have examined any signs of putrefaction, but, on the contrary, a tendency to oxidise. The absence of sulphuretted hydrogen in samples of unfiltered effluent, which have been kept for about six weeks in stoppered bottles, is also a fact of importance. The settled sewage was not in this condition, as it rapidly underwent putrefaction, even when in contact with air, in two or three days. The effluents as collected from the tanks were found to contain a small amount of iron in solution in the ferrous state, and on exposure to air ferric hydrate was precipitated. In order to remove this iron, it was arranged to pass the effluent from the precipitating tank over and through a filter of sand, and by this means oxidation and precipitation of the soluble ferrous salt was brought about, and effluents were obtained, free from iron, of bright appearance, but in no degree chemically purer as regards organic matter than in the unfiltered effluent. The results of this chemical investigation show that the chief advantages of this system of putrefaction are: 1. The active agent, hydrated ferrous oxide, is prepared within the sewage itself as a flocculent precipitate (it is scarcely necessary to add that the inorganic salts in solution are not increased, as in the case where chemicals in solution are added to the sewage). Not only does it act as a mechanical precipitant, but it possesses the property of combining chemically with some of the soluble organic matter, and carrying it down in an insoluble form. 2. Hydrated ferrous oxide is a deodoriser. 3. By this process the soluble organic matter is reduced to a condition favourable to the further and complete purification by natural agencies. 4. The effluent is not liable to secondary putrefaction, and it would fertilise land. Filtration is therefore entirely superfluous, except to remove sentimental objections, or where an effluent has to be discharged into a trout stream, the unfiltered effluents being, as shown by the above report, as chemically pure as the effluent when filtered."

The following table of Sir Henry Roscoe, F.R.S., of Analyses of the Raw Sewages and Effluents, expressed in grains per gallon, are extremely interesting and instructive.

Analyses of Raw Sewages and Effluents,
Expressed in grains per gallon.

Experiment.	Date.		Oxygen absorbed from Permanganate of Potash.	Albuminoid Ammonia.
	1889.			
1	August 13	Raw Sewage	5·64	·87
		Unfiltered Effluent	1·92	·45
2	August 16	Raw Sewage	4·40	1·17
		Unfiltered Effluent	1·41	·46
3	August 21	Raw Sewage	9·03	1·17
		Unfiltered Effluent	3·18	·50
4	August 26	Raw Sewage	6·34	1·12
		Unfiltered Effluent	1·45	·28
5	August 29	Raw Sewage	4·81	1·12
		Unfiltered Effluent	1·32	·56
6	August 30	Raw Sewage	5·30	1·30
		Unfiltered Effluent	1·85	·60
7	September 3	Raw Sewage	5·10	·84
		Unfiltered Effluent	1·20	·105
8	September 5	Raw Sewage	5·49	1·12
		Unfiltered Effluent	1·48	·14

It is a little perplexing that some chemists express their results in different ways, some in grains per gallon, others in parts per 100,000, and others in parts per million. Sir Henry Roscoe's table is no exception to this, and it may help to simplify comparisons if I give the formula for converting grains per gallon into parts per 100,000. A gallon weighs 70,000 grains; therefore grains per gallon is really an extension of parts per 70,000, and to transform grains per gallon—*i.e.*, parts per 70,000 into parts per 100,000—divide by seven and multiply by ten, and conversely, to bring parts per 100,000 into grains per gallon, divide by ten and multiply by seven. For example: Raw sewage, 5·64 grains per gallon, multiply by 10 = 56·40, divide by 7 = 8·057 parts per 100,000.

ANALYSES OF RAW SEWAGES AND EFFLUENTS,

Expressed in grains per gallon, parts per 100,000, and in parts per million.

Experiment.	Date.		Oxygen absorbed from Permanganate of Potash.			Albuminoid Ammonia.		
			Grains per gallon.	Parts per 100,000.	Parts per million.	Grains per gallon.	Parts per 100,000.	Parts per million.
	1889.							
1	August 13	Raw Sewage	5·64	8·057	80·570	·87	1·243	12·430
		Unfiltered Effluent ...	1·92	2·743	27 430	·46	0·743	6·430
2	August 16	Raw Sewage	4·40	6·286	62·860	1·17	1·671	16·710
		Unfiltered Effluent ...	1·41	2·014	20·140	·46	C·657	5·570
3	August 21	Raw Sewage	9·03	12·00	129·00	1·17	1·671	16·710
		Unfiltered Effluent ...	3·18	4·543	45·430	·50	0·714	7·140
4	August 26	Raw Sewage	4 81	6 871	68·710	1·12	1·60	16·00
		Unfiltered Effluent ...	1·45	2·071	20·710	·28	0·40	4·00
5	August 29	Raw Sewage	1·81	6·871	68·710	1·12	1·60	16·00
		Unfiltered Effluent ...	1·32	1·886	18·860	·56	0·80	8·00
6	August 30	Raw Sewage	5·30	7 571	75·710	1·30	1·90	19·00
		Unfiltered Effluent ...	1·85	2·657	26·570	·60	0·857	8·570
7	Sept. 3	Raw Sewage	5·10	7·286	72·860	·64	1·20	12·00
		Unfiltered Effluent ...	1·20	1·714	17·140	·105	0·150	1·500
8	Sept. 5	Raw Sewage	5·49	7·843	78·430	1·12	1·60	16·00
		Unfiltered Effluent ..	1·48	2·114	21·140	·14	0·20	2·00

This system has also been reported upon by Mr. Alfred E. Fletcher, F.C.S., F.I.C., H.M. Inspector of Alkali Works, and Inspector under the Rivers Pollution Prevention Act for Scotland, in November, 1889, from which I gathered the following points and information: The shoot conveying the sewage and holding the iron plates is of wood, 2 feet deep by 18 inches broad, and 400 feet in length, and leads to a tank 3 feet 4 inches deep, containing 25,000 gallons. This long wooden shoot is filled with wrought-iron plates one-eighth of an inch thick, and one inch apart. They are connected alternately with the positive and negative conductors from an electric generator or dynamo, driven by a steam engine. In thus operating with sewage, two agencies are at work—the action of electrolysis, or precipitation and decomposition of organic matter by electricity, and the formation of a hydrated oxide of iron. One of Mr. Fletcher's experiments was to subject the raw sewage to a lime precipitating treatment with various salts of iron, so as to compare the results with that of the Webster electrical treatment. He states that the results obtained were not nearly so good as by the electrical method. This would prove, what must be extremely probable, that reactions take place by the forcible electric current, which cannot occur in the quiescent lime and iron salts methods, and makes the action more complete and absolute, and prevents further decomposition.

Mr. Fletcher sums up his remarks as follows: "The result of my examination of this process has been to convince me of its efficiency in clarifying sewage, of removing smell, and in preventing putrefaction of the effluent. I am of opinion that such an effluent as I saw at Crossness can be discharged into a river, or after passing through a thin layer of sand even into a stream, without causing any nuisance."

It was noticed that while the raw sewage filters very slowly, 500cc. requiring 96 hours to pass through a paper filter, the electrically treated sewage settled well and filtered rapidly.

Where irrigation is carried out, it is obvious that this rapidity of filtration is of incalculable value.

Samples of the raw sewage, having but little smell when fresh, stunk strongly on the third day. The treated samples, however, had no smell originally, and remained sweet, without putrefactive change.

The treatment causes a reduction in the oxidisable matter in the sewage varying from 60 to 80 per cent. The practical result of the process is a very rapid and complete clarification of the sewage, which enables the sludge to separate freely. After the treated sewage has been allowed to settle for two hours, the effluent may be passed through a filter of sand, &c., and discharged bright in appearance and free from offensive smell.

The following table shows the results of his analyses after treatment with Webster's electrical process :—

Description of Sample.	Oxygen consumed.	Free Ammonia	Albuminoid Ammonia		Chlorides	Notes
(A) Mixture of 3 samples of raw sewage collected during time of treatment	22·920	19·0	6·50	Probably low, as the samples had been kept.	444·0	Stunk badly on 3rd day
(B) Mixture of 3 samples of treated sewage flowing into tank before settling	9·10	7·0	6·0		372·0	Faint smell on 5th day
(C) Same as sample A after settling 24 hours			4·0			Stunk badly on 3rd day
(D) Same as sample B after settling 24 hours			1·20			Faint smell on 5th day
No. 7 Average sample of raw sewage under treatment	25·0	18·0	6·50		412·0	Stunk badly on 3rd day
No. 8 Treated sewage after 2 hours settlement in tank...........	5·360	10·80	0·70		358·0	Perfectly sweet after 10 days
No. 9 Same as No. 8 after filtration through 6in. of sand.........	5·360	10·00	0·60		356·0	ditto
No. 10 Treated sewage after 3½ hours settlement in tank	5·0	12·80	0·50		356·0	ditto
No. 11 Same as No. 10 after filtration through 6in. of sand...... ..	5·0	9·10	0·10		356·0	ditto

I fitted up an experimental apparatus in my laboratory, consisting of a six-cell bichromate battery and a glass jar containing three thin plates of hoop iron connecting the anode and cathode with their respective battery poles, and have perfectly succeeded in proving the efficacy of this novel method of precipitating the organic matter of sewage. The Corporation of Salford decided to put this process to the test in a practical way, and called upon the company to provide appliances for that purpose at their works at Mode Wheel. This has been carried into effect, and the process thoroughly tested, and although these practical and costly tests will not be able to show us the results extended over years of working, they will be most valuable guides to us, I think, in showing us the results on such a thorough and comprehensive scale as to help so important a Corporation as that of Salford to avoid the mistakes of the past and to adopt the best method. The Electrical Association inform me that they consider the sewage there treated to be one of the worst combinations which a corporation could have to deal with. Various other processes were also treated at Mode Wheel, namely, that of the International Water and Sewage Purification Company Limited, Messrs. Spence and Co.'s Alumino-ferric processes, and Messrs. Barry's process.* The whole of the work was under the direct control of A. Jacob, B.A., &c., borough engineer, and J. Carter Bell, A.R.S.M., county analyst, and was, at the same time, independently investigated on behalf of the Corporation by Messrs. John Newton and Son, civil engineers, Mr. Charles A. Burghardt, Ph.D., scientific chemist, and, for the purpose of checking the electrical cost, Messrs. Mather and Platt, electrical engineers. Dr. Burghardt, who reported on all the processes tested, concludes as follows:—

"After studying closely the results of some laboratory experiments, and the uniformity of its working, I am of opinion, as a

* Other processes have also been subsequently tried, and will be described further on.

final result of my investigation, that the most reliable process is that of the Electrical Purification Association Limited."

Messrs. Mather and Platt state, after giving the figures obtained during the trials, that the indicated horse power required for treating the foul Salford and Pendleton sewage is 40 indicated horse power per million gallons. They add that during the period the trial was made there was no appearance of anything indicating a practical difficulty in the method as regards supply of electric current, or anything to suggest undue wear and tear of plant.

I refer the reader to the Report of the Salford Corporation for the experimental results of this system at Mode Wheel Sewage Works.

CHAPTER VIII.

THE PROCESS OF THE INTERNATIONAL WATER AND SEWAGE PURIFICATION COMPANY, OF 7, VICTORIA STREET, LONDON, S.W.

THE last of the three systems particularised on p. 47 comes next for description. I have had the satisfaction of seeing it in successful practical operation on a considerable scale, and have therefore been placed in such an advantageous position as to enable me to describe it in its relationship to the needs of a population; so that this chapter could have for its title either "The Ferozone and Polarite System," or "A Description of the Acton Sewage Works, near London."

I visited the works in question, by permission of the Acton Local Board, on Thursday, the 11th of December, 1890 (the only day in the week set apart for their inspection), I having received by post a card of admission a few days previously. I met Mr. Ebbetts, the engineer and surveyor to the Local Board, who assisted me in answering the questions on page 174. These works are by far the most complete I have yet visited, and, should the analysis of their effluent give good results, permanently and regularly, certainly the most effective. The Local Board calculate their population—a residential one mainly—at six persons per house, and in this way arrive at a total of nearly 7,000 persons.

Their system of sewage purification is applicable to a much larger population; their present pair of boilers, each 21 feet by 6 feet, and their two engines, each of 30 horse power, with a vertical pump to each, forming part of the engine next to the flywheel, being capable of doing the work for 20,000 to 25,000 persons. The system being purely a chemical one, or rather that of precipitation and filtration, their Local Board have no need to trouble themselves with any kind of irrigation further than to sell their compressed sludge for agricultural purposes. This is arranged by contract to one farmer, who takes the whole, which amounts to eight or nine tons per week of pressed cake. The sludge also can be, and a little of it is, dried from cakes and ground into powder for grass land, the cakes being ploughed into corn land. They deal with 425,000 gallons per day of 24 hours, which is an average of 63½ gallons per inhabitant. This average being very much larger than that of most towns per person is accounted for by the fact of there being upwards of sixty laundry establishments, large and small, in Acton, which use a large quantity of water.

Briefly, their system is this. The low level sewage is made to arrive into a large well, from which it is pumped to a height of 30 feet, the level of the tanks, into a square brick chamber with such force that it is dashed about so as to break it well up. It runs in brick open channels, where the remaining sewage, which runs down from the high level sewers by gravitation to meet it, and where two 1 inch jets convey ferrozone, ground in water, into them as it streams past into the tanks. There it is left to precipitate and settle for 1½ hours to 2 hours only. Then the top water is drawn off into the adjoining filter bed, and admitted by a mushroom-shaped admission, one in the middle of each of the two separate divisions in the filter bed being placed at the top of the filtering media. It passes through the filter bed and runs off perfectly clear into a sewer, and eventually finds its way into the Thames near Kew, where it is not objected to. The filter bed is contained in a large tank of blue brickwork, 72 feet long by 26 feet broad, and is thus arranged: At the bottom of the tank is placed a series of 4 inch agricultural drain pipes laid in parallel rows 4½ inches apart in width. The pipes are filled in between and just covered by rough balled gravel. Upon this is placed a 4 inch

Plate 7. SKETCH PLAN OF THE ACTON SEWAGE WORKS AND SECTION OF THE POLARITE FILTER BEDS.

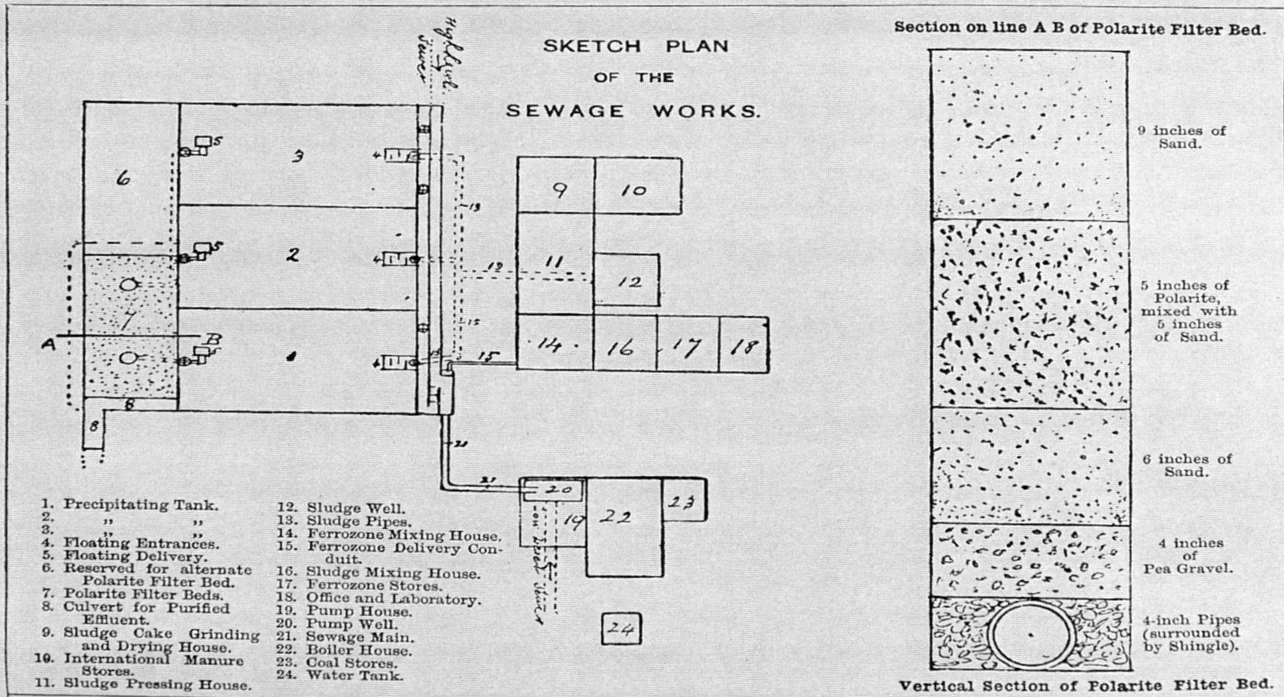

bed of pea or flint gravel, then a 6 inch bed of sand, then a bed of half sand and half polarite mixed intimately together, and then a bed of 9 inches of sand, upon which the liquor flows from the precipitating tanks and filters through rapidly. The filtering tank is divided in the middle, the two compartments being used alternately, one being reserved for cleaning, which is done every other day in a very simple and effectual manner by skimming off a thin layer of blackish deposit and a little sand, to the depth only of ¼ inch to ½ inch, with a skimmer gauged to take off only a stratum of that thickness. The accompanying plan of the Acton sewage works, kindly lent to me for this work, will show the arrangements and the system more clearly.

SLUDGE.

The treatment of the sludge is as follows: After the supernatant liquid has been run off to the filter bed, there remains the precipitate, or sludge at the bottom of the tanks, two of which tanks are used at once, and one is being cleaned. The sludge is pushed into the culvert by men using indiarubber sweepers, where it runs down into a sludge well, from which it is pumped by a double plunger pump into an iron agitating tank, to mix it well together and to make it flow better. It then runs to another pump, a forcing pump, which forces it into and through a large filter press, which forms the more solid parts into cakes, and the water pressed out of it runs away into the sewage well, and is pumped up again with the sewage. Six hundred pounds of sludge lose three hundred of water. There are two filter presses.

The ferozone used to precipitate the organic matter is a peculiar iron ore from Wales treated with sulphuric acid. It is ground up in a mill with water or sewage into a creamy paste. There are duplicate grinding mills, one in use and a spare one. The proportion of ferozone used is 7 lbs. to each ton of sludge and 1½ cwt. of pressing powder, the composition of which I am unacquainted with. It is supplied by the same company, who supply both the ferozone and the polarite from their Welsh works.

The pressing powder was at first used to retain the ammonia in the sludge. The company now find that a special powder for sludge pressing *is* unnecessary, as the ferozone treatment at the tanks is quite sufficient. The Acton Board, when I was there, had just begun to try Buxton newly-burnt lime with the pressed sludge instead of this pressing powder, on account of its greater cheapness. The pressing powder cost £2 5s. per ton delivered, the lime £1 1s. per ton delivered. Besides, the cost of pressing the sludge into cakes is only 1s. 4d. per ton, against 4s. per ton in the old way of pressing powder.

Out of 425,000 gallons of daily sewage at Acton 350,000 have to be pumped, 75,000 gallons running into the tanks by gravitation from the high level sewers. The total cost of pumping this quantity is 2s. 6¾d. per head of population. The district rate is 1s. 10d., and covers everything, police, &c., except poor rate. The total annual expenditure or cost of maintenance of £797 includes wages, chemicals, oils, fuels, wood, repairs, poor rate, income tax, insurance, water, tithe-rent charge, sand, and every other cost.

There is no real nuisance arising from these works. There was a little smell where the sewage emerged from the pumping, but none at the tanks. The manager said there was no smell in summer, and the system was perfectly satisfactory, and there was no nuisance. The polarite used in the filter beds, the analysis of which I give further on, is a ferric mineral containing, or being mixed with, carbon. This is manufactured by the Company at their works in South Wales, where the ferozone and pressing powder is also prepared. The present filter beds would not be large enough for a larger population, but the precipitating tanks are large enough for a much larger population. They are considering the construction of an additional filter bed.

The cost of adapting the works, originally designed for the A B C Company's system, to the International Company's system was very small indeed, amounting in fact to less than £500 for the polarite bed.

The initiation of this system included cost of land for works, the works themselves, tanks, machinery, and building, besides an entirely new sewerage of the town. The old system of sewerage now takes the surface water—that is, the rain water, and any other surface or street water.

It may now be well to show what the International Water and Sewage Purification Company, established in 1884 (whose system in its entirety is adopted and worked at Acton), claim for it, and in further explanation to show of what it consists. It will also be of interest to quote Sir Henry Roscoe's and Dr. Frankland's opinions, each separately given, of this system, which, in consideration of their importance, I will give first:—

In July, 1890, Professor E. Frankland, Ph.D., D.C.L., LL.D., M.D., F.R.S. (one of Her Majesty's Commissioners to inquire into the pollution of rivers), reported to Major Tulloch, R.E. (the chief engineering inspector of the Local Government Board), the results of this investigation of the above-named company's process as carried out at the Sewage Works, near Acton, where the crude sewage is first treated with ferozone in subsidence tanks for precipitation, and the effluent therefrom is passed through polarite filter beds for final purification. His report ran as follows:—

"These results show that the raw sewage contained a very large proportion of highly polluting suspended matter, and an unusually large amount of foul organic matter in solution; and further, that the effluents from the subsidence tank and filter were derived from sewage of about equal polluting power as regards dissolved organic matter. In the subsidence tank, the suspended matter was reduced from 240·80 parts per 100,000 of raw sewage to 5·92 parts per 100,000 of tank effluent, whilst the effluent from the filter was free from suspended matter. It was clear and transparent. This is a satisfactory result. The effect upon the dissolved organic matter in the subsidence tank is very remarkable, its amount being reduced to little more than one-tenth of that present in the original sewage. In its subsequent passage through the filter, the dissolved organic matter is still further reduced to nearly one sixteenth of that present in the original sewage. It is now in a state of purity greatly exceeding that prescribed by the standards of the Rivers Pollution Commissioners. No chemical process of purifying sewage has ever, in my experience, approached this in efficiency; and if the results obtained at Acton can be accomplished in other places, a most important advance will be made in the purification of the sewage of towns. I need scarcely add that the effluent from the filter is not only clear, but inodorous and inoffensive. It is, of course, not fit for dietetic purposes, but it may be admitted in large volumes into running water without any nuisance."

The following is a condensed version of the report made to the company by Sir Henry E. Roscoe, M.P., F.R.S., LL.D., Ph.D., Professor of Chemistry, &c.:—

"August 20th, 1888.—I have, at your request, investigated, and now beg to report upon your process for the purification of sewage as carried out at Acton. For this purpose I visited the works at Acton, on Monday, July 9th, and found that the sewage ran along a channel, into which was flowing a solution of your precipitant ferozone. This mixture flows with the raw sewage into a large precipitating tank of 138,000 gallons capacity. Of these there are three side by side. Subsidence is allowed to take place for a period of from one to three hours, after which the top liquid is run off from the surface by means of a floating arm directly on to the filter bed, which is made up of sand as a top layer, then a layer of your preparation known as polarite, the bottom being made up of coarse gravel. I found the resulting filtered effluent to be bright, clear, and colourless, and free from any objectionable smell. Many samples of the tank liquid and the filtered effluent were taken by Mr.

Lailey, the engineer to the Acton Local Board, and also by my assistant, Mr. Joseph Lunt, B.Sc., and these were analysed. I have always found that the filtered effluent was equal in appearance to the sample taken on the day of my visit, viz., bright, clear, colourless, and free from objectionable smell. Many samples of this filtered effluent have been analysed. The average results obtained are given in the following figures:—

ANALYSIS OF FILTERED EFFLUENT.

	Grains per gallon.		Parts per 100,000
Chlorine	4·480	...	6·400
Free ammonia	0·284	...	0·407
Albuminoid ammonia	0·018	...	0·022
Oxygen absorbed from permanganate in four hours	0·228	...	0·325

"Samples of the effluent have been kept in absence of air for nearly two months, and have still retained their good qualities. They exhibit no trace of putrefactive change, but have rather undergone improvement, inasmuch as in most cases a slight green growth has taken place, showing that the effluent is in a condition to be acted upon by the natural influences which will tend further to improve the water whilst flowing along a stream.

"But it is to be observed that the amount of organic matter, as measured by the quantities of the two forms of ammonia, is above the limit of a safe drinking water. Hence, while the effluent is an excellent sewage effluent, and may be sent into the river at the present outflow without danger, it is obviously unfit for drinking purposes.

"I consider the removal of all the suspended matter to be a very good point in favour of the system. Another is the intermittent nature of the filtration. This is assisted, and the filter kept in good order, by frequent raking over the surface, removing the surface layer of sand, and replacing it by washed and clean material. The porous nature of the oxide, which is used in the filter, its complete insolubility, and its freedom from rusting, constitute, in my opinion, its claim to be considered a valuable filtering material. A sample of this filtering material, taken at Acton, gives the following results:—

ANALYSIS OF POLARITE.

Magnetic oxide of iron	53·85
Silica	25·50
Lime	2·01
Alumina	5·68
Magnesia	7·55
Carbonaceous matters and moisture	5·41
	100·000

"The material called ferozone, used for mixing with the sewage, contains mainly sulphate and magnetic oxide of iron. About eight grains of ferozone per gallon of sewage is added during the flow into the precipitating tank, and after from one to three hours' subsidence the liquid is drawn off. The sludge remaining is run into a well, whence it is pumped into a press, and the pressed cakes dried in a current of hot air from a coke fire. The sludge thus dried is not liable to offensive decomposition. I have come to the conclusion that the precipitation process alone is not capable of producing a bright, clear effluent, nor complete deodorisation. This is especially the case where the sewage is strong, or is undergoing incipient putrefaction. Hence filtration is necessary for obtaining a satisfactory effluent. As the sewage at Acton is received in the fresh state, the liquid has not yet lost its dissolved oxygen, nor has the greater part of the organic matter been broken up into soluble, and therefore more hurtful, products. The larger portion of this organic matter is, therefore, removed partly by natural subsidence and partly by the precipitate formed by the ferrous sulphate. The filtration removes all the suspended matter, and a further purification is effected by the porous magnetic oxide, which, on absorption of organic matter in its pores, oxidises it by help of the dissolved oxygen existing in the liquid. The filter beds of Acton have now been in use for fourteen months, the only cleansing having been the removal of the surface layer of sand.—I am, gentlemen, yours truly (signed), HENRY E. ROSCOE, F.R.S."

Referring to Sir Henry Roscoe's analysis of the filtered effluent, I would observe that the figures given are in parts per 100,000. It is also customary to estimate the organic matter, &c., in parts per million. These figures would, therefore, require to be multiplied by 10, giving the following results:—

Chlorine	64·00
Free ammonia	4·06
Albuminoid ammonia	·250

The Sanitary Committee of the Borough of Salford visited Acton, in August, 1889, and took a sample of effluent. They sent it to my friend, Mr. J. Carter-Bell, A.R.S.M., F.I.C., &c., County Analyst for Chester, Salford, &c., with the following result and report:—

	Parts per million.		Parts per 100,000.
Free ammonia	0·160	......	0·016
Albuminoid ammonia	0·150	.. ...	0·015

"This is a first-class effluent; many well waters in this country are not so pure as this." It must have been collected at a favourable moment. A year later, another and more exhaustive analysis was made by Mr. Carter Bell with samples he took himself from the Acton Sewage Works on Saturday, August 30th, 1890. "The effluent was flowing exactly like clear spring water, and when the samples were put into the bottle there was not the slightest odour of any kind, and it would have been impossible for anyone to have said, judging by the appearance, that it was not good drinking water, and the analysis also shows that the reduction of the putrescent matter has been very great. Such an effluent as this may be run into any stream with perfect safety. I am surprised to find that after more than three years' working, the filter beds can produce such a good effluent.—Signed, J. CARTER-BELL."

It is noteworthy to observe that the albuminoid ammonia in the sewage being 7·000 parts in a million and only ·150 in the effluent, a reduction has been effected by these processes of 6·650 parts in a million—that is, all the putrescible matter has been removed save one-twentieth part only; *i.e.*, one-seventh part only of the amount suggested as permissible by the Royal Commission of 1868. This degree of purification far exceeds the suggestions of the Rivers Pollution Commission of 1868, which are two parts of organic carbon per million, and three parts of organic nitrogen per million, or their equivalent. It is effected in the most needed as well as in the most difficult direction, producing an effluent purer than the legal standard, and fit to be turned into a river without any fear of causing a nuisance.

It will be observed in the replies given to the queries at the end of my report, that the cost of the main drainage, sewerage and outfall works at Acton has been £150,000. The £150,000 includes the main sewers put down throughout the district—a very large and long one—in which, on account of an injunction obtained against the Board by the Metropolitan Board of Works, they had to run sewers a very long way to pick up a small number of new houses. Roughly speaking, 20,000 people drain into the Metropolitan system, 6,000 drain into their works, and new houses as erected are joined on to their works. The expenditure on the sewage disposal works and machinery, which is, with the exception of the polarite filter beds, largely in excess of their present requirements, has been covered by the more moderate sum of £15,000, the cost of the ferozone and polarite system being less than five hundred pounds.

The following is an abstract of a very interesting and important paper read before the Manchester Section of the Society of Chemical Industry on Friday, November 7th, 1890, by Mr. J. Carter-Bell, A.R.S.M., F.I.C. :—

" Mr. Carter-Bell said that he had been making experiments upon the various processes which for the last two months had been in operation at the sewage works belonging to the Salford Corporation. After describing the electrical, porous-carbon, alumina-ferric, and the Barry process, he alluded to the polarite or international method in the following terms: 'The International Water and Sewage Purification Company's method consists in adding to the sewage a substance called ferozone, and then allowing the treated sewage to settle in tanks. From these tanks it is passed on to the filter beds, which are made up of sand and polarite, and it certainly was very surprising to see a liquid running from the filter as clear as spring water, and when one considers the foul nature of the Salford sewage, one cannot but be impressed with such results. Would not a sand filter answer as well as the polarite? To test this he had two filters fitted up in his garden; one was polarite and the other was sand. Gallons of sewage after being treated with ferozone were passed through the two filters. The effluents were compared and analysed. At first the sand produced a good effluent, reducing the albuminoid ammonia by about 70 per cent; but soon the effluent began to be cloudy, and the purification power was reduced. The polarite filter gave clear and bright effluents, and when the filter was given sufficient rest the albuminoid ammonia was reduced to a minimum. In one instance it was as low as ·02 in the 100,000. This is actually as good as many well waters in country places. Mr. Carter-Bell said he was at first rather sceptical as to the length of time these polarite filters could last, but he had paid a visit to Acton and seen the filter beds, which had been in operation nearly four years, and the effluent which was flowing from the filter beds was as bright and clear as those which were obtained from the filters in Salford, and from the experimental filters fitted up under his own superintendence. He said that he fully confirmed what Sir Henry Roscoe had reported upon this process. Mr. Carter-Bell said he had taken samples from the Irwell, Irk, and Medlock—three of the foulest rivers in England—had passed these samples through filters of polarite, and had produced, comparatively speaking, pure water. The Irk and the Medlock are rather worse than the ordinary Salford sewage, and, therefore, the test upon the polarite was very severe. But when ordinary well waters, which are not fit for domestic use, are passed through filters of polarite they are immediately purified and made fit for ordinary use. Therefore, judging from his many experiments, he thought local boards, who had reason to doubt the purity of their water supply, could not do better than pass such water through filters of polarite.'"

At the cost of repetition on some points I quote how the International Water and Sewage Purification Company explain their system: 'Our system is carried on in two stages, viz.: (1) By precipitating and deodorising the sewage in settling tanks by the aid of a magnetic precipitant and deodorant called ferozone. (2) By passing the partly purified sewage effluent from the settling tanks through polarite filter beds, which arrest any solids remaining in suspension, and oxidise and render innocuous putrescible matters held in solution. Ferozone is the trade name of the material successfully used in deodorising sewage and precipitating the solids therefrom. It is a cheap preparation, containing salts of iron and alumina, acknowledged to be the best and most powerful precipitants and deodorants for sewage purification. It is rich in ferrous iron, alumina, and magnetic oxide of iron in a very spongy and absorbent condition. By virtue of its soluble iron and alumina salts it soon causes subsidence of the suspended solids. The iron oxide, being porous and magnetic, parts with its polarised oxygen, and thereby helps to disinfect and deodorise the sewage and sludge. It attacks the molecular constitution of that portion of the organic matters in solution which cannot be removed by precipitation in the settling tanks. It so alters the constitution of these putrescible matters that further oxidation thereof by filtration through polarite becomes certain and under complete control at all times. The sludge precipitated is richer in ammonia than that produced by the lime process, and is therefore of considerable value as a fertiliser. Polarite is the trade name of the material successfully used for filtering and further purifying the ferozoned sewage water. It is a specially-prepared rustless and magnetic oxide of iron, insoluble in water, non-poisonous, and practically everlasting. It is very hard, porous, and absorbent. It is the only permanent oxidiser sufficiently cheap and durable for practical use in filter beds on a large scale for the purification of either drinking water or sewage effluent. It is a powerful deodoriser and decolouriser by virtue of the polarised oxygen contained within its microscopic pores. Although it consists chiefly of iron, it does not rust, and is therefore much superior to spongy iron. In carrying out the process of purification, the sewage on reaching the outfall works is run through strainers to arrest floating solids, such as corks, rags, &c., and then flows quietly through a floating trough into a settling tank, the floor of which inclines towards the centre, so that a gutter may convey the sludge to the outlet valve. Before entering the tank the crude sewage receives a dose of ferozone, which costs about one farthing for every 1,000 gallons of sewage treated. The flow of sewage through the tanks should not be continuous if it can be avoided. Quiescent tanks are better for precipitation purposes, whatever kind of precipitant be used. When a tank is full its contents should be left standing for about two hours, so that the ferozone may have time to act thoroughly. The ferozone will cause deodorisation and precipitation to take place in the sewage, and a considerable portion of the albuminoids in solution will coagulate and be precipitated with the solids. The supernatant sewage water thus clarified would then be drawn off and run through polarite filter beds, which produce a higher degree of purification at less cost than can be obtained by any other known process. The polarite filter beds consist of six inches of broken stone in which four-inch agricultural drain tiles are embedded, three inches of gravel, six inches of sand, twelve inches of polarite and sand mixed in equal proportions, and at the top a layer of nine inches of fine sand, making a total depth of three feet of filtering material. Polarite filters purify tank effluents which have been treated with ferozone at the rate of 1,000 gallons per square yard in twenty-four hours. They give better results than generally produced by land filtration, and at less cost and without nuisance. Land filters about 1½ gallons per square yard in twenty-four hours; therefore one acre of polarite filter bed will do more effective work than 666 acres of land. The cost is therefore much less than that of a sewage farm. The polarite filter bed only requires a few hours rest occasionally for aeration and for cleansing the surface sand, which is done automatically at a nominal cost. The polarite never requires to be removed or replaced. Hence it is better to lay down several small beds rather than one or two large ones, so that they may rest a few hours alternately for aeration and cleansing. After the needful time for precipitation of the solids in the ferozone tank has elapsed, the supernatant sewage water is drawn off through a floating arm surrounded by a floating cill to prevent any solid matter from passing into it. It flows from the tank through an iron pipe to polarite beds for filtration. After passing through the polarite the effluent produced satisfies the demands of the most advanced sanitary critics. It is neither acid nor alkaline. It will not decompose in hot seasons, but remains inodorous, non-putrescible, clear, and tasteless. The higher forms of green aquatic vegetation thrive in it. It will not kill fish, but rather help to purify a river by discharging into it water free from any tendency to undergo secondary decomposition. The sludge deposited in the settling tanks, having been deodorised by ferozone, is comparatively inoffensive—unlike that produced by lime and other processes. The quantity, too, is less, being only half that produced by the lime process; hence it is richer in manurial value, and requires less labour and machinery to treat it. In many places it may be drained into sludge pits covered by a light roof, and arranged so as to send the surface liquor back to the tanks for further ferozone treatment. The sludge can then be mixed with town refuse and sold to farmers. In large works it is probably better to press it by improved sludge pressing machinery, such as used by the International Company. The pressed ferozone sludge will dry spontaneously if placed under cover, and can be so easily and cheaply ground into powder. In this state it can be kept without nuisance in a dry store until the season comes round for selling it as a fertiliser. The sewage manure produced by the International process is shown by analysis to contain nitrogen equal to from 1 to 2 per cent of ammonia, according to the quality of the sewage, besides phosphates. This system is in operation at Hendon, and the following examination has been made by Dr. Angell.

"Analysis No. 1 of effluent obtained by the lime and land process. Sample taken by the foreman to the Hendon Local Board:—

	Dry weather, grains per gallon.		Parts per million.
Free ammonia	3·325	......	47·500
Albuminoid ammonia	1·05	......	15·00
Chlorine......................	6·03	......	—

Very turbid stinking foul fluid. The residue on ignition gave off a foul sewage odour. This is an extremely bad effluent; in fact, it contains practically all the foul matter of the original sewage.—Signed, ARTHUR ANGELL, Ph.D., F.I.C."

"Analysis No. 2 of Hendon sewage, purified by the International process:—

	Grains per gallon.
Free ammonia	0·0218
Albuminoid ammonia	0·0056
Chlorine	6·0200
Suspended matter	none.

Physical properties of sample clean and odourless.—Signed, ARTHUR ANGELL, Ph.D., F.I.C."

SEWAGE MANURE FROM THE HENDON SEWAGE WORKS BY THE INTERNATIONAL PROCESS.

"Dried sludge (not pressed) collected from average daily samples produced during a period of two months. Analysis No. 3 of sewage sludge from Hendon (air dried):—

Moisture	12·62
*Organic matter	46·60
Oxide of iron	3·60
Alumina	2·96
Phosphates of alumina	2·21
Magnesia	none
Potash	0·29
Sand and clay	29·20
Sulphuric acid (S.O.)	1·46
	98·94

"This sludge is in a fine uniform state of division, and is of good quality in manurial properties.—Signed, ARTHUR ANGELL, Ph.D., F.I.C., Public Analyst."

*Containing nitrogen equal to 2·28 per cent of ammonia.

The 2·28 per cent of ammonia, at the present market value of 12s. per unit, would, with other constituents, be worth about 30s. per ton of sludge dried to 12 per cent of moisture; the quantity of such dried sludge obtained from the strong Hendon sewage and laundry refuse is found to amount to four tons per million gallons of sewage of the value of £6. The method of treating the sludge at Acton is quite exceptional, and not necessarily a part of our system of purification. It is necessary to use pressing powder for the sludge beyond the 7lb. of ferozone and 1½ cwt. of lime per ton. The 7lb. of deoderising ferozone is added at Acton to prevent the sewage cake from becoming offensive when stored for a long time. The 1½ cwt. of lime is added per ton solely to help to separate the water from the sludge, which it does by its curdling power. If the manure is not to be stored, deodorising ferozone need not be used. For ordinary purposes this sludge is sufficiently deodorising by the ferozone used in the precipitation process. Now, here is a plan which obviates all land difficulties as to disposal without having to irrigate a yard of land, as its effluent is such that in its discharge into a river or stream it can defy any county council contention, as it is within the limits of contamination prescribed by the Rivers Pollution Act. In saying this I must not be understood to be advocating this method as the best remedy for what we have to do with the Leek sewage. I merely wish it to be considered as part of my inquiry into the relative merits and defects of each system of sewage disposal, whether in operation or projected. Further, I am most thoroughly convinced that no examinations of methods or systems of disposal can be at all adequately considered without giving due prominence to the chemical analysis of their effluents, and to bacteriology, or to the bacteriological inferences which can be fairly deduced from the results of such analysis, and I wish to point out that the Acton report is interesting in these respects, particularly with regard to the chemical state of the effluent as to organic cleansing, by the examinations having been made at stated periods annually since the adoption of their system, and these made by very eminent and most competent chemists, whose results may be unhesitatingly accepted. First, we have the analysis of Sir Henry Roscoe, in the month of August, 1888, after one year's working of the system. Then follows the analysis of Mr. Carter-Bell, in August, 1889; after that, that of Professor Frankland, in July, 1890; and, lastly, another and more exhaustive one of Mr. Carter-Bell, in August, 1890.

The Swinton New Works on this System.

Since I wrote the foregoing I have received the published report of the Swinton Purification Works, the salient points of which are as follow: Swinton, a small suburb five miles from Manchester, with a population of 24,000 people, has adopted the ferozone and polarite system. Their sewage farm is 32 acres in extent. The sewage flow is about 240,000 gallons daily, or 15 gallons per head of population. It is conveyed by low and high level mains; there is no pumping. The solids are arrested and the sewage flow is made to turn a water wheel for working the ferozone mixture. The quantity of ferozone used is 10 to 12 grains per gallon, costing 9s. 6d. per day. There are four settling tanks, holding 250,000 gallons. Three hours after the ferozone has been mixed the effluent is clear, and is run in the filter beds, which have an area of 350 square yards, divided into four chambers. The filter beds are arranged as at Acton, with four pipes, loose broken stone, gravel sand, and polarite. Each bed, when at work, is in operation 18 hours per day, the remaining six hours being occupied in cleaning and aerating at intervals. There is provision for the filtration of 650 gallons of effluent per 24 hours. The effluent passes to the brook down earthenware pipes 300ft., and about 25ft. lower than the filter outlet, the fall being available to work a turbine if necessary, or it might be used to irrigate the farm. The sludge is not at present pressed, but is conveyed to the farm and dug into trenches. The arrangements for the farm are such that either crude sewage effluent or sludge can be used upon it at will, or at any intervals. Four men are employed at a cost of £4 10s. per week. The total working expenses amount to £420 per annum, or about 6½d. per head of the population, which is a very low rate, as it is well known generally that the sewage can be treated at a total inclusive cost of 1s. per head. The Swinton sewage purification works, now completed, have cost about £3,800. This sum includes all new work, but does not include anything for land, law, or engineering expenses.

Notwithstanding the excellent results recorded in this chapter I feel bound to state that the weak part of the system is in the use of ferozone as the precipitating agent. With a better precipitant I consider the results, after passing through the polarite filter-beds, would be as near perfection as could be desired. The strong part of the system is undoubtedly the polarite filtration.

CHAPTER IX.

A Day at Birmingham with the Lime Process and Irrigation. The Works of the Birmingham Tame and Rea District Drainage Board at Saltley and Tyburn, November 5th, 1890.

Should this account of my visit prove of no practical use, it will be sure to possess some points of interest which may help us to form a sounder judgment and derive some lessons of guidance on a question of such great importance to communities as to outlay, and to the purifying of our rivers and of the air we have to breathe from sewage contamination. Being very much struck by a statement of Dr. Reid, the Medical Officer of the Staffordshire County Council, at the Stoke meeting of the deputations from the local authorities, invited to meet the Sanitary Committee of the County Council of Staffordshire, that the effluent water of Birmingham after it left the land contained only ·24 parts of ammonia in a million, and ·22 parts of albuminoid ammonia per million, I determined to go and see the *modus operandi* of this interesting elimination of organic matter, certainly, if correctly reported, one of the most perfect in the kingdom, if not in the world, so complete as to contain less organic matter than some bad drinking waters, notably that of the well at Leek Workhouse, which Mr. Wanklyn found to contain ·34 per million of albuminoid ammonia. Water which contains no more albuminoid ammonia than from ·02 to ·05 per million is considered a pure enough potable water.

Throughout the country generally we find opinions on the treatment and disposal of sewage of a most varied and conflicting character. I do not regard this as a bad sign. I think it shows that everyone is seeking a remedy for an unhealthy state of things, and we shall have to come to a common agreement upon this, as upon every other question which comes before us. What I wish to state here is, that I am not seeking to give an opinion as to the relative merits of the many systems of dealing with sewage, whether now in operation or projected, nor am I presuming to offer any opinion of my own as to what is best to do, at least in the present state of this inquiry. I wish to describe what I have seen, adding a few remarks by way of criticism and explanation. As far back as 1853 Birmingham constructed main intercepting sewers to an outlet at Saltley, about two miles below Birmingham, leaving the sewage of the districts above and around Birmingham only partially dealt with, but in 1877 there was formed a united district for the purposes of sewage disposal, comprising the Borough of Birmingham, the Local Government Districts of Smethwick, Harborne, Balsall Heath, Saltley, Aston Manor, Handsworth, Aston Union, King's Norton Union, West Bromwich Union, and the Borough of Sutton Coldfield, comprising an acreage of 47,275, a population of 619,693, and a rateable value of £2,401,093. The Joint Board numbers twenty-four representatives of this large area, under the chairmanship of Mr. Alderman Avery. The cost is divided *pro rata*, and the duties of the board are to acquire land and to construct such outfall works as will so purify the whole sewage as not to make a breach of the Rivers Pollution Act of 1876. The outfall at Saltley discharges nine-tenths of the sewage of the whole district.

I will give in the words of Mr. Till, C.E., the engineer to these extensive works, the method of treating the sewage. I quote from his report, written in 1886; the figures refer to the previous year; the figures given are therefore liable to some modification at the present time. "The sewage on arriving near the liming sheds at the upper end of the works at the Saltley outfall is mixed with lime, both to neutralise the acids (present to an unusual extent in Birmingham sewage) and also to assist in precipitation, which, however, is not now necessary to so great an extent as formerly; the sewage then passes through the large or roughing tanks, where the grosser impurities settle down, and from thence it is conveyed by the main conduit to the land and disposed of by ordinary irrigation. The sixteen small tanks required at one time for completing the precipitating process are still used under certain circumstances, and are a valuable auxiliary when rainfall has increased the normal quantity of sewage. The sludge from the tanks settles to a depth of about 6 feet, and is removed every ten days, and is elevated by bucket dredgers and pumps into movable wooden carriers, and flows into beds formed in the land at the Saltley or western end of the farm. The sludge contains about 90 per cent of water as it comes from the tanks, but after lying on the ground for about fourteen days, much of this water drains away or is evaporated, leaving the sludge in a layer about 10 inches thick and of a consistency that admits of its being trenched into the land. Crops are then planted, and after a time the sludge becomes pulverised and capable of being irrigated. About 40 acres were required for the sludge last year, and the same land may receive a coating of sludge every two or three years. A few words may be said as to the difficulty at one time experienced in dealing with this mud or sludge from the tanks. After the construction of the two large tanks in 1859, the mud therein deposited was dredged out and run on to the adjacent land, where it accumulated for some years, forming at one time a large mass of foul matter about seven acres in area, and over 4 feet deep. In consequence of the nuisance arising therefrom proceedings were taken about 1871 by the residents in the vicinity, and an injunction obtained restraining the Corporation from depositing the mud so as to cause a nuisance. Great efforts were made by the Corporation to reduce the amount of mud, large quantities being conveyed away in boats, but it was not until the experiment had been tried of trenching the mud into the land, and found perfectly satisfactory, that the present system was adopted, about the end of 1872, and the difficulty finally overcome. Practically, the whole of the sewage of the drainage district, amounting to sixteen million gallons per day, flows by gravitation to the outfall works. Only a very small area requires its sewage to be lifted by pumping, the cost of such pumping being £104 per annum. The Board farms the whole of the land, no portion of it being sublet. Of the produce milk is a large and increasing item, 128,995 gallons, realising £4,406, being sold last year. During the present year about 280 acres of land are devoted to mangolds, swedes, and kohl rabi, 250 acres to market garden produce, 100 acres to Italian rye grass, 130 acres to cereals, and about 340 acres are pasture. The total amount realised last year from the sale of stock and produce was £20,008. During the same time stock was purchased to the extent of £7,760. With regard to the financial aspect of the Board's work, it is perhaps needless to say that a considerable sum of money has annually to be obtained from the rates. The total amount raised by the Board's precept last year was £33,089, of which interest of loans and repayment of loans absorbed £17,516; management expenses, rents, rates, taxes, &c., £5,534; the balance of £9,979 representing the loss on the year's working on the farm. There are 1,227 acres available for works of sewage disposal in a flat valley extending about five miles in length, from Saltley to below Tyburn. Some of it is leased for £4 per acre, and the rest is freehold, having cost £152 per acre. This price included tenant rights, compensation, and law charges. The nature of the land is very favourable for the purification of sewage, the natural surface of the ground being, as a rule, even and unbroken, and the level such as to admit of the irrigation of the whole by gravitation, with the exception of about 100 acres. The subsoil is gravel and sand, varying from 6 feet to 10 feet thick. For conveying the sewage to the land a conduit 8 feet in diameter and about 2¾ miles long has been constructed, capable of discharging thirty-eight million gallons per day when running half full, or double that quantity when running full, the fall being about 2 feet per mile. The land is drained to an average depth of 4 feet 6 inches by 3 inch or 4 inch agricultural drain pipes placed about 11 feet apart, and discharging into main drains into the outfall or effluent channels. The total cost of the land and works to the present has been £403,695, of which the original land and works is £170,544, new land £110,800, new works £113,299, farming stock and implements for new land £9,052.

Mr. Till informs me that some of the constituent authorities, including Birmingham, have constructed considerable lengths of storm-water sewers, but some of the storm water mixed with the sewage, and what cannot be dealt with at Saltley has to go bodily into the river Tame, but, as I have already mentioned, there is capacity in the pipes for dealing with twice as much as the normal quantity. Notwithstanding this, a considerable quantity of the Birmingham sewage must get into the Tame, which at Tamworth is now never clear. In hot weather the stench there is very bad, and in flood times the lower parts of the meadows are black as possible. Formerly the Tame at Tamworth was transparent; now it is always muddy.

The water of the Tame five miles above Tamworth and eleven miles below Birmingham was found by Dr. Hill to contain in parts per million—

Free ammonia	11,000
Albuminoid ammonia	2,800

showing that there is sewage impurity entering the river from the Birmingham, Tame, and Rea District Drainage Board. It should be remembered that this was three years ago.

To return to the Saltley works. After precipitation and removal of the sludge, the liquid sewage is run upon the land below Saltley, in the direction of and past Tyburn, where, after fertilising the crops, it filters through 4 feet 6 inches of subsoil, composed of gravel and sand, to the drain pipes I have already described. It then flows out of these pipes at Tyburn as the effluent product, and finds its way into the Tame and Rea, a clear and colourless fluid.

The results of my analysis of this purified effluent product are—

80 grains of solids per gallon
Large quantity of chlorine
An almost absolute neutrality
1·660 free ammonia per million
·438 albuminoid ammonia per million.

The difference between Dr. Reid's figures and mine are shown in the following table:—

My analysis—free ammonia per million parts ...	1·660
Dr. Reid's figures ,, ,, ,, ...	·240
Difference	1·420
My analysis—albuminoid ammonia	·438
Dr. Reid's figures ,, ,, ...	·220
Difference	·218

These two last figures, compared with Dr. Reid's statement at Stoke, show an effluent very much more highly charged with organic matter than is found in the analysis from which Dr. Reid quoted.

I acquainted Dr. Reid with my results, and he informs me the figures he quoted were communicated to him by Dr. Hill, the Warwickshire County Analyst, as the last analysis made by that gentleman. Dr. Reid adds that one could hardly expect that such an excellent result as that which he had quoted could uniformly be maintained, and that my figures, although pointing to a very much worse state than his, are still greatly superior to any of those effluents taken by himself in North Staffordshire.

I was also told by Alderman Barrow of a very drastic test of the state of the effluent, viz., that the members of the Sanitary Committee go periodically down to Tyburn to test its purity—I will not say by drinking, but by tasting. This is somewhat analogous to ensuring the safety of a railway train by making a director ride on the engine.

By consent of Mr. Till, I am allowed to publish a more recent examination of Dr. Hill's. It is as follows:—

"City of Birmingham,
"Health Department, The Council House,
"November 3rd, 1890.

"Dear Sir,—I beg to hand you my report on the two samples of effluent, which you submitted to me on the 28th ultimo, described—1, Sample of effluent taken from a town in the north of England for comparison. 2, sample of effluent taken from Tyburn outfall, near Castle Bromwich Station, at 3 p.m., October 27th, 1890. The following are the analytical results:—

	1	2
Total solid matter	71·4	111·1
Free ammonia	1·520	0·400
Organic ammonia	0·540	0·084
Nitrogen as nitrates and nitrites	none	1·76
Previous sewage contamination	12·18	20·580
Chlorine	9·6	17·6
Turbidity	Opalescent, slightly dirty white flocculent deposit.	Opalescent, very trifling dirty white flo·culent deposit.
Colour of liquid	colourless	colourless
Reaction	strongly alkaline	neutral
Odour	strongly urinous	none

"The amount of suspended matter in sample No. 1 was slight, and in No. 2 very slight. The above results show that as regards organic constituents No. 1 is very greatly inferior to No. 2. It contains nearly four times as much free ammonia and nearly seven times as much organic ammonia as No. 2 does, and half as much free ammonia as raw sewage. It, moreover, has a filthy urinous odour, clearly proving the presence of unaltered sewage, and that it has not been submitted to an efficient means of purification. These characters prove its unfitness to be discharged into a river, while those of No. 2 show it to be perfectly fit to be so discharged.—I remain, yours faithfully,

"W. S. Till, Esq., C.E.,
"City Surveyor,
"Council House."

"ALFRED HILL, M.D.,
Medical Officer of Health.

I have not had time to analyse the sewage as it goes on the land at Tyburn—that is, before filtration, but it may probably be taken to contain approximately organic matter indicated as follows—say

Free ammonia	10,000 to 15,000	per million
Albuminoid ammonia	3,000 to 4,000	" "

It may not be without interest in showing how nearly this Birmingham effluent comes up to potable water, to give a few tables of the organic matter in several drinking waters supplied to various populations, taken from Dr. Wanklyn's tables.

	Per million parts		
	Free Ammonia.	Albuminoid Ammonia.	Degree of hardness.
Leek town water	·00	·02	3·8
Manchester "	·01	·06	3·0
Glasgow, Loch Katrine	·00	·08	2·0
Southwark and Vauxhall ...	·03	·016	—
Lambeth Water Company ...	·02	·14	—
New River Company, London	·02	·08	15·0
Chelmsford town water	·08	·02	13·3
River Thames at Hampton Court..	·04	·18	—
River Thames, high tide......	1·02	·59	—
London Bridge..................	—	—	—
The well at Leek Workhouse, which produced diarrhœa..	·02	·34	5·2

In addition to these, I append the results of a few analyses of Dr. John Thresh, formerly of Buxton, now Medical Officer of Health in the Chelmsford and Malden Rural Sanitary Districts. His report is most valuable, and he has been at enormous trouble in examining the potable waters of no fewer than upwards of 60 parishes.

I will only quote a few of the parishes, good and bad, giving his organic results in parts per million, with his remarks:

Name of Parish.	Free Ammonia.	Albuminoid Ammonia.	Remarks.
Great Baddow	·00	·05	good
Galleywood	·20	·40	bad
Hanningfield	·07	·14	unsafe
Runwell (rain-water tank)	·80	·90	very impure
Runwell Pond	·64	·96	sewage
Danbury Spring	·02	·02	very good
Little Waltham Pump	·09	·06	good
Southminster	·16	·24	bad

I think the examination of the Birmingham system of precipitation and after-purification of the sewage by land filtration shows us that it has the twofold advantage of high agricultural value, and the production of an effluent remarkable for its small amount of organic matter. The difference between my analysis and those of Dr. Hill shows conclusively that the effluent varies according to circumstances and the time of day at which it is taken. It seems to me that, if a larger area of land could be obtained, this system would be perfect as far as irrigation and filtration are concerned. With regard to the sludge, I should think the authorities must consider it more or less a serious nuisance, and that it would be a good thing if a more distant locality could utilise it, or some other means be found of disposing of it, such as burning it or making it into portable and saleable manure. But a better precipitant is greatly needed.

CHAPTER X.

Manchester and Rochdale: The Pail System.

The Manchester Sewage Works are under the direction of Mr. Hy. Whiley, superintendent, who kindly gave me the following information:—

The sanitary arrangements of the city of Manchester are—(1) sewage which is simply discharged into the river Irwell, and into which human excreta does not enter; (2) the disposal of the excrement by the pail system.

There are in old Manchester 70,000 pail closets or privies, the contents of only 35,000 of these are utilised for the preparation of their artificial manure.

The method of treating the excrement from the pails to make it into the artificial manure is to convey it to the works in four-wheeled collecting vans, holding a number of pails and their contents. It is then passed through riddles, being in a more or less fluid state (solid excreta contains as much as 80 per cent of water); then it is put into immense furnaces and evaporated to dryness, the water and gases passing into a large chimney shaft. It is, in fact, cremated. The dry residuum is then mixed with sulphuric acid to fix the ammonia. Blood and bones are then added, and the whole boiled and ground.

The remaining 35,000 go to Carrington Moss, an enormous expanse of surface, with a depth of some fourteen feet of bog, the surface of which in the parts manured has been prepared by digging and breaking up to receive the excremental manure, which is added without any treatment. There is no soil on the bog, nor is any soil or marl added to it in its preparation for agricultural purposes. This moss has become very fertile since it has been so manured. Excellent crops of oats, turnips, mangolds, carrots, clover, and grass are grown upon it. Wheat and beans are not cultivated, and no human food crops except potatoes, of which 900 tons were sold last year.

A Manchester friend informed me a short time ago, that a considerable area of the moss unexpectedly cropped tomatoes, no preparation having been made for their growth, and the following year the increased crop was so remarkable as to attract many people to see it. The explanation is that tomatoes had been previously very largely used as food, the seeds had passed uninjured through the body into the fæces and had germinated in the fertile soil.

The difficulty of dealing with this immense mass of excretal matter is colossal, and is now so much increased by the addition of seven more townships, which have chiefly midden systems, and a death-rate for the whole of 27 per thousand, which is about that of Salford, that the Corporation are now making a new drainage for the whole of Manchester, which will go to Davyhulme, to be treated and disposed of in some way or other not yet decided upon. Another illustration of the gigantic difficulty of fixing upon a system which shall be the most comprehensive and effective.

Carrington Moss has also of late been found inadequate to the great disposal requirements of their excretal pail system. The land has become surfeited with such wholesale and prolonged applications, and additional land or the adoption of some other system has become necessary.

Complaints of unhealthiness have been presented by the Medical Officer of the Altrincham Union Rural Sanitary Authority, in which Carrington Moss is situated, and the Corporation of Manchester have endeavoured to make arrangements for the purchase and utilisation of 1,800 acres of land at Rampton Manor, Nottinghamshire, to continue their present system, and no doubt to avoid the enormous expense of converting the pail system into that of water-closets, and to provide an adequate output for a time for the disposal of the excretal matter. But owing to the protests of the authorities of the Nottingham County Council, the Sanitary Committee of the Retford Town Council, the East Retford Union, and the Gainsborough Local Board, and probably on account of the manifest permanent shortcomings of the pail system as compared with more modern ideas of sewage treatment and disposal, the Local Government Board have refused the application of the Corporation.

It remains to be seen whether the better method of the water-carrying and precipitation systems will be adopted by the Corporation. It may well be hoped that, notwithstanding the great initial expense which is involved in the conversion of the pail system into that of water-closets, the more sanitary plan will obtain, and the enormously high death-rate be lowered.

The pail system was utterly condemned by the Committee appointed by the Local Government Board in 1875, but reported upon in 1876 by the Committee appointed by the Society of Arts to inquire into various subjects connected with the health of towns as follows: "That the pail system under proper regulations for early and frequent removal is greatly superior to all privies, cesspools, ashpits, and middens, and possesses manifold advantages in regard to health and cleanliness, whilst its results in economy and facility of utilisation often compare favourably with those of water-carried sewage." I do not endorse this too flattering opinion. I regard the pail system as manifestly inferior to the water-carriage disposal, and I regard with surprise so favourable a view. At Manchester a few cinders are thrown into the pails, that is all the defecating matter assistance possible. The cartage of this foul residuum from house to house to its destination must be fraught with much danger and nuisance.

I have before me a circular issued by the Manchester Corporation, offering to the public a fertilising product which they call "concentrated manure." It deserves some passing notice alongside those of the other systems I have described, although we shall only learn from it how a difficulty is dealt with in other circumstances than those of most towns, it being not a notice of sewage disposal, but rather of excreta treatment. The advertisement states that at a cost of £3 per ton can be obtained a manure containing from 3½ to 4 per cent of ammonia, existing in the manure in the same form as guano, over 8½ per cent of phosphate, with potash salts, and 38½ per cent of organic matter. The manure has the appearance of fine mould, and contains all the constituents necessary to render soil fertile. Mr. Heaton, of Endon, informs me he uses it with success on his land. The circular states that "the removal from the city of the material from which the manure is made being an absolute necessity, profit is not considered in fixing the price." The circular states that the Corporation have been unable to execute all the orders they receive, and are putting down additional machinery to enable them to supply the demand. The following quotation gives the source and nature of this manure: In making their concentrated manure they utilise all the urine and fæcal matter from the pail closets of the city, the offal from their extensive markets, and hundreds of tons weekly of the blood, entrails, and fleshy matter from the city slaughter-houses. When it is remembered that there are upwards of 200,000 head of beasts, calves, and others animals killed annually at the city abbatoirs alone, it will be readily perceived that a large quantity of valuable offal for manure is obtained; besides which hundreds of tons of dried blood and bones are added, the result being a perfectly natural manure suitable for every kind of crop, containing 3½ to 4 per cent of ammonia, and 8½ to 10 per cent bone phosphates, these two ingredients alone at the present market value being equal to more than the price charged for the manure:

	£	s.	d.
3½ per cent ammonia (guaranteed) at 12s. 6d. per unit	£2	3	9
8½ per cent bone (guaranteed) at 2s. per unit	0	17	0
	£3	0	9

If farmers were aware of the value of this manure, they would readily avail themselves of such a cheap fertiliser.

In fixing the price at £3 per ton delivered at the nearest railway stations in bags, profit has not been considered, the object of the Corporation being to rid the town of the objectionable matter referred to, thus accomplishing important sanitary reform, and providing a valuable fertiliser from the best and most natural raw material.

The following directions are given for its use in fertilisation:—

Wheat.—Apply 6 cwt. to 7 cwt. per acre in the autumn, and well harrow or drill in with the grain.

Oats and Barley.—6 cwt. to 7 cwt. per acre at the time of sowing.

Turnips, Swedes, and Mangolds.—10 cwt. to 15 cwt. per acre, broadcast or drilled, or 5 cwt. with half the usual dressing of farmyard manure, or mixed with about 2½ of super-phosphates or bone dust.

Grass.—6 cwt. to 8 cwt. per acre, applied early in the spring.

Clover.—6 cwt. to 8 cwt. per acre, applied either in the autumn or very early in the spring.

Vetches, Beans, Peas, &c.—10 cwt. per acre at the time of sowing.

Potatoes.—10 cwt. to 12 cwt. per acre, or 5 cwt. with half the usual dressing of farmyard manure, applied in the rows.

Carrots and Cabbages.—10 cwt. to 12 cwt. per acre, well dug into the soil at the time of sowing.

The chemical composition of this manure is shown in the following analysis made by Mr. Alfred Smethan, F.C.S., consulting chemist to the Royal Manchester, Liverpool, and North Lancashire Society, in October, 1888:—

Organic matter	35·25	yielding ammonia	4·14
Alkaline salts	6·55	yielding potash	0·84
Phosphoric acid	3·75	yielding phosphates	8·20

The Rochdale Pail System and Manure Works.

The pail system has perhaps its most perfect development in Rochdale, and I propose to give a further illustration of the system.

Owing to the courtesy of Mr. S. S. Platt, the Borough Surveyor of Rochdale, I am enabled to do so. He has sent me an exhaustive account of their system, and has given full replies to the series of questions I asked him. I cannot improve on his own description, and insert it *verbatim et literatim*:—

"The works are situated about 660 yards from the centre of the town. The closets are on what is known as the Rochdale pail system, with a roofed ash-place at the back, containing tub for the reception of the house refuse.

"The excreta pails are tubs made from petroleum oil casks, divided into two equal parts, each 18 inches wide, and are removed weekly in specially designed vans, each van containing 24 pails; the full ones on being removed from the closets are hermetically sealed by a special cover, and are replaced by clean pails, each having had placed therein a small quantity of deodorant.

"The tubs containing the house refuse are removed weekly, or as required; 8,645 tons of excreta, and 15,271 tons of house refuse had been collected this last year; the staff required being 2 inspectors, 34 drivers, 22 conductors, 34 horses.

"The ashes and refuse are tipped into a shed containing riddling machines, which remove all the fine ash, which is eagerly fetched by farmers to use as an absorbent in their shippons and for mixing with liquid manure. The tins, cans, and similar articles are taken out and disposed of from time to time. The cinders, vegetable and other refuse are used solely for generating heat and steam throughout the premises, the clinkers resulting therefrom being ground in a mill with lime into mortar, which has a ready sale and is also used by other departments of the Corporation. On entering the works, the vans, after being weighed, proceed into a large shed, which is hermetically sealed, the two doors only being opened when required for the admission and exit of the vans. The pails are removed from the vans and placed on a large table of a hydraulic lift, which raises them to a higher floor, where they are emptied into two tanks, when they are returned and washed and deodorised before being placed in the vans, which are also similarly treated after every journey. The excreta is run from the tanks into revolving cylinders 12 feet long by 6 feet 6 inches to 6 feet taper diameter. These cylinders or machines have been designed and patented by Mr. Harescough (the late manager), and are the outcome of several years' experience at these works in attempting to convert the excreta into a marketable manure. The cylinder revolves on four friction rollers, there being fixed inside the cylinder a movable arm or scraper for keeping the mass constantly in a state of motion, so that the surface of the excreta will be constantly exposed to the hot air, thereby thoroughly drying it. The drying is effected by connection of a large pipe to a boiler, specially constructed for burning the cinders and vegetable refuse referred to, the heat as hot air passing through the pipe and through the revolving cylinder or machine, and absorbing the water of the excreta, the effect of which is that a charge of 70 cwt. of excreta is reduced in about 11½ hours to 5 cwt. 1 qr. 18 lb. of material, very similar in appearance to hard dry clay.

"To each ton of the excreta there is added 24 lb. sulphuric acid, and thus the ammonia is fixed and not driven off by the heat. The heat passes from the machines through flues, is utilised for completing the drying on a drying floor, and conveyed to condensers, which reduce the temperature and drive off the vapory particles; it is then forced by a blower through a furnace, and finds its way up the chimney, and no complaints from nuisance now arise. The manufactured product is taken from the machines, placed in an adjoining shed, and ground in a pugmill to a 'poudrette.' It is sold to the farmers of the district at £6 6s. per ton.

"*Copy of Analysis made by Dr. A. Voelcher.*

"Composition of a sample of Rochdale manure, sent by Mr. W. Holt, manager of the Manure Works, Rochdale:—

Moisture	10·03
*Organic matter and salts of ammonia	66·26
†Soluble phosphoric acid	1·62
‡Insoluble phosphoric acid	2·29
Lime	2·59
Magnesia	0·82
Oxide of iron, alumina, &c.	2·48
§Alkaline salts	12·02
Insoluble siliceous matter	1·99
	100·00

*Containing		7·06
Equal to ammonia		8·57
Including nitrogen present as salts of ammonia		2·86
Equal to ammonia		3·47
Equal to sulphate of ammonia		13·48
†Equal to tribasic phosphate of lime	3·35	8·52
‡Equal to tribasic phosphate of lime	4·99	
§Containing potash		2·96
Equal to sulphate of potash		5·48

AUGUSTUS VOELCHER.

"A very similar analysis was made by the Public Analyst for the Borough of Rochdale in 1837." For further particulars see the answers to the questions on p. 216.

CHAPTER XI.

THE BUXTON SEWAGE TREATMENT AND WORKS.

UP to 1885, the sewage of Buxton was allowed to fall into the river Wye, a short distance below the town, thereby polluting the river to such an increasing extent as to render the purification of the sewage imperative. By a happy thought, the result of a number of experiments, Mr. Thresh, D.Sc., of that town, found that the purification of the sewage might be effected by diverting and adding it to a ferruginous water which emerges from an old disused coal-mine adit in the lower coal strata of the Millstone-Grit or Yoredale measures, about two miles above Buxton, and which contains 2 grains per gallon of iron in solution as ferrous carbonate, or most probably as sesquicarbonate or bicarbonate of iron dissolved in carbonic acid, and which on exposure to the air absorbs oxygen and falls as ferric oxide, coating the surface of the brook, as was formerly well seen in the brook-course and ornamental water of the Buxton Gardens with a copious precipitate of an ochreous deposit. It is a chalybeate water, containing besides a little sulphate of alumina and sulphate of magnesia. This ferruginous water has been conveyed by pipes to the sewage-tanks about half a mile below Buxton, on the river Wye. It is there made to fall into the sewage, to which a small proportion of lime, about 12 grains per gallon, has been previously added, and the more solid portions of the sewage having been removed by iron sieves, to be used as manure. A dense flaky precipitate is formed, the heavier portions of which fall to the bottom of the first brick tank. After that the sewage passes in succession through a series of tanks, the outflow from each being from the top. By the time it has arrived at the last tank it has become almost clear, and the precipitate very fine and sparse. Alongside these tanks is another series of equal size, both sets of tanks being worked together, except when cleansing is required. The sewage after leaving the last tank is made to flow over an artificial waterfall, where further oxidation occurs, and then, passing over a shallow artificial serpentine brook-bed, it issues into the river Wye perfectly clear and without sediment.

The first trial, at which I was present, appeared to be very satisfactory, as the following analysis made in 1890 shows. The sewage, which is entirely domestic, before purification contained free ammonia 11·74 per million parts, albuminoid ammonia 1·00 per million. The purified effluent just before it enters the river contained free ammonia 4·00 parts per million, albuminoid ammonia 0·30 parts per million. The iron water contains 1·2 grains per gallon of iron, which as ferrous sesquicarbonate or bicarbonate amounts to 2·5 grains per gallon. Both the sewage and the purified effluent are distinctly alkaline. The hardness of the effluent is by Clark's method 12·2 degrees.

The works have been constructed from the plans of Mr. J. Hague, the able surveyor of the town of Buxton, and are models of handsome and economical constructive work for this useful purpose. The whole cost of the works, including the sludge-pressing machinery and the building of three houses, was only £6,000 within a few pounds. Should this new method continue to prove to be as satisfactory throughout the year as it is at the present time, great credit will be due to Dr. Thresh, Mr. J. Hague, and the Buxton town authorities, and it points to the useful application of chalybeate waters for this purpose wherever they may be found, as at Leek and other places, in Coal-measure and Yoredale formations.

Some of the springs from the Yoredale rocks around Leek also yield the same ferruginous water, as may be seen in almost any valley where this formation comes to the surface, noticeably below Abbey-Green, where ochreous deposits, which are evident enough in the ditches, drainage-outfalls, and rivulets, are always occurring. It is quite within the reach of practical sewage-problem inquiry that this water could be collected, and the Buxton system applied at Leek, which is only 12 miles distant, and similarly situated geologically, throughout an extensive moorland region to the north, west, and south of Buxton.

The sludge-presses were supplied by Messrs. Johnson, Stratford, London. They are capable of pressing 30 tons of crude or deposited sludge in ten hours, by which they abstract 75 per cent of the moisture, rendering the cake quite hard and portable, and removable from the works in an ordinary cart at any time of the day. There has been no attempt to sell the pressed sludge; the farmers remove it free of cost, and it does not accumulate. It gives very good results in growing grass, turnips, cabbage, &c. The Sewage-works adjoin the Gas-works, and it is quite possible that a further economy might be tried here by using up the refuse gas-lime, either by itself or mixed with the fresh lime now used. The limestone is obtained from an adjoining quarry belonging to the Local Board, and is burnt and slaked at the sewage works. By the side of the Gas Works is a destructor, which burns up all the town garbage, fish, and other offal. It is found to be a most useful acquisition to the town.

It seems to me that instead of continuously running the effluent into the river, the whole or a portion of it might be beneficially used for fertilising the table-lands on each side the valley. It could easily be raised by water-power pumps to the required heights, and might be intermittently applied over a considerable area.

Complaints having been made in 1891 of fouling the river, an examination was demanded by the Board, when it was found that with the exception of a temporary maladministration of the system, which was at once rectified, there was no cause for complaint. Dr. Barwise, the Medical Officer of Health for the County Council, Derbyshire, made a chemical examination of the effluent water as it flowed into the Wye, with the following exceptionally satisfactory results:—

	Parts per Million.
Total solids	3·50
Chloride	25·00
Free ammonia	1·50
Albuminoid ammonia	0·25

Dr. Barwise stated that an effluent showing a quantity of albuminoid ammonia not exceeding 1·00 part per million would be a most satisfactory standard of purity.

Mr. Hague has obligingly furnished me with a plan of the sewage works, drawn to scale, which I reproduce to show the completeness of the method and the wonderfully small cost of an efficient treatment of sewage by precipitation alone.

Glasgow.

Last year I received a letter from a Glasgow friend, a gentleman who takes great interest in sewage questions. His observations are so interesting and suggestive, that I quote them here to show how circumstances and localities affect the question of sewage treatment and disposal, and also to consider if the difficulties which occur at Glasgow are not remediable, and if the questions raised in his letter cannot in some degree at least be answered from an outside point of view. I believe by bringing to the front all the difficulties, faults, and dangers of sewage disposal we shall be doing something towards solving them. He writes:—

1. "We in Glasgow have been not unmindful of the sewage problem. But our difficulties lie here—the mischief to health is not done by our unsightly and unsavoury big drain, the river Clyde, but it is the exhalation of noxious vapours in foul drains between the houses and the street sewers where the real evil arises. Traps and such appliances are all very well as far as they go, but the most fatal diseases we are subject to are

PLATE 8.

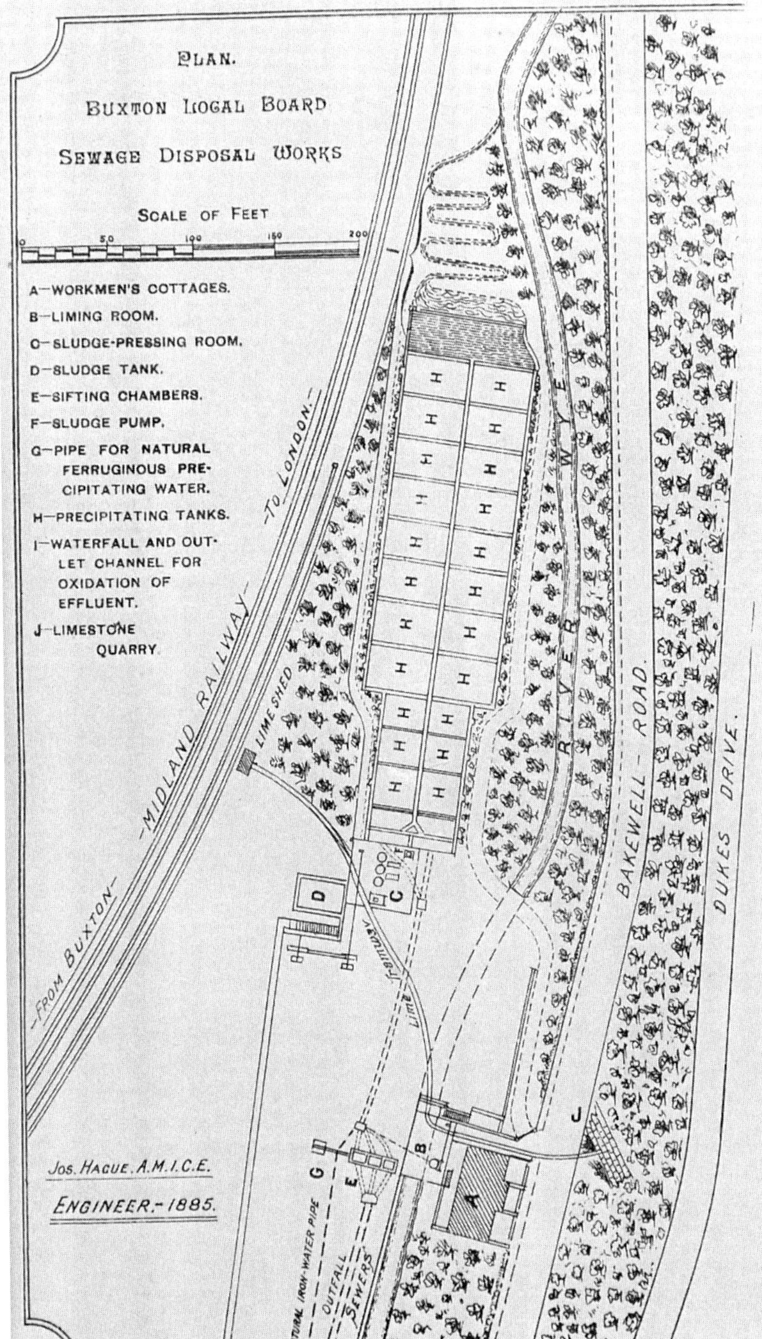
PLAN.
BUXTON LOCAL BOARD
SEWAGE DISPOSAL WORKS
SCALE OF FEET
0
50
100
150
200
A—WORKMEN'S COTTAGES.
B—LIMING ROOM.
C—SLUDGE-PRESSING ROOM.
D—SLUDGE TANK.
E—SIFTING CHAMBERS.
F—SLUDGE PUMP.
G—PIPE FOR NATURAL FERRUGINOUS PRECIPITATING WATER.
H—PRECIPITATING TANKS.
I—WATERFALL AND OUTLET CHANNEL FOR OXIDATION OF EFFLUENT.
J—LIMESTONE QUARRY.
—TO LONDON.—
—FROM BUXTON— —MIDLAND RAILWAY—
LIME SHED
RIVER WYE
BAKEWELL - ROAD.
DUKES DRIVE.
NATURAL IRON-WATER PIPE
OUTFALL SEWERS
JOS. HAGUE. A.M.I.C.E.
ENGINEER.-1885.

invariably, or at least most frequently, traceable to the house draining. However perfect, therefore, our system of treatment and disposal of sewage may be, if it is taken at the street drains, it would have practically no effect in diminishing the death-rate of a town.

2. "We have over and over again proved that the districts on the banks of our abominable river are not the most unhealthy. Indeed, they are healthy districts, and the men who are constantly employed on the river are an exceptionally healthy class of workmen. To keep the sewage out of the river would certainly make it much more beautiful and agreeable to the senses, but that is nearly all. We pause at the enormous expense for such a small result.

3. "Again, even if it were done, the towns up the river would soon pollute it as badly. The valley is large, and there are many growing towns above Glasgow. These have been our difficulties, and I thought you might like to know of them. You will doubtless come to treat of this view before you finish your papers. The treatment of sewage is simply a matter of chemistry, and the removal of it is a matter of engineering.

4. "Glasgow has spent enormous sums on getting reports from specialists on these points, and we all know the thing can be done if we will pay for it. But if anyone can show us a perfect way of preventing the midnight entrance of foul air into our sleeping rooms from the house drains, he would confer a lasting benefit.

5. "For myself, I fear it ever must be a matter of individual action. Every man must be brought to understand his sanitary surroundings, and act in accordance with the laws of health, and keep his own person, his house, and his drains clean.

6. "Still it is a grand thing for a municipality to make their town sweet and clean, and agreeable to visit or to reside in. We are very far behind in that respect in Glasgow. Leek has exquisite natural benefits surrounding it, and I do not wonder at the desire to treat the sewage and remove it, and your

labours will probably get this done in an efficient and economical way. But I doubt if the more serious matter of the death-rate will be affected thereby."

My friend also says he is informed by Dr. Russell, Medical Officer for Glasgow, that the death-rate of Glasgow for the last three years has been as follows:—

1892	23·6
1891	25·3
1890	23·8

The lowest death-rate ever recorded for Glasgow was in 1888, when it was 21·2.

I have put his letter in paragraphs, and will so discuss them.

Nos. 1 and 4.—The danger and nuisance of noxious exhalations from drains I take to be simply bad engineering, which could be most effectively remedied at moderate cost. Most likely all the drains from the houses to the river want relaying, or at anyrate the low part of the city, which is really very foul. The Scotch town system of living in flats, and also having business offices and workshops in them, many of them on the third, fourth, or fifth storey, increases the danger; for it must have occurred to all those whose business requires them to call there, that these toilsome staircases in the older buildings are often reeking with sewage smells, no doubt arising from imperfect trapping or defective pipes. The dissolving power of sewage and sewage vapours are well known, and no doubt the leakages in Glasgow both in the down pipes and in the house and ground drains are enormous.

One has only to see the corroded pipes which have been removed from a fever-stricken house to be convinced of the necessity for strict and constant attention to this part of sanitation, and one at once ceases to wonder why diphtheria or typhoid-fever made that house its home and its occupants its victims. For some years I have been convinced that all town drainage is incomplete without a system of proved acting and effective ventilation; and for that the same may be urged for houses. I should suggest for Glasgow a plan which would be certain to be effectual, and it is that of the ventilation of coalmines—pure air to be drawn into the sewers to fires in various parts of the drainage area, connected with chimney shafts built on purpose for the work. They need not be high, because a properly managed fire would draw the foul vapours through it, decomposing them and passing them into the atmosphere in a harmless state. Nor need the fires be large or their maintenance expensive. They could be placed wherever the sewage gases are known to accumulate.

The difference between an arrangement whereby foul gases are constantly forcing their way out of pipes, drains, and even through traps, and that whereby pure air is having a continuous downcast draught into the sewer, catching the impure air and cremating it, is obvious, and this, I take it, would cure the evil complained of by my Glasgow friend, and very materially raise the tone of health and lower the rate of mortality.

No. 2.—No doubt the Clyde being as far as Glasgow a tidal water, by the curative admixture of salt water neutralises the evil properties of the sewage poured into it, and renders habitation on and near the banks not unhealthy. Still the Glasgow sewage ought not to flow into the Clyde, and the high efficiency of Scotch agricultural knowledge would speedily deal with Glasgow sewage if its engineers have skill enough to distribute it so as to bring it under its powerful sway, and so avoid such wasteful disposal.

No. 3.—The question of "towns up the river" must be answered in the same way. The sewage of one town or village must not be allowed to contaminate those lower down the watershed. Agricultural utilisation, in its various ways, or chemical treatment, or both, must and will be eventually the solution.

What interesting difficulties are coming to the front now the old and costly systems have been tried and found wanting, and the newer ones put on their trial, without having yet served their apprenticeship!

Macclesfield is in this dilemma. The Mayor of Macclesfield, who was chairman of the Drainage Committee, told me that he and his colleagues had inspected many systems in various parts of the country, and had not yet been able to make up their minds as to which system to adopt, and it must be acknowledged that of all places needing good sewage disposal, Macclesfield cries aloud in that respect. So with Manchester, so with Salford.

The following notes on sewage considerations and propositions are not without interest; they relate to Worcester, Cambridge, and Lyons, where the subject is being seriously considered by the several Scientific Engineering and Municipal Authorities.

The Sewage Question at Worcester.

Mr. T. Hawksley, in his evidence before a Local Government Inquiry in 1891, said he was in his 84th year, and was a past President of the Institute of Civil Engineers, and a Fellow of the Royal Society, and had been foremost in sanitary engineering in all his professional experience. The present system of sewerage at Worcester worked well. The Corporation proposed to erect new sewage works away from the city. Population, 50,000.

The dry weather yield of sewage, including the water supply, would not exceed 40 gallons daily per head. The proposed works were therefore to be established to deal in dry weather with two million gallons per day. The sewage would have to be pumped from a depth of 45ft., and 60 horse power would be required. The sewage would be dealt with so as to produce from the residue a solid merchantable cake. There would be twelve tanks, nine to be in use at one time, and three out of use for the purpose of being cleansed. Those tanks would hold 100,000 gallons each, and the ordinary supply would, of course, take about seven hours to work its way through them, but four hours were quite enough for the purpose of desiccation. Only the effluent would go into the river.

The sewage would be pumped up from an outfall well, previously treated with lime and other matters in the well. It would then be brought up by pumps and put in the roughing tanks; there it would deposit the greater part of both the suspended and soluble matter. It would then proceed to the finishing tank, where it would be treated, unless the Local Government Board established some different method, with alum. Then the effluent would go through a covered pipe direct to the river, and no nuisance of any kind would be occasioned in the neighbourhood.

The proposed cost would be from £50,000 to £70,000, and would constitute a rate of 9d. in the pound.

Cambridge Sewerage and Sewage Disposal.

This subject is now occupying the University and Town authorities. Mr. Mansergh, C.E., has been called upon to adjudicate on the respective merits of the two schemes submitted to the local authority for adoption, one by Mr. Anson, the other by Mr. Wood.

Mr. Anson advocates chemical precipitation and filtration through 50 acres of land.

Mr. Wood proposes to utilise the sewage by broad irrigation on a wide area of land.

It is to be regretted that Cambridge should not have put forward a more practical and sounder scheme than, judging from Mr. Mansergh's criticism, is either of the two submitted for his opinion. In the plainest English Mr. Mansergh thinks that Mr. Anson's scheme for the sewerage of the town disregards the principles and primary rules of sanitary science; that it possesses an entire lack of proportion and adaptability, and, therefore, that it is wrong in design. Whilst with regard to Mr. Wood's scheme, he says that, although it is laid down generally upon right lines, it does not strictly adhere to them; adding, in somewhat guarded language, that as the alteration in detail can be effected, he is bound to approve of the latter scheme rather than the former.

In Mr. Mansergh's opinion (which cannot be well gainsaid), the former's estimate is some £23,000 less than it should be, whilst the latter is £15,000 under the mark—a very appreciable omission in a drainage scheme which will approximately cost about £100,000, without taking into account the cost of the land for the disposal works of its preparation.

Mr. Mansergh suggests that 640 acres or so at Shesterton Fen should be laid out for a sewage farm, and the *British Medical Journal* thinks this is a good plan. I think it would eventually prove a nuisance and a mistake.

Lyons.

Lyons distributes its sewage down the Rhone Valley for irrigation purposes.

Kettering.

Population about 20,000. The staple trade is the manufacture of boots and leather.

The sewage disposal works were established in 1884, and consisted of five small tanks, coke-filters, engine, boilers, and lime mixers, the system adopted being similar to that in use at Bradford.

In 1890, the works being inadequate for the population, they were reconstructed and enlarged, and three large tanks, 80 by 30 feet, constructed, together with additional buildings for stores and workshops, sludge-pressing machinery, agitators for properly mixing sewage and chemicals, new storm-water outlet, and effluent-channels.

The process was altered to lime and alumina, and the coke filters abolished. The cost of lime and aluminoferric delivered at the Kettering Railway Station is—lime, 12s. per ton; aluminoferric, £3 per ton.

The storm-water, which could not be accommodated in the tanks, used formerly to be discharged direct into the brook; by the new arrangement, no storm-water can pass as formerly, the

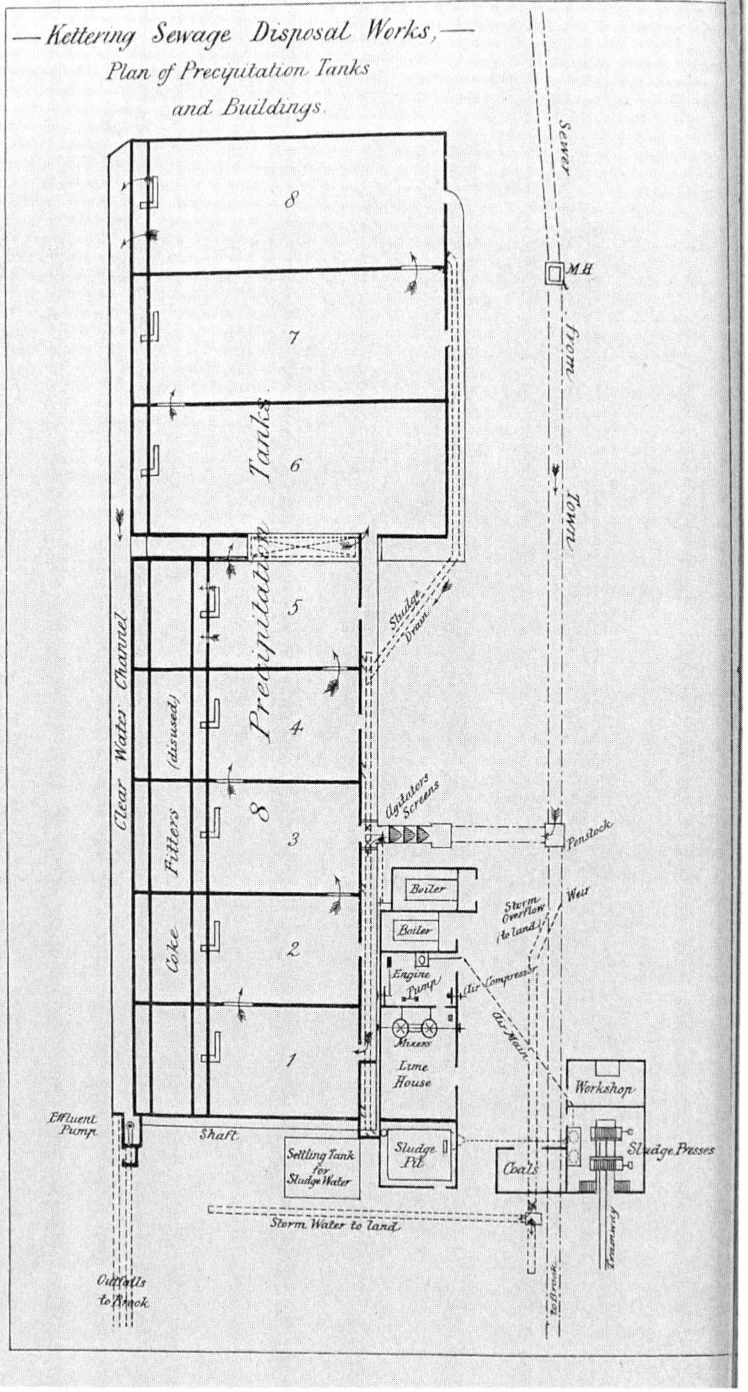
—Kettering Sewage Disposal Works,—
Plan of Precipitation Tanks
and Buildings.
Sewer
from
Town
M.H
Precipitation Tanks
8
7
6
5
4
3
2
1
Clear Water Channel
Filters
(disused)
Coke
Sludge Drain
Agitators
Screens
Penstock
Storm Overflow (to land)
Weir
Boiler
Boiler
Engine
Pump
Air Compressor
Air Main
Mixers
Lime House
Workshop
Sludge Presses
Coals
Sludge Pit
Settling Tank for Sludge Water
Shaft
Effluent Pump
Storm Water to land
Outfalls to Brook
to Brook
Tramway

new outlets delivering it on to the filtering area planted with osiers, through which it is rough-filtered.

The tanks are fitted with overflow weirs laid with white glazed brick, and can be worked either continuously or intermittently.

The works of extension, alteration, and the pressing-plant were designed and carried out by Mr. William Fairley, Assoc. M.Inst.C.E., F.G.S., engineer to the Board, by whose kindness I am enabled to reproduce the accompanying plan of the works, which may prove of assistance to urban authorities, and as an example of a very efficient arrangement for sewage treatment and disposal by precipitation and sludge-pressing.

The cost of these works amounts to £700 a year.

The Nuneaton Sewage Outfall-works.

These works are a model of a carefully-thought-out arrangements for the treatment of sewage by precipitation and filtration. The system of precipitation at present in operation is my own, which is acting admirably, and under extremely difficult conditions, owing to manufacturing refuse wash from Fellmongers' and Woolscourers' works, which ought and must be treated severally at the works, and their effluents only be allowed to enter the public sewers.

The average daily flow of sewage is about 400,000 gallons, about 25 per cent of which is composed of waste liquors from Fell-mongers, Wool-scourers, and Felt-hat works.

Upon entering the outfall works, the sewage is received into a screening chamber B of the accompanying plan, where any foreign matter and articles likely to damage the pumps are intercepted by means of an iron screen. A penstock in the chamber A is used for shutting off the sewage during the removal of road detritus and other heavy matter, from a sump in the screening chamber. Passing to the penstock chamber C, the sewage may either be conveyed along the outfall sewer to the pump-wells in the engine-houses D and E, or delivered into an underground storage tank. This storage tank is generally used for the night and Sunday flow of sewage, which is pumped out the following day. The two pump wells are connected by a heading 5 feet in diameter, and are 46 feet and 50 feet deep respectively. There are four pumps, which together are capable of raising 80,000 gallons per hour. The sewage is pumped into the delivery chamber F, and, passing over a weir, receives the chemical precipitant from a serrated iron trough. Having received the precipitant, the sewage may be delivered to either of the four settling tanks, each of which has a capacity of 120,000 gallons, or, if it is desired to work on the continuous-flow principle, it may be discharged into tank No. 1 at the point G, passing over a weir at the other end, along the channel J, into tank No. 2, and along the channel K, through tanks No. 3 and 4, to the floating outlet.

Additional chemicals may be applied, if found necessary, at the entrance to either tank. By an adjustment of the sluices, and the use of the channels H and V, either tank may be cleansed while the remaining three are in use. Each settling tank is furnished with a floating valve, through which the supernatant liquor flows into the respective conduits, which converge into one main channel at the point N. The pipe sewer P S is used for returning the contents of either settling tank to the pump wells for re-treatment when necessary.

The tank effluent flows along the channel R to the "polarite" filters. These are arranged in two sets of four, each filter having an area of 100 square yards. The process of filtration completes the treatment, and the filtered effluent is discharged into the river through the channels S and T. The bye-pass channel W is used in times of heavy rainfall, when the tank effluent is discharged direct from the settling tanks to the river. The flocculent matter from the surface of the filters is removed by a system of upward washing, and is conveyed by the carrier Y to a plot of drained land. The tank Z receives a quantity of the filtered effluent, which is lifted by a small pump in the press-house for the use of the boilers. The sludge from the settling

PLATE 9.

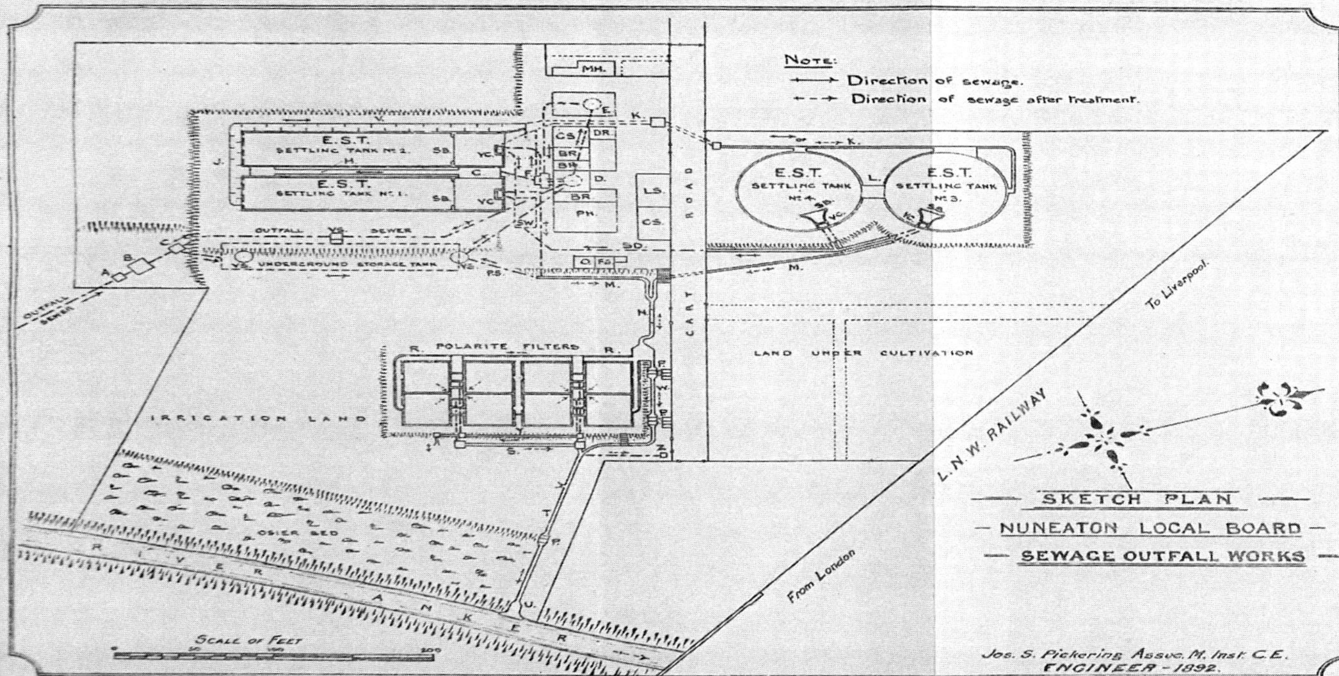

Note:
Direction of sewage.
Direction of sewage after treatment.
E.S.T. SETTLING TANK No. 2.
E.S.T. SETTLING TANK No. 1.
E.S.T. SETTLING TANK No. 4.
E.S.T. SETTLING TANK No. 3.
OUTFALL SEWER
UNDERGROUND STORAGE TANK
CART ROAD
POLARITE FILTERS
IRRIGATION LAND
LAND UNDER CULTIVATION
OSIER BED
RIVER ANKER
L. N. W. RAILWAY
To Liverpool.
From London
SCALE OF FEET
SKETCH PLAN
NUNEATON LOCAL BOARD
SEWAGE OUTFALL WORKS
Jos. S. Pickering Assoc. M. Inst. C.E.
ENGINEER - 1892.

tanks gravitates to the sludge well S W. After the top liquor has been discharged into the pump well through a floating arm, the sludge is pumped to the press-house, where it is made into cakes by two of Johnson's sludge presses, the cakes being given or sold to farmers as manure. I am indebted to Mr. J. S. Pickering, Asso.M.Inst. C.E., Engineer and Surveyor, Local Board of Health, Nuneaton, for this description and plan.

Reference to Plan of the Nuneaton Sewage Works.

A.C.	Penstock chambers.	Z.	Effluent water-tank.
B.	Screening chamber.	U.	Outfall.
D.E.	Engine houses.	P.S.	Pipe sewer.
F.	Pump delivery chamber.	S.D.	Sludge drain.
G.	Screen.	V.S.	Ventilating shafts.
S.B.	Scum boards.	M.H.	Manager's house.
V.C.	Valve chambers.	D.R.	Steam disinfector.
H.J. / V.K.L.	Sewage channels.	G.S.	General stores.
		B.R.	Boilers.
M.N.R.W.	Tank effluent channels.	P.H.	Press house.
E.S.T.	Four effluent settling tanks.	L.S.	Lime shed.
S.T.	Filtered effluent channels.	C.S.	Chemical stores.
V.	Scum carrier.	O.	Office.
P.	Cascades.	F.S.	Fitting shop.

The Richmond Main Drainage Precipitation Works.

Amongst the many attempts to solve the sewage question, the Richmond main drainage scheme stands conspicuous both as a well-considered measure and for the practical assertion of the superiority of the treatment of sewage by precipitation over all others. The works were opened in September, 1891, by Mr. Richie, then President of the Local Government Board, and are probably the most complete and extensive sewage treatment and disposal works, as well as the most recent, in the kingdom. An interesting and exhaustive official description has been published by the Main Sewerage Board; it is entitled "The Richmond Main Sewerage and Purification Works." It contains two illustrations, one a map of the whole of the drainage districts, and the other a descriptive plan of the sewage works. I have reproduced the latter as a really valuable guide to the construction of precipitation works. The following description is compiled from the report which has been kindly furnished me by Mr. W. Fairley, A.M.I.C.E., the resident engineer. I have also availed myself of quotations from an excellent description of the works on the opening day in the *Daily Graphic* of September 29th, 1891.

These works were constructed for the purpose of collecting and disposing of the sewage of Richmond, Mortlake, Barnes, Kew, and Petersham, a district forming the Richmond Poor Law Union.

The sewage of these places had hitherto been partly discharged in its crude state direct into the river Thames, and partly retained in cesspools, to the detriment of the health of the inhabitants. Both these evils have been remedied by the present works, which are due to pressure put upon the localities named by the Local Government Board and the Conservators of the Thames Conservancy Act, who, fourteen years ago, gave notice for the discontinuation of the flow of the sewage into the Thames.

The continued non-compliance was not due to neglect, but, on the contrary, there had been great anxiety by the local authorities to cope with the difficulty.

In 1874 the Thames Conservancy Board took legal proceedings for polluting the Thames, and obtained penalties amounting to £151 and costs, with a further loss in penalties and expenses by an arbitration of £700.

In 1870, the general opinion was in favour of dealing with sewage by irrigation, and the vestry, under the advice of Mr. Bailey Denton, provisionally purchased 230 acres of land at Malden for a sewage farm. After six days' local inquiry by Mr. J. Thornhill Harrison, C.E., on behalf of the Local Government Board, and at which the Duke of Cambridge, the Metropolitan Board of Works, Mr. Garth, and others were strong opponents, the scheme was rejected, and it became clear that no irrigation scheme was possible for Richmond.

Eight separate schemes have been submitted to the Local Government Board, and all have been rejected. Besides these,

RICHMOND MAIN SEWERAGE

Purification Works situated at Mortlake.

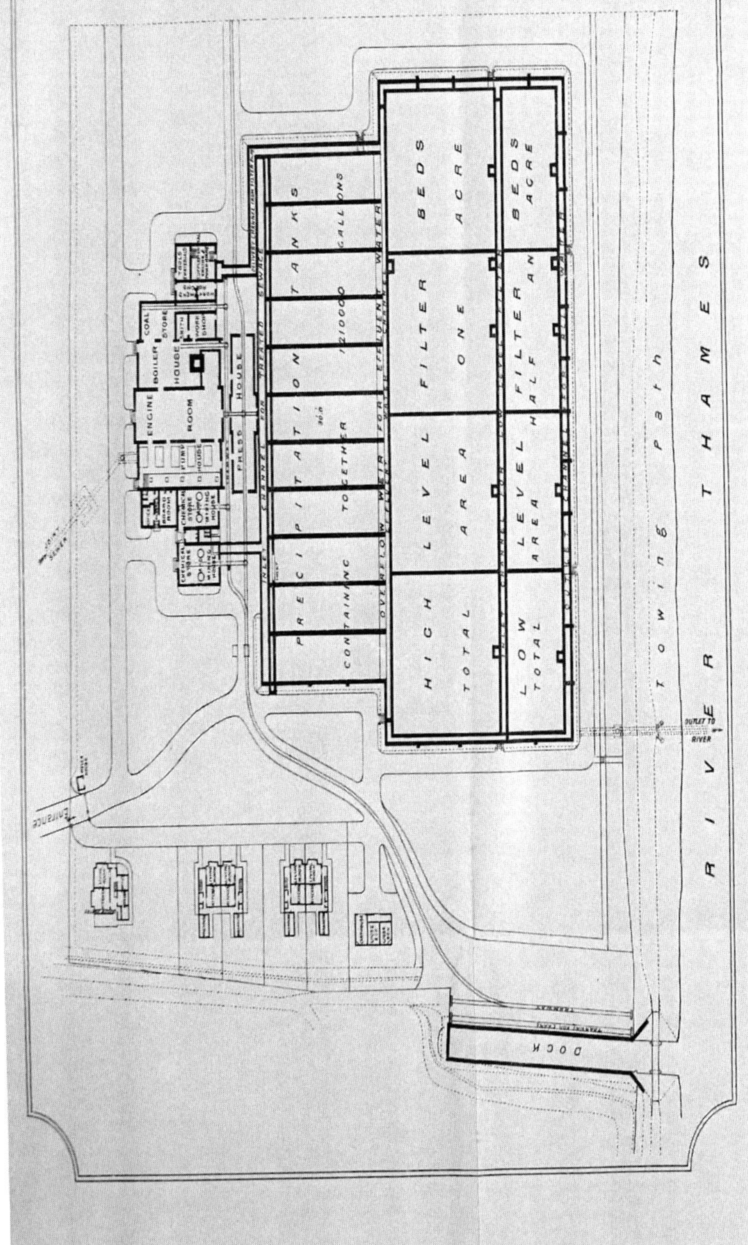

many others have been examined, and the most eminent engineers have been consulted. The costs of these abortive efforts have been nearly £45,000, and to such costs there should be added the costs of the persons and public bodies appearing in opposition to them.

In 1885 Mr. J. C. Melliss, of London, proposed a ninth scheme, which is the present one, and which was confirmed by Parliament in 1887. The works were three years in construction, and cost £120,000. The drainage area comprises a population of about 41,600 persons, but the works are large enough for a future population estimated at 80,000 persons; the outfall sewer having a discharging capacity of 11 million of gallons per 24 hours. The present quantity of sewage for treatment is 2½ million gallons per day; in wet weather, 5¾ million gallons. The present area is 4,983 acres, and the rateable value £316,056.

Flushing of the sewers is provided for by the admission of the Thames from a higher level. Ventilation of the sewers is provided for by Holman Keeling's Patent Sewer Gas Destructors, and by connection with the Works chimney they serve to exhaust all the foul air from the sewers; open gratings in the manhole covers acting as fresh air inlets.

The site comprises 11 acres, of which only seven are at present used.

The accompanying drawing shows the arrangement of the works, which comprise the following: Straining chamber; pump chamber and pump house; engine room; boiler house; coal store; chemical stores and mixing houses; press house and sludge chamber; board-room, office, and laboratory; smiths' and work shops; workmen's day-rooms; store rooms; supplemental engine and pump house; eleven precipitating tanks together holding 1,210,000 gallons; one and a half acres of filtering beds; dock, wharf, and tramways; foreman's cottage and four cottages for labourers; stable, cart-house, &c.

The following particulars give a description of the machinery employed:—

A straining apparatus for the removal of large solid matters.

Three sets of single-acting plunger pumps for lifting the sewage 43 feet, with three 50 horse-power compound, horizontal, intermediate receiver, condensing engines for working the pumps. Each set of pumps, with its engine, is capable of raising about four million gallons of sewage in a day, the three together being capable of lifting 12 million gallons per day. Each set of pumps can be worked separately, one of them being sufficient for dry weather, a second set being brought into use in wet weather, and the third held in reserve in case of accident to either of the others. Space is provided in the buildings for the accommodation of two more similar sets of pumps and engines, if in the future they should be required.

Three of Galloway's boilers, each 18ft. long by 6ft. 6in. diameter, any one of which will supply sufficient steam for doing the fine weather work, the other two being available when wanted, as in the case of the pumps and engines, there being space for an additional boiler provided for future requirements. These boilers also supply steam to the other machinery.

A pair of 14 horse power vertical engines for mixing chemicals, driving tools in workshop, &c., one engine being sufficient to do the work. A Worthington steam-pump for lifting water from wells, for the boilers and other requirements on the works.

A pair of large Worthington steam-pumps for use in case at any time a continuance of high tide and flood in the Thames should prevent the discharge of the effluent water by gravitation. These pumps would then lift it to flood level.

Four vats for dissolving chemicals with agitators; also an agitator for thoroughly mixing the sewage and chemicals together.

Three of Messrs. Johnson and Hutchinson's patent 36in. square sludge presses, each containing 30 plates and chambers, together with all requisite sludge pumps, forcing rams, liming and skimming apparatus necessary to reduce sludge from a condition containing 90 per cent of water to a portable and inoffensive "sludge cake" containing only 50 per cent of moisture. The various parts of this sludge-pressing machinery are provided in duplicate, in case of accident.

Shafting, belting, together with workshop tools, including lathe, &c.; also tramway, weigh-bridge, trucks, and a steam crane for unloading coals and chemicals at the dock and wharf, and conveying them into the buildings, are all provided.

When the sewage rises to a certain height in the sewers and the pump chamber, the fact is communicated by an automatic arrangement of electric bells to the foreman's residence and the engine-room, and attention called thereto, thus avoiding any chance of overcharging the sewers.

The sewage arrives at the works in a fresh state before any decomposition has had time to take place. It is strained through an iron bar strainer before entering the pump chamber; cabbage stalks, rags, and similar articles being thus removed, and daily disposed of by being dug into the gardens. In quantity this will be about a hundredweight per day.

The strained sewege then flows continuously into the pump chamber, where it receives a small dose of milk of lime from the vats in one of the chemical mixing rooms. It is then pumped to the surface, the operation of pumping also serving to thoroughly mix the sewage and the milk of lime together. The partially treated sewage then flows from the delivery pipes of the pumps by a covered channel along the side of the pump-house. It then flows continuously to the inlet channel for treated sewage, passing in its course through a close iron bar strainer, which removes any large matter that may have passed the first strainer; also through the second chemical mixing room, where it receives a dose of a solution of sulphate of alumina, carbon, and iron, and is thoroughly mixed therewith by an agitator placed in one corner of the chemical room.

I may observe here that the precipitation process of the Standard Sewage and Water Purification Company would serve as a most effective precipitant, and, being a solution, is most easily applied.

The treated sewage is admitted continuously into one or more of the precipitation tanks, as required, by the opening of valves, and flows continuously over weirs extending the entire width of the tanks. As the tanks fill precipitation takes place, the precipitate rapidly falling to the bottom of the tanks, while the effluent water, now deprived of the whole of its polluting matter in suspension and nearly all of that in solution, flows continuously over weirs at the other end of the tanks. This effluent water is sufficiently pure to be discharged direct into the river, and is passed thereto by the outlet provided at the north-east corner of the works.

The effluent water can be raised to a higher degree of purity whenever required by filtration through filter beds, constructed of layers of various sized gravel, sand, and carbon, the surfaces being covered with a thin layer of agricultural earth sown with grass.

The precipitation tanks are emptied of their entire contents once a day, and when a tank has to be emptied it is shut off from the inlet treated sewage channel by valves, and allowed a short time for repose; the water is then drawn off from the tank by a floating valve, the upper portion of the water, owing to the levels, being discharged on to the high level filter-beds, and the lower portion on to the low level filter beds.

The precipitate (commonly termed "sludge") at the bottom of each tank averages from two to three inches deep, being about 90 per cent water. The water drawn off from the tanks and passed away to the filters, or the river direct, is not, however lowered quite down to the precipitate, the water immediately overlying the sludge for a depth of six or seven inches being retained with the sludge.

The bottoms of the tanks slope so as to form a channel, and the sludge and water last mentioned together gravitate, with the aid of sweeping, from the tanks into a covered sludge passage passing along their upper ends, and thence into the sludge chamber beneath the press house.

In this sludge chamber precipitation still rapidly takes place, and the water which rises to the top of the sludge is drawn off, but not allowed to pass to the filter beds or the river, but flows back again into the pump chamber, mixes with the sewage, and is therewith treated over again. The quantity of water thus dealt with is comparatively very small.

The sludge, having thus been deprived of some of its water in the sludge chamber and brought into a favourable condition for pressing, is lifted by pumps into iron receivers placed in the press house. These receivers each contain a charge for one press. In these it is limed—that is, a small quantity of lime is added to it, so as to facilitate the operation of pressing, and it is forced by rams into the presses. The water pressed out of it (again a comparatively small quantity) passes to the pump chamber and mixes with the sewage to be treated over again, and the solid "sludge cakes," measuring 3 feet square by 1½ inches in thickness, are removed from the presses.

Sludge in this pressed condition occasions no nuisance whatever, and may be stacked and stored without creating even an annoyance.

For agricultural purposes it has a manurial value rather higher than that of the best farmyard manure. The quantity that will be produced from the present population will be about ten tons daily, and considering the great demand for manure of all kinds by the market gardens surrounding the locality, there should be no difficulty in disposing of it.

Pressed sewage "sludge-cake," which contains about 50 per cent of water, and readily parts with such water, is very suitable for making up and raising low-lying land, and the Board will be able to utilise much of it for this purpose of raising to a higher level, and thus materially improving, their own land.

The entire cost of the works, including the sewers, will amount to between £115,000 and £120,000. The exact outlay has not yet been, and cannot at present be, ascertained.

The annual expense of carrying on the works, including

pumping the sewage, is estimated at £3,500, and this, together with the repayment of capital and interest, will not, it is expected, exceed a rate of 7½d. in the pound for the entire cost.

The charge is borne in due proportions by the urban and rural parishes, according to their respective rateable values from year to year.

Mr. J. C. Melliss has been the engineer for carrying out the works.

Mr. James McKie, Assoc.M.Inst.C.E., acted as resident engineer during their construction, and Mr. Fairley is the present resident engineer.

The contractors who have completed the sewers are Messrs. Nowell and Robson, of Kensington; and the pumping station and precipitation works have been completed by Mr. William Webster, of St. Martin's Place, London. The contractor who commenced the undertaking was Mr. C. Dickinson. The pumps, engines, boilers, and mixing machinery have been supplied and erected by Messrs. James Simpson and Co., of Pimlico. The pressing plant and machinery have been supplied and erected by Messrs. S. H. Johnson and Co., of Stratford. The cottages have been erected by Mr. George Wade, of Chelsea; and the fencing has been supplied by Mr. Parham, of Bath.

CHAPTER XII.

SPENCE'S ALUMINOFERRIC PROCESS.

THE Spence aluminoferric process is the invention of the late Mr. Peter Spence, of the firm of Messrs. Peter Spence and Sons, Manchester Alum Works, Manchester. The author of the process asserts with truth there is now a general recognition that purification by precipitation, either with or without subsequent land filtration, is the only practical solution of the problem in the vast majority of cases, and asserts that, for this purpose, the substance he names "aluminoferric" is the most efficient and economical precipitating agent yet introduced. This substance contains 46·7 of sulphate of alumina. He affirms that one ton of aluminoferric will purify three million gallons of average sewage. The cost of it is £2 10s. to £3 per ton. The advantages claimed for it are:—

1. To render the albuminoid matters of the sewage insoluble, thereby enhancing the value of the sludge as a nitrogenous manure, the decolourising of the effluent, and the rendering both effluent and sludge practically free from smell.

2. That, it being entirely soluble in water, the weight of sludge it produces is practically no more than that of the impurities it precipitates.

3. There is scarcely any labour-expense incurred in the addition of aluminoferric to the sewage; any ordinary labourer will, in a few minutes per day, do all the work it involves. It requires simply to be put into the channel or conduit (in a perforated wooden box, or cage made of iron bars), and sufficiently immersed in the sewage, to admit of the necessary proportion being dissolved to effect purification.

4. That it is the most economical and efficient precipitating agent yet introduced.

5. That it is the cheapest known form of soluble alumina and iron for precipitation purposes.

6. That it completely obviates the two serious objections to the use of lime, viz., the great weight of sludge produced, and the coloured, highly alkaline, and consequently putrescent character of the lime effluent.

It is produced in solid cakes or slabs measuring about 21in. by 12in. by 5in., and weighing about 56lb. each. It can be used alone or in conjunction with lime. This compound is also used to purify refuse waters from manufactories, but without lime. With some kind of sewage lime is used with advantage. Appended to the aluminoferric prospectus of date 1888 are letters from the surveyors of the following towns, all praising its efficiency: Wimbledon, Hinckley, Swinton, and Pendlebury, Brentford, Aldershot, Long Eaton Rural Sanitary Authority, Atherstone Union, Saffron-Walden. The letter from Atherstone is interesting in a twofold point of view, namely, the purification of sewage and the decolourising of dyewater. The surveyor remarks: "As the sewage contains the dyewash from six hat factories, and is continually black, it has been a source of difficulty to deal with it, for a long time having used lime and other recommended precipitants, which have failed. But the effluent is now perfectly clear, and has been tested and found chemically purer than the water in the river into which it flows. The works are simple, and open for inspection, and it is stated that results can be seen."

If this method for decolourising dyewater is really efficient, and would not destroy fish after its being so used, there would be no need of our rivers being any longer discoloured, as the slabs of aluminoferric might be placed in the sewers so that the dyewash could run over them and be decolourised.

The aluminoferric process is in operation in the manufacturing village of Whitefield, near Manchester. The village contains several dyeworks, and is frequently heavily charged with dye-water, but I am informed by Messrs. Peter Spence and Sons that after treatment with the aluminoferric a perfectly clear effluent is obtained. They inform me that their process is alone used at Eccleshill and Guiseley, in Yorkshire.

From a letter written by Messrs. Spence and Sons to the *Contract Journal*, in 1891, I extract the following statements:—

"After experimenting on every possible description of sewage with alumina, iron, and other ingredients, and perceiving that engineers and sewage committees could not possibly use a high-priced material, we were led to invent our now well-know aluminoferric.

"MODE OF APPLICATION.—Nothing can exceed the simplicity of our process. The aluminoferric is produced in solid cakes, measuring 21in. by 12in. by 5in., and weighing about 56lb. each. These are placed either in a stout wire cage, or a cask or box freely perforated with large holes. This is immersed in the conduit or stream of sewage leading to the precipitating tanks. As the cakes dissolve, fresh ones are added. This is all the labour required, and need not occupy an ordinary labourer more than a few minutes per day. No machinery or plant of any kind is necessary beyond what we have named. We recommend in many cases the use of a small amount of lime along with the aluminoferric. As the proportion required depends upon the character of the sewage, a few experiments with an average sample of the sewage will indicate the proper quantity. We do not recommend the quiescent system of using the tanks—*i.e.*, the plan of filling up a tank and then shutting off the flow till its sludge has settled—but the continuous system, the raw sewage with its proportion of aluminoferric and lime going in at one end of the series of tanks, and the purified effluent flowing out over the last lip at the other. If there is adequate tank capacity, complete precipitation will take place during the passage through the tanks.

"RESULTS OBTAINED.—What is really necessary in sewage purification is the production of an effluent that is clear, colourless, and, above all, permanently non-putrescent. The Local

Government Board have for the present, we think somewhat unwisely, insisted that plans for sewage treatment submitted for their sanction must be either land filtration schemes, or must include provision for passing the effluent through land. As, however, land filtration, even in those cases where it is possible to procure land, almost always involves a very heavy outlay, and is, moreover, liable to cause a nuisance if the sewage has not been previously treated with aluminoferric, and as the aluminoferric precipitation process alone, without filtration of any kind, produces an effluent as clear and colourless as natural river water, and permanently non-putrescent, there is, it seems to us, no real necessity for land filtration.

"ACTION OF THE ALUMINOFERRIC ON SEWAGE.—When a cake of aluminoferric is immersed in a stream of sewage, a remarkable change takes place in the liquid, which is quite evident on comparing a portion of it after passing the cake with a sample taken before. The sewage before passing the cake is dark and muddy, and on standing does not improve. After passing the cake it is seen to be separated into two portions—a clear, bright, inodorous effluent, and a curdy precipitate which rapidly settles. This is equally the result whether the sewage is a purely domestic one, or contains also a large proportion of dyers', tanners', and other manufacturers' refuse. The aluminoferric, by the action both of its bases and acid, throws down the suspended matters and destroys the smell.

"The power of alumina in combining with dissolved organic matter, and so removing it from solution, is taken advantage of in many commercial processes." The report of the Royal Commission on Metropolitan Sewage Discharge, 1884, says: "The artificial precipitate, or coagulum, produced in the fluid, mechanically entangles and carries down the organisms into the sludge." It is well known to chemists that the phosphates present in sewage tend powerfully to promote putrefaction; and the fact that the aluminoferric removes from the sewage every trace of phosphoric acid, may, to some extent, explain the remarkable freedom of its effluent from after-putrescence.

"One very great advantage of the aluminoferric process is that it solves, for all practical purposes, the problem hitherto left unsolved by all other processes, viz., the treatment of storm water. No existing land filtration area, artificial filter, or series of precipitating tanks, can grapple with the enormous volume of storm water which at times rushes down the sewers and inundates all ordinary arrangements. With the aluminoferric process, however, all that is necessary to be done is to give the storm overflow, which cannot be treated in the tanks, its proper proportion of aluminoferric. The result is that the impurities are rendered insoluble, the oxygen of the sewage is left intact, and fish life is probably no more affected by the insoluble precipitate than it is by the clay and earth suspended in the storm water. Any portion of this precipitate, not carried to sea by the flood, is in a condition to be disposed of by the organisms in the water, which would be killed by the soluble, and therefore oxygen-removing, impurities of the untreated sewage.

"There is no difficulty whatever in adding, night or day, the proper proportion of aluminoferric to the storm water, as we compass this by an automatic arrangement without moving parts, and which cannot get out of order.

"THE SLUDGE.—A fact of great practical importance is that the sludge produced is of extremely little weight, is entirely free from offensive smell, and when air-dried is found on practical trial to be, weight for weight, equal in crop-giving value to farmyard manure. This is borne out practically by the statement of Lieut.-Colonel Jones, V.C.: "He had tried sewage sludge side by side with farmyard manure, and with 60 per cent of moisture it was, bulk for bulk, rather superior to the manure."

"COST OF THE ALUMINOFERRIC PROCESS.—Experience proves that one ton will purify quantitie varying from one to three million gallons, according to the degree of impurity of the particular sewage. As the price of aluminoferric is £2 10s. free on rails Manchester, and as it is always supplied of uniform strength, the cost of treatment can be readily calculated in any given case after one or two experiments with the sewage. In all cases it will be found to be a very small amount per head of the population.

"There is one serious practical difficulty which we must frankly admit the aluminoferric process will not overcome. It will not produce any effect whatever upon the sewage in those cases where a mistaken economy leads sewage committees to save the rates by keeping the precipitating material locked up in their stores whenever they think their effluents are not under observation."

Messrs. Spence and Sons go on to say: "There is, we believe, one and only one way of dealing with both these classes of cases. The sanitary authorities have practically unlimited power over manufacturers who send polluted matter into their sewers, but towards the authorities themselves the Local Government Board have extended almost unlimited indulgence.

"Let the Local Government Board now require every sanitary authority to produce (by whatever method) a permanently non-putrescent effluent, and appoint permanent inspectors to see this requirement carried into effect.

"All chemists agree that a permanently non-putrescent effluent will withstand the following simple test: Retained for any length of time in a light room at ordinary atmospheric temperature (60 deg. F.) in a bottle half filled with it, and tightly stoppered, it will not, when the stopper is taken out, give off a putrid smell.

The Local Government Board would find this system much more simple in operation than their administration of the Alkali Works Regulation Act, under which their inspectors have to visit each separate works. Every chemical works sends out its gaseous sewage direct into the air, instead of—as in the case of the liquid sewage of local sanitary districts—into one or two common sewers.

In the latter case, therefore, two or three outfalls, at most, would have to be sampled by the inspector in each district; every outfall, of course, being so placed by the local authority as to give free access to it by all persons at all times of the day and night.

If this course were adopted by the Local Government Board, we have no doubt that every river and stream in the United Kingdom would be entirely and permanently freed from sewage impurity within twelve months.

Messrs. Spence and Son's process was tried at the Salford Corporation Sewage Works. The following results are recorded in the tables of the two chemists who reported upon it:—

	Mr. Carter-Bell.	Dr. Burghardt.
Effluent from Tanks.	Parts per Million.	
Free ammonia	17·55 ...	8·11
Albuminoid ammonia	3·00 ...	4·54

Mr. Carter-Bell's figures are the mean of two analyses, and he reported that he feared secondary decomposition when the effluent got into the Ship Canal.

Dr. Burghardt's are the mean of nine analyses. He reported that a partial secondary decomposition took place after some time. It will be necessary to wait for the result of trials on a larger scale which Messrs. Spence and Co. are anxious to make before forming a correct conclusion as to the merits of this process. They were dissatisfied with these trials, and threatened to take action against the Corporation, alleging erroneous conclusions and inadequate trial.

MR. BIRCH'S SYSTEM OF FILTRATION.

For sometime past there has appeared in the Manchester papers the following advertisement: "Rivers Pollution Act. 10,000 gallons per hour of sewage is made as brilliant as well-water at the Sewage Outlet, Soho Street, Radcliffe, from 9 to 5 daily, by W. Birch's System, of Milton Street Ironworks, Lower Broughton, Manchester. Members of Local Boards are invited to inspect."

This attractive statement led me to visit the scene of operations. Radcliffe is a town of about 20,000 inhabitants.

The river Irwell here is broad and rapid, with high banks; on the top of the left bank, close by the river side, is Mr. Birch's system in constant operation.

Mr. Sharples, Clerk to the Radcliffe Local Board, was present, and kindly explained the system to me. Everything in the advertisement is correctly but not fully stated; but as the operations are very interesting, a detailed account may be useful. Those who desire to see how a foul black sewage can be perfectly decolourised should visit these temporary works without loss of time. Twenty yards below them the whole of the sewage of Radcliffe, except that portion being operated upon by Mr. Birch's system, rushes into the river in a black torrent, whilst at the works is poured into the river the sewage in a clarified and decoloured state. Nothing can be more striking, and this remarkable change is accomplished in a few minutes by what the patentees call Birch's system. Although I have nothing to say in hostile criticism of this so-called system, my duty and effort being to describe and examine accurately each method which puts forward claims for attention, it is one which shows the necessity of taking *cum grano salis* the way by which its claims are stated. Two things struck me in examination: First, it is not a process for sewage treatment—*i.e.*, precipitation; the advertisement does not say so. There is precipitation, but that is simply the application of Spence's aluminoferric process, which I have fully described. The sewage so treated, after being allowed to settle, passes into a filtering machine, which I will describe presently. The claim, therefore, is limited to a filtration of the effluent, and this filtration is perfectly accomplished. First the black sewage is pumped from the sewer into a cistern, from which it runs through a series of wooden troughs, say one foot wide, with here and there alternate intercepting strips of wood, to break up the current and to cause a thorough mixing of the sewage and the precipitant. At the upper end of the trough is placed a block of Spence's aluminoferric cake, perhaps 6 to 10 lbs. in weight; over this the sewage flows and passes on to a large iron settling tank, where precipitation immediately takes place, the bulk of the precipitant falling down, but leaving the effluent only partially clear. It is at this point that Mr. Birch's system comes into play. The system is described by him in his circular, "For filtering supernatant water from settling tanks," and that is what it is, and that only. The supernatant water passes on from the settling tank to a trough, over which passes an endless sheet of very thick porous flannel or woollen cloth, when wet more than half an inch thick, and very spongy. Over the top of this cloth the effluent is made to pass; it passes through very quickly; the 10,000 gallons per hour percolates this filter in a space of about four yards in length by from 40 to 50 inches in breadth. The cloth is constantly revolving over and under a series of four cylinders or drums, the upper or dirty side of the cloth being cleaned as it passes the second cylinder by an automatic roller attached to the cylinder underneath the pulley, which squeezes out the dirty particles. The effluent then passes to an overshot wheel of about 8 feet diameter, the purpose of which is to provide motive power to work the filtering machine and to keep its operation in constant motion.

There is no cessation, no time allowed in the tank for settling; a constant stream goes in and out. The filtering machine is evidently not intended for sewage filtration, but for effluent filtration; the waterwheel discharges the effluent into the river, and nothing is more striking than to see two volumes of liquid being poured into the river within a few yards of each other, the filtered effluent pellucid, the other, the sewage, like ink. It is necessary to state, from the information I obtained, that the

WM. BIRCH'S AUTOMATIC SEWAGE CLEANSING, FILTERING, AND SLUDGE COMPRESSING PLANT,
FOR ALL CLASSES OF POLLUTION WHEN NOT TOO ALKALIED.
Treating 7,000 Gallons per hour and upwards to 20,000 according to class of pollution.
SLUDGE COMPRESSOR (AUTOMATIC)
SLUDGE WELL
FILTERING MACHINE
6 Ft BROAD 18 Ft LONG
SELF CLEANING
SLUDGE MUCH THICKER THAN ORDINARY TANK SLUDGE
PUMP WATER FOR CLEANING FILTERING MEDIUM
These Machines can be had Sepcrately
Wm BIRCH,
ENGINEER,
MILTON St IRON WORKS,
LOWER BROUGHTON,
MANCHESTER
ENTIRELY•AUTOMATIC WITH WATER WHEEL,
OR ONE QUARTER HORSE POWER WILL DRIVE THIS PLANT WITHOUT THE WATER WHEEL.
Fig. 45.

sewage of Radcliffe is mainly refuse-wash from manufactories, the principal part being dyers', printers', and bleachers' wash; this being so, there is ample ocular demonstration of the power of the aluminoferric cake to decolourise dyewash.

This leads me to a necessary digression and repetition.

It is necessary to state with regard to the Radcliffe sewage that it is chiefly manufacturers' refuse, the whole town being a privy arrangement, with but very few water-closet exceptions. They have the ordinary closed midden system, and the excreta is removed and used for agricultural purposes; so we have nothing to guide us in the experimental system at Radcliffe how it would operate if the sewage were water-closet sewage, and not manufacturing.

The following is a copy of Mr. Birch's circular, to which is attached the engraving of his filtering machine, which he has kindly permitted me to copy:—

"Milton Street Iron works,

"Lower Broughton, Manchester.

"GENTLEMEN,—By the kindness of the gentlemen composing the Radcliffe Local Board, I am working for a short time only my patent tank and filtering machine combined, cleaning sewage which varies in colour and pollution as often as two or three times in one hour. This pollution comes from dye houses, soap works, and manufactories in the town of Radcliffe, besides household pollution. This is the dirtiest sewage I have ever seen, and I am glad to say the effluent is as transparent as well-water, and more brilliant than the town's water.

"The machinery is situated in Mount Sion Road, Radcliffe, near Manchester, and I shall be very glad to show you and your friends the machine, which is the result of more than two years' work. I am now treating at present about 10,000 gallons per hour. The filtering machine is automatic, being driven by the filtered effluent driving a water wheel, which again drives the filtering machine.—Yours very respectfully,

"WILLIAM BIRCH."

I have not had opportunity to analyse either the Radcliffe sewage or effluent, and am consequently unable to pronounce on the state of organic sewage impurity or upon the proportions of free or albuminoid ammonia in the effluent.

It is proposed on the large scale to deal with the precipitate or sludge at Radcliffe in the usual manner, but it can hardly be expected to possess much, if any, fertilising property, being, as I have stated, principally manufacturing refuse sewage.

The Barry Patent Manure Company Process.

This system has also been tried on a practical scale by the Salford Corporation, side by side with the electrical and aluminoferric processes. According to their report, the process consists of adding a solution of perchloride of iron, containing six per cent of iron, and a powder which is refuse lime from the gasworks, containing carbonate of lime and free lime, sulphate of lime, &c.

The iron is obtainable from the spent oxide of iron of gasworks. It does not appear to have been successfully reported upon by Corporation chemists.

Spencer's Magnetic System or Carbide of Iron Process for Purifying Water and Sewage.

The offices of the Company are 32a, Euston Square, London, N.W., and 17, Water Street, Liverpool. This system depends for its efficacy more upon its "magnetic carbide filtration," or a combination of carbon and magnetic oxide of iron filters, than upon its precipitation process; in fact, any good method of precipitation can be advantageously applied to it.

The form of filtration appears to be mainly used for the filtration of potable waters, especially those which contain organic impurities, which it removes by oxidation, or, as the author of the system remarks, "the power of occluding oxygen in the form of ozone, the ozone thus produced seizing on every species of organic impurity present in water, and converts them into carbonic acid, water, and free nitrogen."

The prospectus of this company, which is dated 1892, states that the water of the river Calder, which supplies Wakefield, and which is one of the foulest in England, containing the sewage refuse of dyeworks of over a quarter of a million of a manufacturing population residing above the point from which Wakefield is supplied, has been filtered since 1864 by this system, and that the purification of the water continues to be as perfect as when the works were first established, and that in the course of the intervening years the surface of the filters has not been penetrated for cleansing more than an inch, all below being perfectly sweet and odourless, and, in fact, untouched.

The filter-beds at Wakefield are only 7 ft. deep, including the head water.

The following properties amongst others are claimed for this system of filtration: It deprives water of the colour, taste, and odour which arise from the chemical dissolution of organic matter, such as peat, decayed wood or vegetation, and that arising from animal putrescence. It also completely purifies water contaminated with dye-stuffs. It reduces the hardness of chalk waters, thirty to forty degrees of hardness being reduced to eighteen or twenty degrees, whilst those under twenty degrees are not altered, though equally deprived of their organic matter. The system is stated to be in use at the undermentioned places: Wakefield, purified water from the river Calder, contaminated with sewage; Calcutta, where water from the river Hooghly is by its means rendered fit for drinking purposes; Spalding, in Lincolnshire, where water from the fen lands largely contaminated is rendered pure and drinkable. Besides these instances, it is employed in a number of private manufacturing concerns. But I am informed the system is not yet applied for the treatment of sewage.

It was used, experimentally, at the Salford Sewage Works in

1892, and has been reported upon by Mr. Carter-Bell to the Company in the following terms:—

"This system is very similar to that adopted by the International Company, and also that of Mr. Ernest Bell, of Durham. It has one great advantage, that it does not come forward as an untried process, for it has been in use for many years for purifying water, and also, in a small degree, for purifying sewage.

The experiments commenced on June 29th, 1892, and ended July 13th, 1892. Two precipitants were used. No. 1 consisted of—

Insoluble matter in the water	32·01
Carbon	2·50
Protosulphate of iron	33·26
Persulphate of iron	2·12
Sulphate of alumina	3·41
Water	26·22

No. 2 precipitant consisted of—

Aluminium sulphate	47·92
Insoluble in water	·75
Free acid	·62
Protosulphate of iron	·87
Persulphate of iron	·45
Water	50·30

About 3 grains of No. 1 were used together with about 5 grains of No. 2 to the gallon as a precipitant, or half a ton of the mixture to the million gallons. When the mixture was added a certain period of rest was allowed before the effluent went on to the filter. There were two filters, each six square yards; one was in use one day, while the other was at rest. 50,400 galls. of sewage effluent were passed through the filters. All the effluents, with very few exceptions, were bright and clear.

The filter was made up as follows: 4 in. of fettles, 4 in. pea-gravel, 4 in. fine sand, 12 in. of a mixture of "magnetic carbide" and sand in equal proportions, and on the top of this 8 in. of sand.

The following table gives the appearance of the effluents after keeping them for a few days at varying temperatures:—

No.	1892.		Per cent of Purification.		Appearance of Effluent.
1	June 29	......	72·9	......	clear
2	" 30	......	50·0	......	do.
3	July 1	......	80·0	......	do.
4	" 2	......	58·8	......	sewage fungus
5	" 3	......	15·0	......	do.
6	" 4	......	62·5	......	clear
7	" 5	......	72·3	......	do.
8	" 6	......	40·8	......	do.
9	" 7	......	78·5	......	do.
10	" 8	......	61·1	......	sewage fungus
11	" 9	......	88·4	......	clear
12	" 11	......	75·0	......	do.
13	" 12	......	65·0	......	do.
14	" 13	......	63·3	......	do.

Table giving the amount of Albuminoid Ammonia in fourteen analyses of sewage and effluents in parts per 100,000, with percentage of purification.

No.		Sewage Albuminiod Ammonia.		Effluent after Filtration, Albuminoid Ammonia.		Per cent of Purification.
1	......	·48	......	·13	......	72·9
2	......	·34	......	·17	......	50·0
3	......	·40	......	·08	......	80·0
4	......	·34	......	·14	......	58·8
5	......	·20	......	·17	......	15·0
6	......	·24	......	·09	......	62·5
7	......	·325	......	·09	......	72·3
8	......	·30	......	·17	......	40·8
9	......	·28	......	·06	......	78·5
10	......	·36	......	·14	......	61·1
11	......	·52	......	·06	......	88·4
12	......	·40	......	·10	......	75·0
13	......	·40	......	·14	......	65·0
14	......	·30	......	·11	......	63·3

Taking the 14 samples, the average purification is 63 per cent; but if No. 5 sample is excluded, then the average is nearly 67 per cent.

The process was not, as it ought to have been, continuous. At the very least, one hundred thousand gallons (a day of twenty-four hours) should have been purified, and from what I have seen of this process, and the experiments I have performed, I consider that it is capable of producing first-class effluent, but before I or anyone else could recommend the adoption of this process a most searching investigation should be carried out."

(Signed) J. Carter-Bell, A.R.S.M., F.I.C., &c.,
August 30th, 1892. County Analyst.

The following analysis by Dr. Burghardt was made during the trials of this process upon the Salford sewage. The results will be seen to be almost like those obtained by Mr. Carter-Bell.

	Parts per 1,000,000.	
	Raw Sewage.	Filtered Effluent.
Mineral matter	72·50	66·50
Loss on ignition	25·00	20·00
Total solids at 212° Fah.	97·50	86·5
Mineral matter	3·00	nil.
Organic matter	10·50	nil.
Total suspended matter	13·50	nil.
Mineral matter	69·50	66·50
Organic matter	14·50	*20·00
Total solid matter in solution	84·00	86·5
Oxygen required to oxydise the organic matter in—		
3 minutes	2·92	1·32
3 hours	5·44	1·60
Free ammonia	1·60	1·60
Albuminoid ammonia	0·55	0·25

* Dr. Burghardt remarks that the increase here is due to oxidation of organic matter by the filter, the product being practically non-putrescent.

PLATE 10.

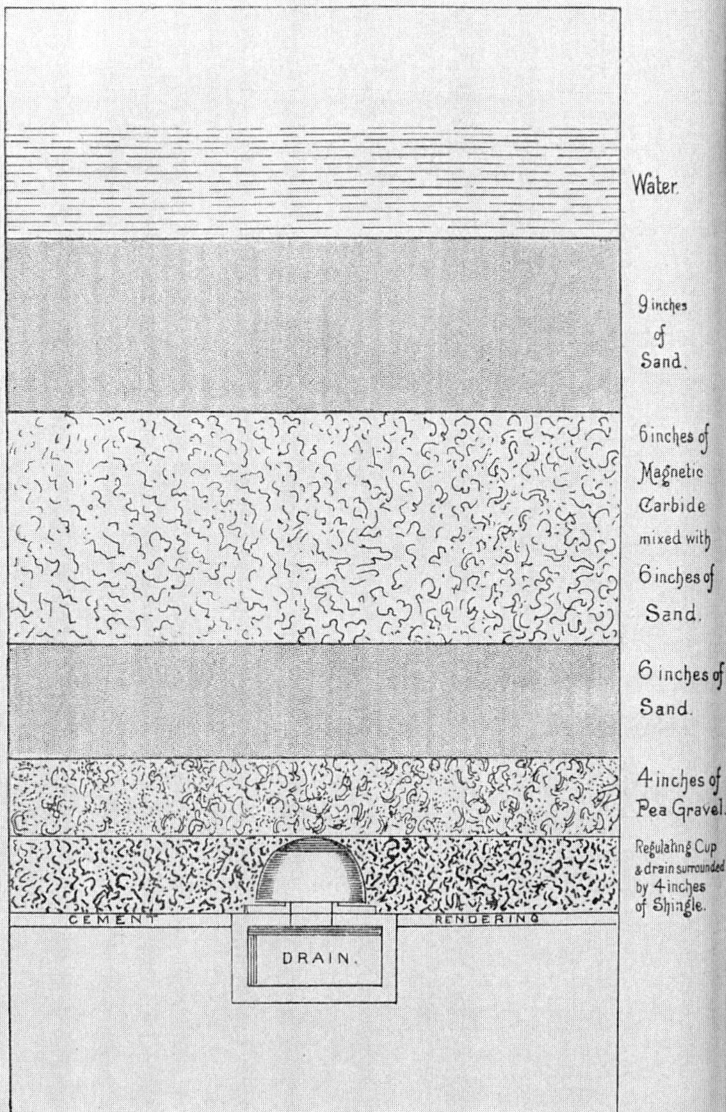

VERTICAL SECTION OF MAGNETIC CARBIDE FILTER BED.

SCALE 1½ INCHES = ONE FOOT

From the above results it is evident that the treatment has produced a very great improvement, both in the amount of the purification of the oxidisable constituents of the sewage, and also the albuminoid bodies.

The filter has produced a powerful oxidation of the oxidisable matter. With more efficient precipitants the albuminoid matter would have been still less in amount.

This system of filtration having proved so effective in the removal of organic matter, I give a sketch of its construction and contents.

Complete analyses of two other trials, showing before and after filtration :—

Parts in 100,000	A Sewage.	B Effluent Before Filtration.	C Effluent After Filtration.
Total solids at 212° Fah....	102	105	106
Suspended matter	14	8·5	nil
Ditto mineral	4	4·5	nil
Ditto organic	10	4·0	nil
Solids in solution	88	96·5	106
Mineral matter in solution	70	80·5	77
Loss on ignition	18	16·0	29
Chlorine	21·5	20·0	13·8
Oxygen required for 15 min.	·8	·5	·4
Ditto for 3 hours	1·3	1·2	·8
Free ammonia	1·8	1·5	·40
Albuminoid ammonia	·28	·16	·07
Percentage of purification ...	—	40	70

Parts in 100,000	D Sewage.	E Effluent Before Filtration.	F Effluent After Filtration.
Total solids at 212° Fah....	95	86	91
Suspended matter...........	8·5	5	nil
Ditto mineral	4·5	4	nil
Ditto organic	4·0	1	nil
Solids in solution	86·5	81	91
Mineral ditto	67·5	63	69
Organic ditto	19·0	18	22
Chlorine......................	20·0	13·5	12·8
Oxygen required for 15 min.	·8	·5	·20
Ditto for 3 hours	1·9	1·5	·48
Free ammonia	1·2	1·5	1·3
Albuminoid ammonia	·20	·12	·06
Percentage of purification	—	42	75

Bradbury's Patent, Worked by the Ferric Sewage Co.

The precipitant is termed Clarine, and is a crude chloride of iron obtained from by-products; it has a specific gravity of 1·411, and has the following composition:—

Protochloride of iron	1·655
Perchloride of iron	21·721
Sulphate of alumina...	2·310
Chloride of aluminium	2·310

This has been tried at Salford, and the official results will shortly be published. From figures given me by Mr. H. Grimshaw, who is interested in this patent, his analyses showed—

Albuminoid Ammonia—Parts per Million. Sewage.	Effluent.	Percentage of Purification.
10·00	6·20	38·0
10·50	5·20	50·5
17·50	4·20	76·0
6·20	4·50	27·4
8·20	5·50	32·9
8·00	3·50	56·2
5·20	2·80	46·1
3·70	2·70	27·0
3·50	2·50	28·5

Mr. Carter-Bell, in an average of 16 analyses, gave the percentage of purification as 16·8.

Mr. Grimshaw's paper on this process will be found in the "Journal of the Society of Chemical Industry of last year."

Mr. Lockwood's Patent.

This method was tried at Salford on the large scale in 1892. I am unable to give the results, owing to their not having yet been published; they will no doubt be given in the fourth part of the Corporation's report.

The process is for using lime and a salt of iron, and the patent consists in the mode of preparing the iron-salt. The oxide of iron, which is ferrous-oxide with a little ferric-oxide, is a waste product; the acid used is also a by-product in the manufacture of benzole.

Instead of using fresh lime the patentee proposes to use the waste lime of gasworks. In the manufacture of the sulphate of ammonia of gasworks vast quantities of lime are used for driving off the fixed ammonia of the gas liquor. When the lime has done its work it has been converted into the carbonate, chloride, and sulphocyanide of calcium, with other calcium compounds; there remains also a certain quantity of free lime. This mixture is a slimy mass, from the large quantity of water which it contains. He proposes to put it into the sewage instead of using fresh lime, using a considerably larger quantity.

Roughly, the proportions of the lime compound in this waste product are—water, about one half, carbonate of lime 33 per cent, free lime 13 to 14 per cent, and the rest other salts of lime.

The System of Mr. Ernest Bell, of Durham.

This is a precipitation and filtration scheme, carried out as stated by the inventor, Mr. C. E. Bell, in the following stages:—

First by precipitation with a specially prepared mixture of sulphate of alumina and protosulphate of iron in such proportions as the quality of the sewage may require. To this mixture the registered trade mark of "Claritone" has been given.

After a period of rest in settling tanks, the sewage enters on the next stage. This consists in passing the partly purified sewage through filters containing a substance which has hitherto been known as Clarence Magnetite, and to which the registered trade name of "Magnetone" has been given. In this way any remaining solids that may be in suspension are arrested, and the oxidation of the decomposable matter in the sewage is completed.

The effluent from the filters is quite clear, and will remain so for months without decomposition undergoing.

Magnetone is an impure magnetic oxide of iron, of an exceedingly porous nature, quite insoluble in water; it has the property of oxidising the noxious matter during the process of filtration through it.

The plant necessary for working the process consists of an

apparatus for mixing the Claritone with the sewage, settling tanks to receive the sewage after the Claritone has been added, and filter beds.

The filter beds are made as follows:—

Sand	12 inches
Magnetone	6 ,,
Sand	6 ,,
Small gravel	6 ,,
Large gravel	6 ,,

A filter bed of this description will filter about 1,000 gallons per square yard of surface in 24 hours.

An experimental trial of this system took place at Salford in June and July last. The results as summarised from the inventor's prospectus are as follows. The results of the official analyses of these trials are not yet published.

Albuminoid ammonia, per million parts, highest ...	3·80
,, ,, ,, lowest ...	·70
,, ,, average of 19 analyses ...	1·60

To this method of filtration any other good process of precipitation can be adopted.

THE GLAUBER-FERRITE SYNDICATE LIMITED.

The prospectus of this Syndicate states that the process is for the purification of sewage on a large or a small scale. Dr. Burghardt is the author of it. It consists of a precipitant composed of ferric chloride, produced in the form of a slab, brick, or cake of any size, thickness, or shape, and containing such percentage, up to 40 or 50 per cent, of ferric chloride the strength of the sewage or manufacturers' foul waste waters may require for their purification, and to this preparation he has given the name of "Glauber-Ferrite."

The cake, slab, or brick is readily soluble in water. Dr. Burghardt states it has been established beyond question that ferric chloride is the most thorough and effective purifier of sewage. This statement is contrary to my experience and experiments. The price is 3d. per 9 lbs., or £3 per ton. The patent is dated the 25th October, 1892. The London offices are 22, Cockspur Street, Pall Mall, S.W.

THE PROCESS OF THE ALUM, CHINA-CLAY AND VITRIOL COMPANY, OR "CARBFERALUM."

This is a patented compound for purifying and deodorising sewage and other polluted waters. The offices of the Company are at 63, Queen Victoria Street, London.

In their published prospectus the Company remark that in various systems of sewage purification sulphate of alumina, protosulphate of iron, or copperas, and carbon, in some form or other, are used and applied singly, in the form of a mechanical mixture, but the Company claim that they have combined these three substances as a chemical compound, and apply them as such to sewage, either in a granulated or slab form, to be dissolved by the sewage, upon which it at once operates, causing a precipitation of both the organic and inorganic matter. They state that by the fine carbon being carried down with the silica, the deodorisation of the sludge and aeration of the effluent liquors are effected more completely and at less cost than has yet been accomplished by the use of any other preparation.

They go on to state that about 7 cwt. of carbferalum is sufficient to treat 1,000,000 gallons of ordinary sewage, but that in some cases it is necessary to use 5 cwt. of lime per 1,000,000 gallons, and that to thoroughly complete the purification of the effluent it is advisable that filter-beds of sand, carbon, and loam should be used in conjunction with their precipitant.

The Company's works are situated at Rainham-on-Thames, Essex, and their precipitant is used at the Richmond Main Drainage Works, at Mortlake. This system has not been tried at Salford, nor, as far as I know, have any analyses been published by the Company; I am therefore unable to give any information as to the degree of purification the process will accomplish, but the quantity recommended to be used seems too small to remove much organic matter.

Messrs. Mather and Platt's Invention.

Messrs. Mather and Platt, of Manchester, have recently introduced an aeration process by which the waste waters of bleachers, dyers, printers, and paper makers become clarified and made fit to enter into a river. It is by a mechanically accelerated precipitation by aeration. The following description of the method and plant is from their recently published prospectus.

The plant consists of two or more small tanks, in which the water to be treated is collected, admitting of its continuous treatment by the precipitation being carried on in one tank while the other is being discharged. In each tank are certain arrangements by which the chemicals intended to act as precipitants are introduced in such a manner as to blend almost instantaneously with the water.

After the introduction of the chemicals, the same arrangements are used for the introduction of air in a special way, the effect of which is to produce a sudden precipitation almost immediately after the chemical action has been completed.

A distinctive feature of the process is that the precipitated impurities which accumulate at the bottom of the tanks form a medium in conjunction with the air to clarify the inflowing foul water. This combined action has also a marked effect on the resultant water, impurities and colours being removed which otherwise would require expensive chemical treatment, or a much longer time for settling in the tanks.

The drawing-off is effected by means of a floating discharge pipe, which is so constructed that the top water is first drawn off, and as the water lowers the discharge pipe follows, and thus only the completely clarified water is drawn from the tank. This arrangement enables the drawing-off to commence as soon as the top layer of water is in a condition to be discharged, and full advantage to be taken of the accelerated precipitation.

The space occupied by the plant is small, about 60 feet by 60 feet only being required to deal with 60,000 gallons per hour. The plant is simple, and there is nothing to get out of order.

Their claims of purification also extend to sewage.

When treating sewage they state that the aeration of the water, which is an essential part of their system, is of immense importance. Any chemicals which form a precipitate of the impurities in the sewage may be used with the system.

The precipitant of the Standard Sewage and Water Purification Company is admirably adapted to this system, as well as to all other precipitation arrangements.

THE KAYE-PARRY SYSTEM, OR PURIFICATION OF SEWAGE BY MICROBES.

This proposed system is fully explained in the October (1892) number of the "Journal of State Medicine." It is preceded by a very interesting paper by Mr. W. E. Adeney, R.C.Sc.I., and Curator in the Royal University of Ireland, on the Chemical Bacteriology of Sewage and its Hygienic Aspect, read at the annual meeting of the British Institute of Public Health, at Dublin, August 18, 1892.

The position taken up is briefly this: that Nature's method of the purification of sewage or of any waste organic matter is one which has for its active agent the influence of saprophytic organisms growing in the presence of atmospheric oxygen in such waste media, or, in other words, the cultivation of bacteria requiring a large amount of oxygen, as the means of the slow combustion and destruction of the organic matter of sewage.

A few thoughts taken from this remarkable and interesting paper will be still further explanatory. Saprophytic organisms are set up in sewage putrefactive fermentation, and emit volatile bodies having offensive odours, but such bodies are not necessarily products of the life processes of these organisms. The writer goes so far as to state that these products are only formed in polluted water when the supply of oxygen is limited, and this he affirms to have proved by careful and long-continued experiments, from which other evolutions were proved to take place; for example, the action of the bacteria varies according to circumstances, producing varying results, such as complete oxidation of the organic matters and ammonia, in some the ammonia is partially or wholly unoxidised, in others complete oxidation of the organic carbon, with or without the production of nitric acid, in addition to the offensive products of putrefactive fermentation. The maximum growth and activity of these organisms being greatest when oxygen is freely supplied to them, it was deduced by the author of this paper that it might be possible and practicable to call in their aid in effectively dealing with the question of sewage disposal. He maintains that this would be effective if the sewage is fed continuously with oxygen throughout its mass during the whole life period of these organisms, or until the organic matter is wholly decomposed. He found on adding nitre to filtered sewage that bacteria rapidly developed and gradually decomposed the organic matter, both in the presence and absence of air and without putrefactive fermentation. Air being present, the decomposition was effected in four or five days, but in 14 or 15 if air was excluded, air being more quickly effective than nitre in developing these aerobic bacteria. He advocates theoretically the breaking up of the sewage into these films of liquid, all its surfaces being fully exposed to air until complete decomposition is effected. In a laboratory filter of coarse pumice stone he completely effected the decomposition of the soluble organic matters in filtered sewage in a few hours.

The solid matters of sewage present a difficulty which seems to militate against the complete practicability of this idea. He says they must be dealt with in an opposite manner by being quickly dried and burnt, as at Ealing.

In a paper read by Mr. W. Kaye-Parry, M.A., at the same

meeting, in which he quotes Sir Henry Roscoe's statement that it was idle to discuss the relative value of different chemicals as agencies for the purification of sewage until the precise part which was played by the micro-organisms in the liquid had been thoroughly investigated, he proposes a plan based upon the lines of his colleague, Mr. Adeney, the outline of which I have just described. He proposes that all the suspended matter be removed from the sewage to be treated before the biological treatment is applied. To this sludge he would add a little oxy-hydrate of manganese as carrier of oxygen to the microbes in the sludge, thereby preventing putrefactive fermentation during the drying of it.

The sewage, after passing through a preliminary straining chamber, flows into a series of tanks seven feet square by sixteen feet deep. The suspended matter subsides, and is pumped from the bottom of the tanks by an air pump on drying floors, and dried.

The effluent is treated with sodium or potassium manganate, or permanganate, not for the purpose of precipitation, but as an oxygen carrier for the bacteria, which increase in the second and third tank, and flourish only so long as all the organic matter has been resolved into gases and inorganic compounds, when they die and fall to the bottom.

This new departure in sewage treatment is now in operation at the Criminal Lunatic Asylum, Dundrum, Ireland.

In the original experiments, as described in Mr. Kaye-Parry's paper, one of the features of the process consisted in the introduction of a blast from an air-propeller or blower into the liquid sewage.

It was at first claimed that the air current oxidised the organic matter of the sewage and destroyed the micro-organisms, but Mr. Adeney soon demonstrated that the direct oxidation of the putrescible matter did not take place, but that the air blast tended to develop micro-organisms in the liquid, and therefore the measure of success was clearly due to the presence and activity of these organisms, and that the partial failure of the process, and the variable character of the effluents were found to arise from unsuitable conditions for the healthy life and vigorous action of these micro-organisms.

Mr. Kaye-Parry's improvements upon this method, and which constitute the Parry-Adeney patent, are these briefly:—

The Sludge.

1. The removal of the suspended matter or sludge by subsidence and straining.
2. Chemically treating it with a little oxyhydrate of manganese, to prevent the formation of putrefactive products.
3. Drying and burning it.

The Sewage in Solution.

1. The partial reduction of the organic matter in solution by direct oxidation, by adding to it a sodium or potassium manganate or permanganate.
2. The sewage so treated sets up indirect oxidation by encouraging a rapid development of bacteria, which complete the reduction and destruction of the organic matter.

The Effluent.

The patentees claim that by this method an effluent is obtained which is absolutely non-putrefactive; that the purification continues after the effluent has left the works, which, being free from chemicals, is not inimical to fish life. The oxygen-carrying chemicals are recovered as insoluble oxyhydrate of manganese, and re-used, by being re-converted into manganese or permanganate of sodium or potassium. The patentees do not give the percentage of purification, or the amounts of free and albuminoid ammonia, in the effluent.

This ingenious method would have more claims to safety if only non-pathogenic micro-organisms could be selected for the indirect oxidation work. To rapidly multiply all sorts and conditions of saprophytic organisms, pathogenic and non-pathogenic, may possibly create a source of greater danger than that which it is the object of the patentees to prevent, but it will be of much biological and sanitary interest to watch the progress of the practical working of the idea at Dundrum.

The Porous Carbon Company's System of Precipitation and Filtration.

Works, Newton Abbot, Devon.

This system, like the other precipitation and filtration ones, relies for its efficacy chiefly on filtration. The Company state their claims in the following terms: The treatment of sewage adopted by the Porous Carbon Company is by precipitation and subsequent filtration, and involves no principles other than those which govern the purification by other methods of like procedure; its success resulting from the care used in selection and treatment of materials, and in the use of the Bovey Tracey Lignite, found in the Newton Abbot district.

The precipitant manufactured by the Company is termed Alfesil. It depends mainly for its success on the use of protosulphate of iron and sulphate of alumina, the balance of proportion between the two being considered with a view to the rapid production of a precipitate which will settle readily, and to ensure a sufficiency of iron to act as a carrier of oxygen to the nitrogenous and carbonaceous matter contained in solution in the sewage.

In the treatment of sewage the use of lime is as far as possible avoided, and it has only been found to be necessary in rare and isolated cases of distinctly acid sewage.

The peculiarities of the system consist in the adjustment of the relative proportions of iron and alumina, the use of materials calculated to contain as low an amount as posssible of insoluble and inactive matter, and in the presence of a proportion of carbonaceous matter, which, derived from the Bovey Tracey Lignite deposits, assists in deodorising and decolourising the effluent, though the Company do not desire to lay much stress on this point, as the small percentage of carbonaceous matter necessarily confines its action within narrow limits. The physical condition in which the precipitant exists as it comes on the market makes it very readily soluble and easily mixed with the sewage.

The precipitation and subsequent settlement are, of course, facilitated if the tanks can be allowed to remain at rest for a time, but good results have been obtained in continuous tanks, the effluent flowing over a weir.

The Company has prepared a very simple apparatus, by which a graduated and regulated flow of water will, with a very small amount of attention, carry the alfesil into the sewage in any desired quantity.

The Porous Carbon which constitutes the filtering medium in the process is quite essential to it, where a high degree of purification is necessary, as here occurs the greater part of the oxidising action—an action well understood and similar to that obtained by the use of animal charcoal.

The material used for its preparation is a peculiar variety of clay occurring in immediate juxtaposition to the Bovey Tracey Lignite beds, and itself containing, or it may be termed impregnated with, Lignite, in a very fine state of division. This clay is carefully carbonised in special retorts to eliminate any volatile matter and sulphur which may be present, and to obtain the required degree of induration and porosity.

The carbon is then ground in a mill constructed so as to avoid crushing and disturbance of the original physical condition. Various grades or degrees of fineness are made to meet the requirements of individual cases, as both the character of the effluent to be filtered, and its quantity and the size of the bed must be taken into consideration in determining the best grade to use; generally it may be said that the larger the bed the coarser the grade desirable. The quantity of carbon required for the filter is 38 lbs. per inch deep per square yard.

The construction of the filter-bed recommended by the Company is shown in Fig. 46.

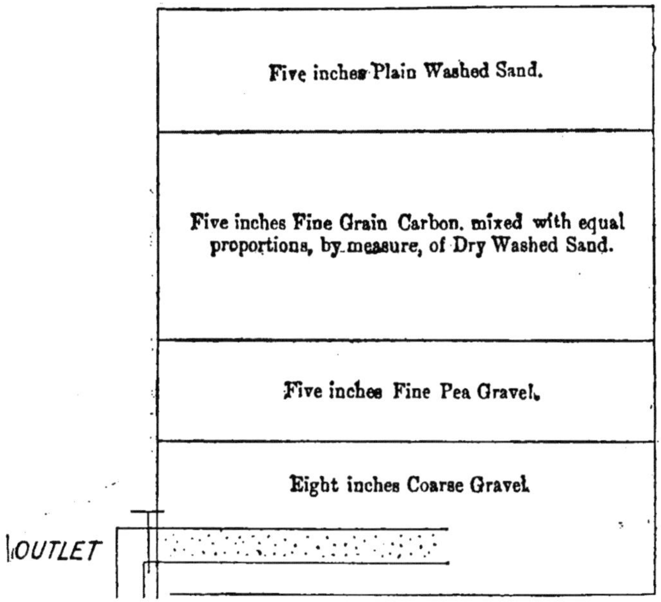

FIG. 46.—SKETCH OF SECTION OF POROUS CARBON FILTER BED.

Its filtering power is 800 to 1,000 gallons per square yard per 24 hours.

This system was tried at Salford; the following results are taken from Mr. Carter-Bell's report issued by the Company:—

"The filtered effluent was beautifully bright and clear, and from the appearance one could not distinguish it from clear spring water.

"The quantity of alfesil used was 16 grains to the gallon.

	Parts per 100,000.	
Total solids	156	93
Suspended matter	62	nil
,, organic	31	nil
,, mineral	31	nil
Solids in solution	94	93
Organic ,,	22	15

	Parts per 100,000.	
Mineral in solution	72	78
Oxygen required for 15 min.	2·3	·22
" 3 hours	3·5	·39
Free ammonia	2·2	1·35
Albuminoid ammonia...	·43	·04

"This precipitation and filtration has given a purification of 90 per cent.

"These results I consider first-class, particularly if they have been effected at a low cost."

The Wardle System, or the Ozonine-Ferricum Process of the Standard Sewage and Water Purification Co.

This is a precipitation process, discovered by the author after several years of experimental research with the persulphates and other salts of iron. The object was the construction of a persulphate of iron in such a basic and delicately existing state as would most readily combine with the organic matter in solution in sewage. This is effected by his patented basic persulphate, to which he has given the name of Ferricum as most indicative of its composition, but on the formation of the Standard Sewage and Water Purification Company to work this patent, the Directors thought the term Ozonine better, and it has been adopted as the Company's trade word. This salt forms the principal precipitation agent, but other substances are also included in the patent, which may be advantageously employed, together or singly with the Ozonine, as varying sewages may require.

After three months' trial at Salford on 100,000 gallons of sewage per day, it was pronounced so satisfactory by the Rivers Conservancy Committee of the Corporation of Salford, that it was preferred, and instructions have been received by the Standard Company from the Salford Corporation to prepare 300 tons of the Ozonine, and proceed with a two months' trial of the purification of the whole volume of their sewage, amounting from 10 to 15 million gallons per day.

Chapter XXVI. (*quod vide*) is wholly devoted to a detailed description of this process and its proved efficacy. As the Fourth Salford Report will not be published before the issue of the present work, I have been enabled, by the kindness of the Salford Corporation, to obtain from Mr. Carter-Bell the results of his analyses of the Salford effluents after the three months' trial of my ozonine, or ferricum precipitant, and the results will be found tabulated at the end of Chapter XXVI.

In addition to this precipitation method, a very good filtration system is incorporated. A waste-product of oxide of iron is used with sand. The name Ozonite has been given to it, but it is not at all a necessary part of the process, because excellent results are obtained without filtration. It is only necessary to incorporate the ozonine with one of the layers of sand in any ordinary filter construction, such as those described in this work.

This ozonite filtration process is incorporated with that of the ozonine precipitation process in the Standard Sewage and Water Purification Co. The offices of the Company are 43, Victoria Buildings, Manchester.

CHAPTER XIII.

The Salford Trials.

To the Salford trials, to which I have several times adverted, may be conceded the most thoughtful and safest procedure in trying to arrive at a sound conclusion as to the adoption of the best method of sewage treatment and disposal.

The large and costly experiments undertaken by this Municipal body are most important; they have extended over a period of about four years, and have been conducted on a large and practical scale, generally upon a flow of 100,000 gallons per 24 hours, tanks and filter beds having been constructed suitable for the treatment of this quantity. These trials are of national interest and importance.

All the best systems have had an opportunity of being tried, and in a short time the final report will be issued. Already three instalments of reports have been published.

The first was issued in July, 1889, but relates only to processes in operation in several other towns.

The second was printed in October, 1889, when the following processes came under review and examination: The Ferrous Carbon Co., the Porous Carbon Co., and the Barry Patent Manure Co.

The third report was issued in 1891, under the joint ægis of the Borough Engineer (Mr. A. Jacob, M.Inst.C.E.), the Borough Analyst (Mr. J. Carter-Bell, F.I.C., A.R.S.M.), Messrs. John Newton (M.Inst.C.E.) and Son, and Dr. C. A. Burghardt, F.M.S., &c. This report dealt with the experiments of the following systems: The Lime process; Messrs. Jagger, Son, and Turley's patent; the Barry Manure Co.'s process (a second trial); Messrs. Spence and Co.'s process; the International Sewage Purification Co.'s system; and the Electrical Purification system.

In the forthcoming and final report the following systems and processes will come under review: The Magnetone and Claritone system of Mr. Ernest Bell, of Durham; Mr. H. Grimshaw's trial of "Clarine," or Bradbury's patent; Mather and Platt's Aeration system; Spencer's Carbide of Iron filtration; Lockwood's process; the Wardle Basic Persulphate of Iron, or Ozonine process, and perhaps one or two other methods.

The conditions aimed at by the Salford authorities of such a system of sewage treatment as would meet their requirements, were the following:—

1st. That the effluent should constantly pass into the Manchester Ship Canal in such a pure state that no nuisance of any kind should arise from secondary decomposition.

2nd. That the suspended solid matter and sludge, when separated from the sewage, should not exceed (when dried at 212° F.) about half the weight of the lime sludge hitherto produced per million gallons at the Sewage Works.

3rd. That the efficacy of the process should be practically permanent in ordinary use; and

4th. That the working expenses and the capital cost of carrying on the works should not be disproportionate to the result to be attained.

The necessity of carefully considering the first of these propositions will be more apparent when we take into consideration that the water of the Ship Canal will flow through a series of levels or pools at not more than the rate of one mile per day, and that such water will consist of one-fourth sewage effluent and three-fourths water containing the refuse from all sorts of manufacturing purposes, having been repeatedly so used.

It is thought by Mr. Carter-Bell that so long as a sewage effluent does not contain more than 2·00 parts per million of albuminoid ammonia, it may be assumed that after-decomposition will not take place.

I propose to give a short resumé of the results of the trials of those processes which have been already published, and refer the reader to the forthcoming final report of the Corporation for those yet unprinted.

THE LIME PROCESS.*

This system was in operation at Salford previous to the adoption of the experiments; the works and tanks were constructed for it, and it may be safely inferred that there were sufficient reasons for wishing it to be superseded by a better system. The death-rate of Salford has for some time been lamentably high, about 30 per thousand, and the lime process by itself is found by Mr. Carter-Bell to be very defective.

The following is a synopsis of his report on this system:—

ANALYSIS OF THE EFFLUENT.

	Parts per million.
Free ammonia	7·20
Albuminoid ammonia	2·70

"The effluent contains an excess of lime, but compares favourably with Messrs. Spence and Co.'s process and also that of the Barry Manure Co., but it is not as good as those of either the Electrical or Polarite systems." The cost per week of the Lime process at Salford, without reckoning interest on capital, depreciation, or other incidental expenses, is as follows:—

Lime treatment	£54 10
Pumping sewage	£36 0
Manipulation of sludge	£29 10
	£120 0

Here, then, as in other places, the Lime process is condemned, and has yielded its claims in favour of other methods and of progressed knowledge.

Lime alone is now by common consent also condemned; its objectionable action may be thus summarised: the quantity

---

* See also pp. 47 and 48, and 94 to 102.

required to be used gives an effluent which is fatal to fish, causes secondary decomposition in rivers and much foul odour of ammonia and sulphur compounds, besides yielding large quantities of sludge, and, as previously stated, it has been officially vetoed both by the Local Government Board and by the Rivers Pollution Commission.

The following is the full analysis of the effluent from the Lime process at Salford by Mr. Carter-Bell :—

		Parts per 100,000.	Parts per million.
Total solids at 212° Fah.		115	
Suspended matter—			
Mineral	2	3	
Organic	1		
Total solids in solution—			
Mineral	84	112	
Organic	28		
Chlorine		48·5	
Oxygen required for 15 minutes		0·12	
Ditto 3 hours		1·66	
Free ammonia		·72	7·20
Albuminoid ammonia		·27	2·70
Percentage of purification		30	—

The average purification of effluent with the lime process is 30 per cent.

The effluent contained an excess of lime, and gave off an ammoniacal odour.

Jagger, Son, and Turley's Process.

This process does not involve more than simple downward filtration of the crude sewage, first through a rough cinder cage or filter, and then through a layer of fine cinders or breeze, which are sifted out of the kiln-burnt ashpit refuse of the town.

No analysis of this process is recorded. The resulting effluent was clear, and the analyst, Dr. J. Burghardt, spoke of it in favourable terms. The engineer states the cinder used in extracting the organic matter is of the size of wheat grains. The only tangible solid refuse left to be dealt with consists of a very thin film of mud in the top of the filters.

THE BARRY PATENT MANURE COMPANY'S PROCESS.*

This process will be found described in Chapter XII., and I need only give the analysis of the effluent obtained in four samples:

	1	2	3	4
Free ammonia in 1,000,000 parts ...	8·00	10·40	9·20	14·60
Albuminoid ammonia „ ...	1·80	2·10	2·30	2·40

For further particulars see pp. 6 and 7 of the third Salford Report, published 19th January, 1891.

THE ALUMINO-FERRIC TREATMENT.†

This process has been fully described in Chapter XII. (*quod vide*). Nine samples of effluents were analysed by Dr. Burghardt, and four by Mr. Carter-Bell, but as Messrs. Spence and Co. did not consider them to have been made under sufficiently favourable circumstances, I omit the results arrived at; they are recorded at pp. 10 and 79 of the third Salford Report.

THE INTERNATIONAL SEWAGE PURIFICATION SYSTEM, OR FEROZONE AND POLARITE.

As before stated, the chief merit of this system lies in its excellent "Polarite" filters, which burn up or oxidise the organic matter in effluents very satisfactorily. A full account of the system, with plan, will be found in Chapter VIII. The experiments with this system were commenced in May, 1889.

The effluent from the Salford Sewage Works contained no more than 2·0 per million of albuminoid ammonia, and there was little fear of secondary decomposition, and such an effluent might safely go into the Ship Canal.

The following observations occur in the Salford Report: that it is extremely difficult to give absolute proof that the polarite

* See page 142. † See pages 131 to 137.

system permanently maintains its oxidising efficacy, or that under all conditions of working the effluent purity is permanently maintained, because it has not been long enough in operation, no works having been established for a sufficiently long time to decide these points, except perhaps at Acton, where the authorities are satisfied with it in these respects, and where it has been working about five years.

Dr. Burghardt's analyses give the following results, in an average of five analyses of both sewage and effluent:—

	Free Ammonia. Parts per million.		Albuminoid Ammonia. Parts per million.
Raw sewage	12·00	......	7·60
Effluent	·70	......	·44

He reports the effluents from this process to be excellent, and in five months had not undergone any decomposition. Owing to the trials being closed in August, he regrets that sufficient time had not elapsed to ascertain whether the filter beds had increased or diminished in efficacy.

The following table shows the difference in weight of the dried sludge produced by the Lime and Ferozone processes. (*Salford Report*, p. 15.)

No. of Experiment.	Weight of Dried Sludge at a temperature of about 212° F.		Relative proportions in weight of Dried Sludge.		
	Ferozone Precipitation.	Lime Process.	Ferozone Precip.	Lime Process.	
	T. cwt. qrs. lbs.	T. cwt. qrs. lbs.			
No. 1	5 19 0 0	11 6 0 0	1·00	1 90	Per million galls. of Sewage.
„ 2	1 15 2 24	3 19 1 12	1·00	2·22	
„ 3	1 1 2 4	3 9 1 24	1·00	3 22	
„ 4	0 17 3 12	2 4 2 16	1·00	2·50	

Average ratio of lime-sludge to sludge from ferozone treatment = 2·46 : 1.

The full analyses will be found in pp. 20 to 30, and pp. 77 of the third Salford Report, compared with those of the Electrical Company's system.

It will be interesting to compare the effluents of the Electrical and the International, and still more so as both companies were treating the same sewage on the same days. To avoid confusion in using a large number of figures, only the albuminoid ammonia in the two effluents will be given, with the percentages of purification.

The average percentages of purification of 18 analyses of Mr. Carter-Bell were—

Analyses of the Effluents of the International Co.'s System Compared with the Electrical System.

	Electrical.			International.	
	Filtered Effluent. Alb. Amm. Parts per Million.	Purification.		Filtered Effluent. Alb. Amm. Parts per Million.	Purification.
1	2·50	76	1	2·00	81
2	1·70	72	2	1·40	72
3	1·10	73	3	1·60	65
4	1·70	65	4	0·50	91
5	1·00	83	5	0·60	90
6	1·10	75	6	0·60	86
7	1·10	75	7	1·20	84
8	1·30	91	8	1·40	90
Not	working.	...	9	1·20	91
9	0·90	93	10	1·00	88
10	1·30	91	11	1·40	90
Not	working.	...	12	1·20	91
11	0·92	93	13	1·00	88
Not	working.	...	14	1·80	86
"	"	...	15	1·50	86
"	"	...	16	0·40	83
"	"	...	17	2·40	45
"	"	..	18	1·80	56
Average Purificat'n	...	80	...	...	81

The Webster Electrical Process.

In Chapter VII. will be found a full account of the process. The results of the Salford trials of this system are recorded at pp. 12, 47, 49, 62, and 66 of the third Salford Report.

The average percentages of purification of analyses by Mr. Carter-Bell are given in the foregoing table.

Mr. Carter-Bell, in his report, remarks that "the effluents of the International and Electrical systems are much the same as regards the reduction of organic matter; the effluent from the Polarite is, however, lower in albuminoid ammonia than in the effluent from the electrical treatment, but the electrical treatment shows a greater reduction in the soluble organic matter, reducing it from 27 to 20. There is very little to choose between the two effluents, for they are both very good; they are bright, clear, colourless, and remained so for weeks, not giving off the slightest odour of decomposition. Ten million gallons a day of water like the above would, no doubt, be an immense boon to the Ship Canal Company."

Of a further analysis Mr. Carter-Bell states: "These two effluents are so close to each other as regards purity, that it is difficult to say which is the better of the two."

It was found necessary to pass the effluent after electrolysis through a light sand and gravel filter bed, to separate a dark flocculent material which floated upon the surface, having, although small in quantity, an objectionable appearance.

The sludge produced by the Electrical system is considerably less than by the International.

Included in the report is a descriptive notice by Mr. Bertram S. Giles, electrical engineer to Messrs. Mather and Platt, on the electrical apparatus employed, and the quantity of electricity used, with other important details. This notice also contains detailed estimates of the cost of applying the electrical system at Salford and of the cost of constructing filter beds, thus summarised:—

	£	s.	d.
Electrical appliances and machinery	36,419	14	8
Filter beds..	9,596	15	0
Total cost for Electrical system for the treatment of 10,000,000 gallons of sewage daily	£46,016	9	8

To this must be added the cost of maintenance, one important item of which would be the waste of iron from the electrical action in the cast-iron plates, which it is estimated would be nearly two tons daily, or 700 tons per annum; which, say at £4 per ton, would cost £2,800. The horse power required would be 336, or 420 indicated.

Messrs. John Newton and Son estimate the cost of applying these systems respectively as follows:—

		£
The International, per million gallons	....	862
The Electrical ”	....	1,172
The Barry Co. ”	....	1,932
The Spence and Co. ”	....	1,292

The capital sums of outlay are estimated as 5 per cent, and not on their respective amounts. (See report.) This estimate does not include the expenses attaching to the disposal of the sludge.

Messrs. Newton say that matters in suspension amount to 25 grains per gallon, which in the case of Salford, with its 10,000,000 gallons per day, would be thus represented:

In dry state.	In pressed cake.	As wet sludge.
16 tons.	32 tons.	160 tons.

And that these weights are increased in proportion to the nature of the precipitant employed. Thus they estimate the annual quantities of the total solid matter—*i.e.*, sludge, in four of the systems tried to be as follows:—

	In dry state.	In pressed state.	As sludge.
	Tons.	Tons.	Tons.
Raw sewage	5,840	11,680	58,400
International	7,500	15,000	75,000
Electrical	6,900	13,800	69,000
Barry Co.	12,000	24,000	120,000
Spence and Co.	12,000	24,000	120,000

Messrs. Newton and Co. estimate the annual cost of dealing with these sludges as cake to be respectively as follows, estimating the value of the cake produced at 3s. per ton:—

	£
International	2,250
Electrical	2,070
Barry and Co.	3,600
Spence and Co.	3,600

This concludes my excerpt of the Third Report of the Salford Trials. The information given I consider to be of great value, as from the results of these trials we may expect to learn more definitely which system gives the best and most reliable results, and we may await their final report with much interest. I cannot help expressing my opinion that these trials greatly lack completeness from the omission of biological experiments. I am compelled to make this statement from an every-day increasing conviction of the necessity of making the consideration of the bacterial element of the first importance. The chief merit claimed for two of the systems tried at Salford is that of sterilisation, and that they should not have been fully tested by some one skilled in bacteriology, appears to me an oversight, and an omission which impairs the completeness of these otherwise most valuable experiments.

The State Board of Health of Massachusetts have considered an inquiry into this branch of the subject so necessary, that eminent biologists have been for a long time engaged in the examination of both the water supply and sewage of Lawrence and other towns in the State. Their reports are of extreme interest and value, and are recorded in two voluminous books, the twenty-second and twenty-third annual reports.

Their researches are still being continued. A summary of their important work will be found in Appendices A and B.

CHAPTER XIV.

English Cities and Towns and their Methods of Sewage Treatment and Disposal.

The next step in the consideration of this subject is to record the working of the methods now used in various parts of the country; in fact, to show the particulars of the practical application of the methods already described.

Whilst I was a member of the Leek Sewage Committee, and before we began any work of visitation, I formulated a series of forty questions, which were sent out by the Town Clerk to many of the principal cities and towns of England. A considerable number of replies were courteously returned, given in answers in whole or in part. They give an enormous amount of information never before published collectively. It is unfortunate that, owing to the exigencies of book-printing and size, they cannot here be conveniently given in a tabulated form, and admitting of an easier comparison of towns with each other.

I propose briefly to incorporate those questions which have been answered with their replies, giving to each answer the number of the question, so as to prevent, in most cases, needless repetition of the questions, which I first give consecutively numbered as follows:—

Queries Sent Out.

1—Name of town?

2—Population?

3—Rateable value?

4—Collecting area of the town's sewage?

5—System of sewage purification and its date?

6—Total cost of sewage scheme?

7—Single or duplicate system of sewers?

8—Cost of construction of laying storm-water system?

9—Annual cost of maintenance of sewage works?
10—Annual loss in the maintenance of sewage works?
11—Any pumping of sewage?
12—How pumped?
13—Capacity and cost of pump?
14—To what height is the sewage raised?
15—Volume of sewage dealt with per 24 hours?
16—How many outfalls?
17—Settling tanks, how many and what size?
18—Cost of settling tanks and their total capacity?
19—Analysis of effluent:—

Free ammonia per million parts?
Albuminoid ammonia ditto?
Chlorine?
Total solids, grains per gallon?

20—What quantity of sludge is formed and how disposed of?
21—Cost of dealing with the sludge?
22—Disposal of night-soil?
23—Proportion of population using privies to that using water closets?
24—Precipitation?
25—Precipitant?
26—Acreage under irrigation?
27—System of irrigation?
28—Depth of drain pipes, their size and distance apart?
29—Nature of land under irrigation?
30—Geological formation?
31—Cost per acre of preparing irrigation land to receive the sewage?
32—Is the land under irrigation freehold or leasehold?
33—Cost per acre of the freehold?
34—Rental per acre paid by Urban Authority to owner if leasehold?
35—Is the management by the Board or by sub-letting?

36—Rental per acre obtained if sub-let.

37—Crops grown ?

38—Annual sum realised by sale of crops ?

39—Annual sum realised by sale of other products ?

40—Do you consider your system satisfactory ?

The following places will come under review :—

Acton, Banbury, Barking, Bedford, Blackburn, Bolton, Bradford, Buxton, Canterbury, Chester, Chiswick, Coventry, Crewe, Dewsbury, Doncaster, Dudley, Dukinfield, Ealing, Enfield, Fenton, Forfar, Friern-Barnet, Grantham, Great Malvern, Hanley, Hitchen, Hyde, Ilford, Kingston-on-Thames, Leeds, Leek Rural, Littleborough, Longton, Manchester, Merton, New Malden, Newcastle-under-Lyme, Nottingham, Oldham, Oxford, Peterborough, Portsmouth, Reigate and Redhill, Rochdale, Sheffield, Swanwick, Swinton and Pendlebury, Stretford, Taunton, Tottenham and Wood-Green, Walton-on-the-Hill, Wednesbury, Wimbledon, Workingham College.

In the following alphabetical list of places such answers to the foregoing questions as the several authorities were good enough to supply will be found recorded.

ACTON.

1. Name of town, Acton, district of Middlesex.

2. Population 6,000 to 7,000, contributing sewage to sewage works; total population about 26,000.

3. Date of sewage scheme, August, 1887.

4. Total cost of sewage scheme, £150,000, sewers, works, and land. [For an explanation into the cause of this great outlay see my report.* The sewage works cost only £15,000.]

5. Single or duplicate system of sewers ? Duplicate.

6. Cost of construction of laying storm-water system ? Used the old sewerage system.

7. Annual cost of maintenance of sewage works, £797. This includes cost of treatment of sewage, pressing, rates, taxes, tithes, &c.

*Chapter VII., p. 84.

8. Any pumping of sewage? Yes; 350,000 gallons out of a total daily delivery of 425,000 gallons.

9. How pumped? Two "butterfly flap" pumps, built by Boulton and Watt, Soho, London.

10. Capacity and cost of pump? Each pump will raise 80,000 gallons per hour at 32 revolutions of engine per minute. The average speed used is about a quarter of this. Cost between £3,000 and £4,000 for duplicate engines, boilers, and pumps, two pumps to each engine.

11. To what height is the sewage raised? 36ft., well to tanks.

12. Volume of sewage dealt with per 24 hours, 400,000 to 500,000 gallons. The present average is 425,000 gallons.

13. How many outfalls? Only one, at the works. Effluent outfall, one, at the Thames.

14. Settling tanks, how many and what size? Three, each 51ft. by 76ft., depth 7ft. Each holds about 150,000 gallons, or a total capacity of 450,000 gallons.

15. What quantity of sludge is formed, and how disposed of? After pressing, about 9½ tons per week; sold at 5s. per ton not dried; 27 tons in a million gallons, but this is a variable quantity.

16. Cost of dealing with the sludge? Not separately ascertained. Sludge is pressed into cakes by filter-presses.

17. Disposal of nightsoil? There is none. It is an entirely water-closet town.

18. Proportion of population using privies to that using water-closets? All water-closets.

19. Precipitation? Precipitation and filtration only.

20. Precipitant? Ferozone precipitation with polarite filtration. A pressing powder is used with the sludge; see notes.

21. Geological formation? London clay.

22. Cost per acre of the freehold? None except sewage works.

23. Annual sum realised by sale of other products? Pressed sludge in cakes undried is sold at 5s. per ton, used for wheat

and barley crops, ploughed in. If sludge is ground and dried for manuring grass land, it is 30s. per ton.

24. Do you consider your system satisfactory? Yes.

BANBURY.

2. The population of Banbury is 16,000.

3. The rateable value is £41,137.

4. The collecting area of the town's sewage is 4,800 acres.

5. The system of sewage disposal is Broad irrigation. The date 1869.

6. The total cost of the sewage scheme is £26,700.

7. The system of sewers is both single and duplicate.

8. The cost of construction of laying storm-water system varies according to the size of pipes and nature of ground.

9. Annual cost of maintenance of sewage works is £250 for pumping, &c.

11. All the sewage is pumped.

12. The sewage is pumped by horizontal condensing pumping engine, in duplicate, of 18 H.P.; cylinders 12in. diameter, with 20in. stroke; two Cornish boilers, 5ft. diameter, 14ft. long, with flues 2ft. 8in. diameter; makes 6,000 strokes per day.

13. The capacity and cost of pump is: Three-fold action plunger pumps in duplicate; each pump having a piston and plunger 23in. by 23in., and 16in. diameter, with a stroke of 20in.; each plunger 30 gallons, or total of 90 gallons per stroke

14. The sewage is raised to a height of 20ft.

15. The volume of sewage dealt with per 24 hours is 320,000 gallons, average.

16. There is one main outfall emptying to tanks with storm overflow *en route.*

17. There are three settling tanks in duplicate. Duplicate measurements: 72ft. by 23ft.; 80ft. by 22ft.; 72ft. by 25ft.; average depth, 8ft.

18. The cost of settling tanks was £3,500. The capacity is 260,800 gallons.

19. The analysis of effluent is as follows:—

	Grains per gallon.		Parts per million.
Free ammonia	0·350	...	·50
Albuminoid ammonia	0·035	...	0·05
Chlorine	3·360	...	4·80
Total solids, grains per gallon	45·700	...	65·28

The effluent finally finds its way into the river Cherwell, which is good and pure, giving general satisfaction.

20. The quantity of sludge formed is about 375 tons every two months. The tanks are ordinary settling tanks, where settlement takes place before reaching pumping well; the sludge accumulating therein is cleaned out once perhaps in two months, according to the weather, which, of course, regulates time of filling.

21. The cost of dealing with the sludge is £12 for cleaning out the tanks each time; the sludge is carted to a farm and spread on land.

23. Banbury is a water-closet town.

26. The acreage under irrigation is 138a. 3r. 38p.

28. The depth of the drain pipes and their size is (average) 4 feet, 3, 4, 6, 9, with open carriers of various sizes, with iron pipes and troughs across ditches. Distance apart 30 feet to 50 feet.

29. The nature of the land under irrigation is light staple soil of gravelly nature, resting on subsoil of clay.

30. The geological formation is alluvia (earth), deposited on bed of clay, &c.

31. The cost per acre of preparing irrigation land to receive the sewage is £50.

32. The land under irrigation is freehold.

34. The rental per acre paid by Urban Authority is £4 10s., paid at first when tenanted.

35. The irrigation, &c., is under own management; the cultivation is now let.

36. The rental per acre obtained by letting is £2 10s., with stipulation to turn on sewage *ad libitum.*

37. The crops grown are Italian rye-grass, mangolds wheat, oats, and beans.

38. The annual sum realised by sale of crops is £550, less £500 expenses when farming ourselves.

39. The annual sum realised by sale of other products varies according to what we have.

40. The system is satisfactory.

Barking.

2. The population of Barking is 15,000.

3. The rateable value is £70,000.

5. The date of the sewage scheme is 1886. Sewage settling-tanks only are used.

6. The total cost of the sewage scheme is £21,000.

7. The system of sewers is duplicate.

8. The annual cost of maintenance is £150.

11. There is no pumping of sewage.

16. There is one outfall.

17. There are three settling tanks, each 260ft. by 40ft. by 4½ft.

16. There is one outfall.

18. The capacity of the settling tanks is 250,000 each.

20. The sludge is pumped from settling tanks into sludge pits 18in. deep, and when dry is carted on to land by farmers, to whom the sludge is given.

24. The effluent discharging into a strong tidal river, no precipitating agent is employed. The sewage in the tanks precipitates naturally with 12 hours' rest.

40. The system of sewage disposal here is considered satisfactory.

Bedford.

2. The population of Bedford is about 2,650.

3. The rateable value is £107,048.

4. The collecting area of the town's sewage is 700 acres.

5. The system is Broad irrigation, dating from 1865.

6. The first cost was about £6,000.

7. The system of sewers is partly duplicate.

10. The annual cost in the maintenance of the sewage works is £800 to £900.

11 and 12. The sewage is pumped by centrifugal pumps.

13. The capacity of pumps is two 12 horse power.

14. The sewage is pumped to a height from 14 to 24 feet.

15. The volume of sewage dealt with per 24 hours is 1,000,000 gallons.

16. There is one outfall.

17. There are no settling tanks.

23. Water-closet system.

26. The acreage under irrigation is 130 acres.

29. The nature of the land under irrigation is loam, gravelly subsoil.

32. The land under irrigation is leasehold.

34. The rental per acre is from £2 13s. 7d. to £4 10s.

35. The management is by the Board.

37. The crops grown are onions, rye-grass, mangolds, and cabbages.

38. The annual sum realised by sale of crops is £2,150.

40. The system of disposal is considered satisfactory.

Blackburn.

2. The population is 120,000.

4. The collecting area of the town's sewage is 6,973a. 1r. 25p.

7. The system of sewers is partially single and duplicate.

9. The cost of management of works is £400 per year.

15. The volume of sewage dealt with per 24 hours is 4½ million gallons.

17. There are five settling-tanks, each of about 100,000 gals.

20. The quantity of sludge formed is about 650 loads. It is dried naturally on a filter-bed, and disposed of to farmers in a semi-dry state.

22. Water-closets are now adopted in all new houses.

24. Rough precipitation lime process and subsequent irrigation.

25. The precipitant is lime, a crude method.

26. The acreage under irrigation is 589 acres.

28. The effluent flows into the river Darwen (a foul river).

40. The system here is quite satisfactory, and no complaints are made of the smell arising from the works.

Bolton.

2. The population of Bolton is 120,000.

3. The rateable value is £413,833.

4. The collecting area of the town's sewage is 11,620 acres.

5. Precipitation chiefly by lime. The date of the sewage scheme is 1872 and 1886—Burnden Works 1872, and Hacken Works 1886.

6. The cost of the works was £48,950.

7. The system of sewers is single.

8. They have no separate storm sewers, but have constructed storm overflows on all their trunk sewers.

9. The annual cost of maintenance of sewage works is about £1,600.

10. The annual loss in the maintenance of the sewage works is £1,500.

11. There is no pumping of sewage.

15. The volume of sewage dealt with per 24 hours is about 3½ millions of gallons, normal flow.

16. There are two outfalls, one at Burnden and one at Hacken.

17. There are four settling tanks at Burnden, and five at Hacken (including detritus tanks), with a total superficies of 11,813 yards.

18. The total capacity of settling tanks is 4¼ millions of gallons.

20. The quantity of sludge formed has not been ascertained, but what is made is given away.

22. Part of the nightsoil is sold as manure, part cremated, and part tipped away.

23. It is a privy and ashpit town; but they have 1,640 closets.

25. The precipitate is chiefly lime.

30. The geological formation is gravel and sand.

35. The works are under the management of the Corporation.

37. The crops grown are carrots and hay, chiefly meadow.

38. The annual sum realised by sale of crops is £100.

40. The system is considered satisfactory.

Bradford.

2. Population, 200,000.

5. Precipitation by lime since 1885.

6. The engineer says the original cost is not known, but he estimates that it would not at the present time exceed £40,000.

9. The annual cost is £3,302 2s. 4d.

11. The sewage is pumped.

15. 8,450,000 gallons daily.

Analysis by Dr. W. Wallace, of Glasgow:—

Crude Sewage.	Grains per gallon.		Parts per million.
Ammonia, free or saline (equivalent to 133 parts per million)	9·3	...	132·80
Ammonia albuminoid (equivalent to 20 parts per million)..................	1·4	...	20·0
Effluents.			
Ammonia, free or saline (equivalent to 8·00 per million)	·56	...	8·00
Ammonia, equal to nitrogen, combined in other forms (equivalent to 4 parts per million)	·28	...	4·00

20. The Corporation has an agreement with a contractor, who receives the sludge gratuitously. He carts it away and sells it to farmers.

23. There are 4,000 water-closets.

24. Precipitation by slaked and sifted lime in tanks on the intermittent plan, with filtration of the clarified fluid. The average proportion of lime used is 12·88 cwt. per million gallons of sewage.

BUXTON.

2. The population of Buxton is about 8,000, including Fairfield district, whose sewage is dealt with at Buxton.

3. The rateable value is, Buxton £53,000, Fairfield £3,000.

4. The collecting area of the town's sewage is 1,089 acres.

5. The system is Dr. Thresh's application of a natural chalybeate water and lime, with settling tanks. No filtration. The date of the sewage scheme is 1884-5.

6. The total cost of sewage scheme was £6,000.

7. The system of sewers is duplicate in all the new streets and in several old ones.

9. The annual cost of maintenance of sewage works is £443, which includes £240 for interest and repayment of principal.

11. There is no pumping of sewage.

15. The volume of sewage dealt with is 700,000 gallons; in the season it is 1,000,000 daily.

16. There is one outfall.

17. The settling tanks are duplicate, 260 feet by 36 feet each, with six divisions in each set.

18. The total capacity of settling tanks is 500,000 gallons; the cost is included in cost of works.

19. The analysis of the effluent is—Free ammonia, per million parts, 2·4; organic, ·5; total solid grains per gallon, 23·9.

20. The sludge formed is about 1 ton per 1,000 of population in 24 hours, which is pressed into cake, the liquid flowing into tanks again. The compressed solids are given to farmers, who eagerly cart them away as fast as they are turned out of the machine.

22. There is very little nightsoil, as Buxton is practically a water-closet town.

23. The proportion of population using privies to that using water-closets is probably not 1 per cent.

25. The precipitant is natural iron-water and milk of lime.*

40. The system here is considered perfectly satisfactory.

Canterbury.

2. The population is 26,400.

4. The collecting area of the town's sewage is 3,968 acres.

5. The sewage scheme has been in work 24 years, and is one of sewage filtration.

7. The system of sewers is duplicate.

9. The annual cost of maintenance of sewage works is £601 8s.

15. The volume of sewage dealt with per 24 hours is one million gallons.

17. The settling tanks hold 190,000 gallons. The works are about half a mile from the nearest dwelling-house or public-road.

20. The contractor pays £40 per annum for the sludge, and carts it away.

24. The sewage is filtered through straw wattles, then through two beds of flint. No chemicals are used as precipitants.

26. The acreage under irrigation is 24.

40. Have had complaints on two occasions, but no proceedings. Have had no complaints made of the smell arising from the works.

Chester.

2. The population of Chester is 40,000.

3. The rateable value is £168,000.

4. The collecting area of the town's sewage is 1,000 acres.

5. Precipitation by lime. The date of the sewage scheme is 1875.

6. The cost of sewage scheme, works only, £25,000, land extra.

7. The system of sewers is single.

---

* See Chapter XI., page 111.

9. The annual cost of maintenance of sewage works is £600.

12. The sewage is pumped by Gwynne's centrifugal pumps.

13. The capacity of each pump is 4·15in. diameter.

14. The sewage is pumped to a height of 12 feet.

15. The volume of sewage dealt with per 12 hours is 1¼ million gallons, 8 millions maximum.

16 There is one outfall.

17. There are two settling tanks, one 200 by 30 feet, one 200 by 50 feet.

19. The analysis of effluent is—Free ammonia, in grains per gallon, 1·68; chlorine, ditto, 4·6; total solid grains per gallon in solution, 30·41.

20. The quantity of sludge formed is 1,200 tons, and it is sold to farmers at 1s. 6d. per load, at works.

21. The cost of dealing with the sludge is not kept separate.

22. The nightsoil is disposed of to farmers.

23. There are 4,428 water-closets, remainder privies, say one half.

24. Milk of lime.

40. The system is considered fairly satisfactory.

Chiswick.

2. The population is 21,000.

3. The rateable value is £115,000.

5. The system of precipitation is by Spence's Aluminoferric process.

7. There is a duplicate system of sewers.

9. The annual cost of maintenance of the sewage works is £1,572 16s. (representing a rate of 4s. 6d. in the pound).

11 & 12. The sewage is pumped by two of Davy's differential pumping engines, each indicating 15 horse power.

13. The pumps are capable of delivering 1,200,000 gallons of sewage in 12 hours.

14. It is raised to a height of 19 feet.

15. The volume of sewage is 550,000 gallons per day.

16. There are two outfall sewers.

17. There are eight settling-tanks of a capacity of 800,000 gallons.

20. The sludge is pressed by two of Messrs. Johnson and Co.'s filter-presses, each 36 inches in diameter, with 24 plates. Air is employed in forcing the sludge into the presses. Lime is then mixed with the sludge in the receivers, in the proportion of 14 lbs. of lime to each cwt. of wet sludge. The cases of sludge are given away to market-gardeners, and is all got rid of in this way.

21. The cost of pressing is 3s. 6d. per ton. This is rather more than sludge ordinarily costs in pressing, but it appears that the sludge contains quantities of size and china clay from a wall-paper printer's works, which block the filter cloths and add very materially to the cost of pressing.

23. There are 4,600 water-closets.

Coventry.

2. The population is 53,000.

3. Rateable value, £131,500.

4. The collecting area of the Board's sewage is 3,126 acres.

5. Precipitation. The date of sewerage scheme is 1852.

6. Cost of sewerage scheme, £37,000.

7. Single or duplicate sewers?—Single.

8. Cost of storm-water sewers?—£300.

15. The volume of sewage dealt with each 24 hours is 2¼ million gallons.

16. Outfalls?—One.

17. There are eight settling tanks, 140ft. by 29ft., nearly 5ft. deep.

18. The capacity of the settling tanks is nearly one million gallons, which is equal to 42 per cent of each day's flow, to allow for settling and rainfall.

20. The quantity of sludge formed is 520 tons, disposed of by pressing.

24. Precipitation by sulphate of alumina and lime.

26. There are eight acres under irrigation.

Crewe.

2. The population of Crewe is estimated at 32,000.

3. The rateable value is £93,292.

4. The collecting area of the town's sewage is about 550 acres.

5. Broad irrigation. The date of the sewage scheme is 1870.

6. The total cost of the sewage scheme is about £46,800.

7. The system of sewers is single.

10. The annual loss in the maintenance of sewage works is about £3,400, including instalments of principal and interest.

12. The sewage is pumped by a ram pump driven by steam engine.

13. The capacity of the pump is 100,000 gallons per hour.

14. The sewage is raised to a height of 93 feet.

15. The volume of sewage dealt with per 24 hours is 1,050,000 gallons (average for the year).

16. There are two outfalls.

17. There are two settling tanks, 210 feet long, 24 feet wide, 14 feet depth of sewage.

18. The capacity of settling tanks is 882,000 gallons.

20. The quantity of sludge formed is about 40 tons, dug into the farm land.

22. The disposal of nightsoil is upon the sewage farm land, as above.

23. There are about 900 water-closets, 4,825 pails, privies, and slop-water closets, &c.

26. The amount of land under irrigation is 269 acres, irrigated alternately.

27. The system of irrigation is broad or surface irrigation.

28. The depths of drain pipes are 6 feet, and 4 feet 6 inch pipes, 20 feet apart.

29. The nature of land under irrigation is stiff clay soil.

30. The geological formation is red marl (Keuper), with drift sand and gravel overlaid in places.

32. The land under irrigation is nearly all leasehold.

33. The cost of the freehold is from £100 to £190 10s. per acre.

34. Rental paid by Urban Authority is from £3 1s. to £4 per acre.

35. The management is by the Blythe Corporation.

37. Crops grown are mangolds, turnips, wheat, oats, but principally rye-grass.

40. The system is considered fairly satisfactory.

Dewsbury.

2. Population, 30,000.

3. The rateable value is £106,767.

5. The system used is the Bailey-Denton system of intermittent downward filtration.

14. The sewage is lifted to two levels.

19. Albuminoid ammonia in average sample of effluent in January 1893 was 7·00 parts per million.

20. The sludge is spread over the level beds or lands on each side at times when there are no crops growing upon them, and then dug or ploughed in.

26. There are ten or twelve acres under irrigation.

27. The sewage is distributed through the filtration areas by means of furrows, so as not to come in contact with the leaves of the growing vegetation, but, spreading laterally through the soil, reaches their roots, and can be applied during almost any stage of growth without disadvantage. The position of the sewage furrows is changed every one or two years, so that as much land as possible may derive benefit from mixture with the sludge, which is taken out of the furrows when the sewage is not under distribution.

29. The soil for the most part is porous and of a sandy character, very suitable for filtration.

37. The crops grown are principally roots and rye-grass, which succeed very well, and have in dry summers proved of great value.

40. In 1889 the weather was not propitious for sewage farming. It was very wet, but general satisfaction was nevertheless expressed by the Council, as in better seasons the farm pays its way.

DONCASTER.

2. Population estimated in 1890 to be 26,300. Including the out-districts whose sewage is dealt with, the population is estimated at 30,000.

3. Rateable value?—The borough, £111,703.

4. The collecting area of the Board's sewage?—The borough 1,690a, 2r. 12p.; area built upon, 456a.

5. The system of sewerage is Broad irrigation. Completed about 1872.

6. Cost of sewerage scheme?—Pumping station, rising main, and laying out farm, £20,000. This is exclusive of cost of the sewerage in the town.

7. Single or duplicate sewers?—In a portion of the town only is a duplicate system of sewers provided.

8. Cost of storm-water sewers cannot be estimated, as many of the old sewers were retained as surface water drains where new sewers were constructed.

9. The annual cost of maintenance of sewage works is £1,500, including interest and repayment of principal, pumping expenses, wear and tear.

10. Annual loss in maintenance of sewage works, £1,030 at present; in 1902, when the capital will have been paid off, there will be a small profit.

11. Pumping of sewage?—All pumped.

12. How pumped?—By means of compound beam engines at the pumping station.

13. Capacity and cost of pump?—35 nominal H.P. Annual

cost of pumping from £300 to £400 per annum. First cost of engine and boilers, £5,405.

14. The sewage is raised 52ft.

15. The volume of sewage dealt with each 24 hours is 500,000 to 600,000 gallons, dry weather flow.

16. Outfalls?—One.

17. Settling tanks?—None.

19. Analysis of effluent?—No analysis. Very little effluent; it is nearly all absorbed in the land. What effluent finds its way into the dykes is perfectly clear and inoffensive.

20. The sludge is all distributed on land.

22. Disposal of nightsoil?—Is let to a contractor, and costs (including inspection) about £622 per annum, or about 1½d. in the pound.

23. The proportion of the population using privies to that using water-closets is about one-third using water-closets.

24. No precipitation.

26. There are 287 acres under irrigation.

28. Depth of drain pipes, 3ft. to 6ft.; various sizes. Distance apart, 66ft. Some parts of farm are so porous as not to require under-drainage.

29. Nature of subsoil under irrigation, light, with gravelly subsoil; surface flat, but slightly undulating.

30. Geological formation, gravel substratum.

31. Cost per acre of preparing irrigation land to receive sewage?—Accounts have not been kept separate.

32. The land is "freehold," and belonged to the Corporation prior to the application of sewage.

35 and 36. The land is sub-let to a farmer at £2 per acre, but prior to the application of sewage it was 15s.

37. The crops grown are beans, peas, oats, barley, wheat, Italian rye grass, mangolds, swedes, white turnips, market garden products.

40. The system is considered satisfactory.

Dudley.

2. The population is 50,000.

4. The collecting area of the town's sewage is 3,600 acres.

5. The system is Broad irrigation.

9. The sewage is received and used by the Earl of Dudley on his farm without cost to the Corporation.

25. There is no precipitation.

26. There is unlimited acreage for irrigation.

40. The system is very satisfactory, and there are no complaints. The success is due to the nature and extent of land available.

Dukinfield.

2. About 18,000 population.

3. The rateable value is £59,000.

4. Collecting area of sewage, 1,263 acres.

5. System of sewerage ?—Broad irrigation. Date 1880.

6. Cost of sewerage scheme ?—Estimated £54,000. Already expended about £20,000.

7. Single or duplicate sewers ?—Single.

9. Annual cost of maintenance of sewage works ?—Not yet in operation.

21. Disposal of nightsoil ?—Cost about £1,000 per annum.

32. The land is "freehold."

38. Cost per acre of freehold, £200.

36. Rental per acre obtained if sub-let, £2 10s.

Ealing.

2. The population is 23,000 (20,000 drain to the sewage works).

3. Rateable value, £152,000.

4. The collecting area of the town's sewage is 3,225 acres.

5. The system is precipitation.

20. The sludge is run into beds formed of ashes allowed to drain for a week or ten days, and then destroyed with house refuse in destructor.

24 and 25. The precipitation is by milk of lime and sulphate of alumina and clay in the following proportions: Lime, ten and a quarter grains, sulphate of alumina two grains, and eleven and a half grains of clay per gallon

40. There are no complaints.

Enfield.

2. The population is 32,000.

4. The collecting area of the town's sewage is 12,600 acres.

5. The sewage system is filtration through land.

7. The system of sewers is duplicate.

9. The working expenses are almost covered by the sale of crops.

15. The volume of sewage dealt with per 24 hours is 500,000 gallons.

17. There are no settling tanks. The works are about 500 yards from public road, and a few houses adjacent to farm.

20. The sludge is ploughed into the land.

26. The acreage under irrigation is 105.

28. The effluent is discharged into the intercepting cut of the East London Water Co., and finally finds its way into the Thames.

40. Do you consider your system satisfactory?—Our farm is generally satisfactory; and we consider that when land in sufficient quantity can be obtained irrigation is the most perfect and inexpensive system for sewage disposal. No complaints are made of the smell arising from the works.

Fenton Sewage Disposal.

The Leek Sewage Committee visited Fenton on November 17th, 1890.

They were received by Mr. Goodall, the Surveyor of the Fenton Local Board, and were conducted to the sewage outfall, the land upon which the sewage falls for irrigation, and to the effluent near Trentham.

The sludge is removed from the tanks alternately about every six weeks, is ploughed into arable land, and is found to be beneficial to the crops.

There is at present only one outfall; it is near Longton Pool, in the meadows, about a mile below Fenton.

There are three sewage levels; at two of them the sewage is lifted into the third by a pneumatic system of pumping, known as Shrone's. The motive power is air compressed by steam-power.

There are three osier beds, of about one acre of well-grown and healthy willows, but the roots were covered with a dense, filthy, and unhealthy deposit of semi-solid sewage mud, varying from 2 feet to 2 inches in depth. The osiers are cut in spring.

The fields nearest the tank are of loam; those further away are gravelly, with thin soil. The gravel is chiefly red sandstone drift-pebbles.

Mr. Goodall considered that under the Broad irrigation system it was necessary to calculate one acre to each 100 people.

The effluent was clear and large in volume, and was discharged into a gutter, and found its way into the Trent, near Trentham. There was evidently much subsoil drainage in addition to the effluent drainage. The subsoil water was slightly ferruginous, and deposited a thin film of yellow ochre along the gutter bottom, and appeared to have brought down some of the organic matter of the effluent, for the sides of the gutter were coated with black organic mud, smelling very offensively.

The valley used for the broad irrigation disposal of Fenton sewage is well adapted by its contour for the purpose, slightly shelving downward on each side, and forming a slight hollow down its longitudinal axis.

The crops grown are principally hay.

This system of irrigation appears to be satisfactory; but with the above-mentioned exception of an effluent very liable to secondary decomposition, and consequent putrescence. There was no analysis of the effluent, and it is therefore impossible to form a correct opinion whether the effluent is generally pure enough to be discharged into the Trent. I think it is very doubtful.

In my opinion, the fertilisation of osier beds by crude sewage is very questionable, and I cannot but regard it with unqualified disfavour, as being eminently fitted for the spread of disease-germs in warm weather. No doubt osiers thrive very well under such conditions, but the accumulation of sewage and fæcal matter under cover of so thick a shrubbing is most favourable to both the growth and spread of bacteria. If the effluent after precipitation were so applied, there could be no objection to it.

The following gives the information in the tabulated form with additions:—

FENTON.

2. The population is 16,000.

4. The collecting area of the Board sewage is 1,599 acres.

5. The system is Broad irrigation.

6. The total cost of the sewerage scheme is £2,000, inclusive of pumping and settling tanks.

7. The sewers are in duplicate.

12. Shone's system of compressed air is used for pumping the sewage.

16. There is one outfall; but provisions are made for two.

17. There are two settling tanks, each 18 yards long by 4 yards wide, with a holding depth of about 6 feet.

23. Cottage privies chiefly prevail.

26 and 28. There are 149 acres pasture and meadow land under irrigation, which is managed by allowing the sewage after having passed through the settling tanks to run upon the land. The depth of the drain pipes is 3 feet 6 inches; distance apart 21 feet.

29. The upper meadows are loam; the lower meadows have light and gravelly soil.

30. The geological formation is: Upper meadows lie on coal

measures, lower meadows on drift gravel, mostly new-red sandstone pebbles.

32 and 34. The land under irrigation is leasehold, at a rental of £5 per acre.

35. The farm is sub-let.

FORFAR.

2. The population is 12,500.

5. The Bailey Denton system of downward intermittent filtration. The date of the sewage scheme is 1878 and 1879.

6. Disposal works cost £1,450, and Orchard Bank Farm £4,000; total cost of sewage scheme £5,450,

15. Volume of sewage dealt with daily, 400,000 to 600,000 gallons.

19.

Analysis made 24th July, 1888.		Natural Sewage.	Sewage after passing through area.
Solids dissolved...	Grains per gallon	37·5	36·6
Chlorine...........	„	5·2	3·7
Free ammonia ...	Parts per million	69·0	0·12
Alb. ammonia ...	„	2·5	0·78
Suspended matter	Organic, grains per gallon	21·0	none.
	Inorganic	23·5	none.
Nitrate	Nitrate of soda, grains per gall.	none.	9·6

26. Forty acres are under irrigation, of which only 24 acres are laid out for treatment.

27. The system of irrigation is surface irrigation, with Bailey Denton's downward intermittent filtration on seven acres, leaving 17 acres for filtration on a series of terraces, intersected by main furrows.

29. The nature of land under irrigation is: The soil is of a free and open character, sandy, and gravelly in parts, though occasionally partaking of a somewhat loamy character.

31. Cost per acre of preparing irrigation land to receive the sewage is £60.

37. The crops grown are rye-grass, cabbages, carrots, turnips, mangolds, swedes, potatoes.

38. The annual sum realised by sale of crops is £392 3s. 1d.

40. The system is considered satisfactory.

Friern Barnet.

2. The population is 6,400 (1888).

3. Rateable value, £22,500.

5. The system of sewage disposal is lime and alumina, and black-ash wash. Date September, 1887.

6. Cost of disposal works, £10,354.

9. Annual cost of maintenance of sewage works, £627 14s. 1d.

11. and 12. The sewage is pumped by an accumulator, giving a normal working pressure of 280 lbs., a pair of high-pressure pumps, a two-inch hydraulic main 1,430 yards long, and Davey's differential hydraulic pumps, automatically controlled at Ely Place.

15. The volume of sewage daily dealt with is 120,000 gallons.

16. There are three outfalls.

17. There are three settling tanks, 97 feet by 21 feet.

18. Their capacity is 9,300 feet.

20 and 21. The sludge is raised by a chain-pump into a lime mixing-pit, from which it is pumped by three-throw ram pumps into Johnson's filter-presses and converted into sludge-cake, or pumped to a portion of land and incorporated with the soil.

25. The precipitant is 1 cwt. of lime and ¼ cwt. sulphate of alumina, or ¼ cwt. of Hanson's black-ash wash.

40. The system is considered satisfactory.

Grantham.

2. Population, 19,000.

3. Rateable value, £54,416.

4. Collecting area of the Board's sewage ?—About 500 acres.

5. System of sewerage and date ?—Broad irrigation. Outfall 1879, Streets 1866 to 1871.

6. Total cost of sewerage scheme ?—Outfall £15,000, street sewers £11,000.

7. Single or duplicate system of sewers ?—We have and are duplicating as we can.

8. Cost of construction of laying storm-water system ? Included in above.

9. Annual cost of maintenance of sewage works, £300.

10. Annual loss in the maintenance of sewage works ?—On farm £200 per annum.

11. Any pumping of sewage ?—None.

15. Volume of sewage dealt with per 24 hours ?—Three-quarters to 1¼ million gallons.

16. Outfalls ?—One.

17. Settling tanks, number, and size ?—None.

23. Proportion of population using privies to that using water closets ?—Only about 50 privies left.

26. Acreage under irrigation ?—105 acres under-drained, but we can and do run it over 500 acres of land.

27. System of irrigation ?—Broad.

28. Depth of drain pipes ?—5 to 7, their size 3 inches, distance apart 22 yards.

29. Nature of subsoil under irrigation ?—Sandy; 97 per cent of silica when we started.

30. Geological formation ?—Alluvial.

31. Cost per acre of preparing irrigation land to receive sewage ?—Draining £22, surface levelling £11.

32. Land under irrigation freehold or leasehold ?—Forty years' lease.

34. Rental per acre paid by Urban Authority to owner if leasehold ?—30s.

35. Farming management by the Board or by sub-letting ? By Board.

37. Crops grown ?—Roots, corn.

38. Annual sum realised by sale of crops?—On irrigated portion £10 to £20 per acre.

40. Do you consider your system satisfactory?—Yes.

Great Malvern.

2. The population is 6,808.

3. Rateable value, £54,000.

4. Collecting area of sewage 884 acres.

5. System of sewage?—Broad irrigation and intermittent filtration.

7. Single or duplicate system of sewers?—Two-thirds of town is duplicate.

8. Cost of construction of laying storm-water system?—£3,000.

9. Volume of sewage dealt with per 24 hours?—From 150,000 to 250,000 gallons.

16. Outfalls?—One.

17. Settling-tanks, number and size?—Two $\begin{cases} 35 \times 15 \times 6 \text{ ft.} \\ 30 \times 12 \times 3 \text{ ft.} \end{cases}$

20. Sludge formed, how disposed of?—On the farm.

23. Proportion of population using privies to that using water-closets?—All water-closets.

26. Acreage under irrigation?—40 acres intermittent filtration.

28. Depth of drain pipes, 4 and 5 feet; size 4, 5, and 6 inches; distance apart 1 chain.

29. Nature of subsoil under irrigation?—Strong loam.

30. Geological formation?—Gravel and clay.

32. Land under irrigation freehold or leasehold?—Freehold.

33. Cost per acre of the freehold?—£200.

35. Farming management by the Board or by sub-letting? By the Board.

37. Crops grown?—Italian rye grass and osiers.

40. Do you consider your system satisfactory?— Yes.

Hanley.

The Leek Sewage Committee visited the Hanley Sewage Works, and were met, by appointment, by Mr. Lobley, the

Borough Surveyor, who conducted them over the outfall station and works near Stoke, where the milk of lime was prepared and run into the tanks. There were two engines of about 14 horse power, to each of which was attached a centrifugal pump, but only one of which was worked, the other being a spare one. They were struck with the large quantity of sludge in the neighbourhood of the works—too much, in my opinion, for the well-being of the large populations so near to it.

The effluent was fairly clarified, but it was not quite clear. There was an absence of smell, and the system seemed to be in good working order.

The answered questions will give all the needful information.

I made rough qualitative tests, and found a considerable amount of organic matter, much more than in the Newcastle effluent, and this is confirmed by Dr. Reid's recent analysis, which Mr. Lobley kindly sent me, and which is entered in the list of questions.

The inference is, that the process was not working so well as it did, and Mr. Lobley, in a letter to me, afterwards informed me that the reasons why the effluent was then not so good as it should be were well understood.

Let me now compare, as far as is possible, these results with the conclusions of the Commissioners appointed in 1868 to inquire into the best means of preventing the Pollution of Rivers.

The conclusions must only be taken as suggestions, as far as acting up to them are concerned, inasmuch as Lord Salisbury, in framing his Act of 1877, did not venture to include them in that Act. No doubt he hesitated to formulate such conclusions into legal obligations, owing to the chronic impure state of the rivers, the then impure state of sewage purification, and the insurmountable difficulties attending the minimising of the impurities to such a comparatively high standard.

I should state that neither the Commissioners of 1868 nor Lord Salisbury's Act made any suggestions as to the separate amounts of ammonia and albuminoid ammonia which might be allowable, but classed them as organic nitrogen; so we can only compare the Hanley analysis with their suggestions as far as they go, which are as follow, the blank spaces in the Commissioners' report column being where no suggestions were made:—

On the effluent from Sewage, parts per million.	Commissioners' standard of purity, 1868. Parts per million.	Grains per gallon.	Hanley effluent, Oct., 1888. Per million.	Hanley effluent, Dr. Reid, 1890. Per mil.
Mineral suspended matter	30,000	2·10	0·000	—
Organic suspended matter	10,000	0·70	0·000	—
Organic carbon in solution	20,000	—	0·000	—
Organic nitrogen...........	3,000	2·1	2,650	—
Free ammonia..............	—	—	4,800	9,760
Albuminoid ammonia ...	—	—	0·444	1,200
The following figures represent grains per gallon of impurities:—				
Total solid matter in solution	1,000	—	56,000	53,000
Chlorine	1,000	—	4,000	7,760
Equal to common salt......	—	—	6,590	—
Soluble silica	—	—	7,300	—
Oxide of iron and alumina	—	—	3,000	—
Lime.........................	—	—	6,550	—
Magnesia	—	—	·260	—
Sulphuric acid (combined)	—	—	20,660	—

Effluent neutral, free from arsenic, lead, sulphuretted hydrogen, and soluble sulphides, absence of smell, taste, and colour.

1. Name of town?—Hanley.
2. Population?—56,293.
3. Rateable value?—£174,770.
4. Collecting area of the town's sewage?—1,800 acres.
5. Precipitation date of sewage scheme?—Completed 1881.

6. Total cost of sewerage scheme, £21,000; total cost of sewage works, £48,000; total to date, £69,000.

7. Single or duplicate system of sewers?—For the most part on the single system; new streets below level of Cauldon Canal being on the duplicate system.

8. Cost of construction of laying storm-water system?—No separate account.

9. Annual cost of maintenance of sewage works?—About £1,200 net.

10. Annual loss in the maintenance of sewage works?—About £1,200 net.

11. Any pumping of sewage?—Yes.

12. How pumped?—By two centrifugal pumps.

13. Capacity and cost of pumps?—Each pump can lift 4,000 gallons per minute; boilers, engines, and pumps cost £1,267.

14. To what height is the sewage raised?—12 to 15 feet.

15. Volume of sewage dealt with per 24 hours?—1¾ to 2¼ million gallons.

16. How many outfalls?—One, except small groups of houses under temporary arrangements.

17. Settling tanks, how many and what size?—Four tanks 200 by 30 feet, and one tank 200 by 60 feet.

18. Cost of settling tanks and their total capacity?—£5,000; 1¼ million gallons.

19. Analyses of effluent—Dr. Reid's analyses:—

	Oct., 1883.		Nov., 1890.
Free ammonia, per million parts	4·80	...	9·76
Albuminoid ammonia ditto	·44	...	1·20
Chlorine	4	...	7·70
Total solids, grains per gallon ...	56	...	53·00

20. What quantity of sludge is formed and how disposed of?—No record.

21. Cost of dealing with the sludge?—No separate account.

22. Disposal of nightsoil?—To the farmers.

23. Proportion of population using privies to that using water-closets?—One half, say, but rapidly decreasing.

24. Precipitation?—Yes.

25. Precipitant?—Lime and alumino-ferric cake.

26. Acreage under irrigation?—About 11 acres.

27. System of irrigation?—Intermittent filtration.

28. Depth of drain pipes?—6 feet, their size 4 inches, distance apart 18 feet.

29. Nature of land under irrigation?—Generally loam.

30. Geological formation?—Alluvial deposit.

31. Cost per acre of preparing irrigation land to receive the sewage?—About £111 per acre; total cost £1,226.

32. Is the land under irrigation freehold or leasehold? Copyhold.

35. Is the management by the Board or by sub-letting?—By the Board.

37. Crops grown?—A few osiers, also grass on land not yet in use as filter.

38. Annual sum realised by sale of crops?—£5 osiers, £30 grass.

40. Do you consider your system satisfactory?—Yes, having regard to the position of Hanley, surrounded by other urban districts.

HITCHIN.

2. Population, 10,000.

3. Rateable value, 27,803.

6. £2,300 was spent in laying out for sewage purposes.

19.

	Grains per gallon.	Parts per million. Free Amm.	Alb. Amm.
Sample of effluent water from the sewage farm at Hitchin, sent by the Croydon Local Board	36·3 33	5·00	1·00

This is not drinking water, but it may be discharged into a stream.

20. The sludge is treated similarly to that practised at Dewsbury.

24. Intermittent filtration.

27. The land is laid out in a similar manner to that at Dewsbury, excepting that the beds are narrow, the under drains more frequent, and the sewage furrows far more numerous. The position of the sewage furrows is changed every winter or early spring, and when doing this care has to be taken that they are not placed immediately over the under-drains, but at some distance from them.

29. The sewage farm consists for the most part of peat mixed with gravelly clunch (lower chalk).

30. Greensand formation.

35. The management of the sewaged land is in the hands of a Committee of the Local Board, and is not farmed at a loss.

37. Mangolds and rye-grass are the principal crops, whilst osiers are grown on a certain portion of the land, which is not closely under-drained. Roses and strawberries prosper under moderate doses of sewage.

HYDE.

2. Population, 32,000.

3. Rateable value, £104,000.

4. Collecting area of the Board's sewage, 3,042 acres.

5. Date of system of sewage?—Commenced to lay down main sewers 20 years ago.

6. Total cost of sewerage scheme?—Cost of main sewer up to now £31,000

7. Single or duplicate system of sewers?—Both.

8. Annual cost of maintenance of sewage works?—Just commencing to construct sewage works.

11. Pumping sewage?—Shall have some little.

12. How pumped?—A centrifugal pump intended.

14. To what height is the sewage raised?—About 18 feet.

15. Volume of sewage dealt with per 24 hours?—Works designed to deal with 900,000 gallons.

16. Outfalls?—One.
17. Settling tanks?—Six.
22. Disposal of nightsoil?—We have very little.
33. Cost per acre of freehold?—Over £1,000 per acre.

ILFORD, CHADWELD, AND DAGENHAM.

3. Houses rated in full, cottages at half, and the land at one-fourth.

11. The sewage is pumped.

20. The sludge is swept from the settling tanks into a well, from which it is pumped into drying beds, and eventually disposed of on land.

24. The sewage is treated with milk of lime in the usual proportions, and after passing through the settling tanks, in which the suspended matters subside, the clarified water is applied to the land, on the intermittent downward filtration principle.

KINGSTON-ON-THAMES.

2. The population of Kingston-on-Thames is 38,068.
3. The rateable value is £195,360.
5. Precipitation by the A B C process.*
6. The cost of the works is £23,000.
11. There is pumping of sewage.

12 and 13. There are three pumping engines, each capable of raising 100,000 gallons of sewage per hour; and three (for Surbiton), each able to raise 45,000 gallons per hour; eight engines.

17. There are eight settling tanks, one 85ft. by 49ft.

20. Compressed cakes of sludge are dried, put into a disintegrator and broken down into a fine manure, and sold at £3 per ton to farmers and gardeners under the name of "native guano."

25. The A B C mixture and a solution of alum.

28. The depth of the drain pipes is 7ft., 4ft. in size, 15ft. apart.

30. Geological formation?—New red sandstone.

35. The works are entirely operated by the Native Guano Co.'s A B C process, who receive a subsidy from the Kingston

* See page 51.

Corporation and from the Surbiton Improvement Commissioners of 3d. in the pound upon the rateable value.

40. The system is considered satisfactory.

Leeds.

2. Population, 375,000.

4. Collecting area of Board's sewage, 21,600 acres.

5. Precipitation.

7. Single or duplicate system of sewers?—Single.

9. Annual cost of maintenance of sewage works?—£4,311 for 1889.

15. Volume of sewage dealt with?—9,000,000 gallons daily.

17. Settling tanks?—Eleven, 100 feet by 60 feet; one, 88 feet by 60 feet; average depth, 6 feet. The works are 134 yards from the nearest dwelling-houses or public road.

20. What quantity of sludge is formed, and how disposed of? 270 tons of fluid sludge, containing 95 per cent water. It is dried in the open air and given away.

24. Lime precipitation.

25. Precipitated lime one ton to 1,000,000 gallons of sewage.

26. Acreage under irrigation, 26a. 1r. 24p.

28. The effluent flows into the river.

40. No complaints. There are also no complaints made of the smell arising from the works.

Leek.

Population, 14,000.

A sewage disposal scheme is under consideration. At present there is only a partial crude sewage or broad irrigation, and sewage-flow into the river Churnet.

Leek, Rural.

2. Population, 34,238. This includes Smallthorne Local Board, which is under the Union, but not in the Rural Sanitary district of Leek.

4. Collecting area of the Board's sewage?—Milton and Norton Green.

5. System of sewage disposal?—Broad irrigation.

7. Single or duplicate system of sewers?—Single.

8. Cost of construction or laying of storm-water system? None.

9. Annual cost of maintenance of sewage works?—About £50.

11. Pumping of sewage?—Yes.

12. How pumped?—The sewage runs into tanks, and is then pumped on to the sand and conveyed by means of wooden carriers to the different parts of the land.

16. Outfalls?—None.

17. Settling tanks, number and size?—Two.

19. Analyses of effluent?—No effluent.

23. Proportion of population using privies to that using water-closets?—All privies.

24. Precipitation?—None.

25. Precipitant?—None. No effluent.

26. Acreage under irrigation?—About 10 acres each in the two fields where sewage tanks are situated.

29. Nature of subsoil under irrigation?—Meadow in one, pasture in the other.

32. Land under irrigation freehold or leasehold?—Leasehold.

34. Rental per acre paid by Urban Authority to owner if leasehold?—Only acknowledgment of 10s. a year.

35. Farming management by the Board, or by sub-letting?

37. Crops grown?—Grass.

40. Do you consider your system satisfactory?—Yes, for the small quantity of sewage dealt with.

Littleborough.

2. Population, 12,000.

3. Rateable value, £45,000.

4. Collecting area of the Board's sewage, 500 acres.

5. Intended precipitation and filtration system?—In course of construction.

6. Total cost of sewage scheme?—Not complete.

7. Single or duplicate system of sewers?—Duplicate.

8. Cost of construction of laying storm-water system?—Not complete.

9. Annual cost of maintenance of sewage works?—Not at work.

11. Pumping of sewage?—None intended.

15. Volume of sewage dealt with per 24 hours?—Estimated at 250,000 gallons.

16. Outfalls?—One.

17. Settling tanks, number and size?—Four, 70ft. by 30ft.

18. Cost of settling tanks and their total capacity?—Not complete, 300,000 gallons.

22. Disposal of nightsoil?—Pail-system, carted on to farm.

23. Proportion of population using privies to that using water-closets?—50 water closets; rest pail-system and ashpits. Ashpits being rapidly converted to pail and tub-system. This latter system very satisfactory.

24. Precipitation?—Intended.

25. Precipitant?—Not yet decided.

26. Acreage under irrigation?—No irrigation intended; precipitation and filtration.

30. Geological formation?—Alluvial deposit.

32. Land under irrigation freehold or leasehold?—Freehold.

33. Cost per acre of freehold?—About £300 per acre; 12 acres.

40. Do you think your system satisfactory?—Not complete.

Longton Sewage.

The system of sewage disposal at Longton is by broad irrigation on land belonging to the Duke of Sutherland, who takes it from the borough boundary, and distributes it by means of costly and efficient pipes over his farms between Longton and Trentham for pasture meadows and green crops, and also for allotment gardens.

The sewage is so efficiently disposed of and distributed that the farmers complain that they have too little of it.

The Borough of Longton is under agreement to pay the Duke of Sutherland for disposing of the sewage. He receives £750 per annum, and disposes of it as he thinks best.

The storm water is taken out of the sewage, except the first washing of the streets; after that the storm water is discharged into the duplicated system of sewers, earthenware storm-water pipes being chiefly used, at a depth of four to five feet. The storm water is discharged into a brook.

There are a number of outlets.

The principal outlet at the borough boundary where the sewage is delivered to the duke is a culvert, 4 feet by 3 feet.

The effluent was said to be bad where it is discharged into the Trent immediately below Trentham Hall.

The Leek Committee had not time to visit more than one farm of the duke's under irrigation, nor the effluent outfalls. They were, however, impressed with the completeness and efficacy of the drainage arrangements made on field with a clayey subsoil of New-Red marl, there was a large crop of cabbages and turnips, some of which were well grown, whilst others showed evident signs of considerable disease.

There is a considerable fall between Longton and the duke's property, especially nearer to Trentham.

The Committee regarded the Longton system as a fair illustration of the advantages of broad irrigation without precipitation, but were unable to give any opinion, in the absence of analysis, of the character of the effluent.

On two subsequent visits, I found the Trent in a very impure state at Trentham, much gas being given off, evidencing the decomposition of a considerable amount of organic matter, and, consequently, a decidedly unsanitary state of the river, which was very dirty-looking.

QUERIES.

1. Name of town?—Borough of Longton.
2. Population?—35,000.
5. System of sewerage?—Broad irrigation, without precipitation
6. Total cost of sewerage scheme?—£30,000; this includes the sewering and storm-water drains.
7. Single or duplicate system of sewers?—Duplicate.
9. Annual cost of maintenance of sewage works?—Sewage

outfall, Longton district, £495; sewage outfall, Dresden district, £220; nightsoil removal, Florence district, £131 16s.; nightsoil removal, Dresden district, £97 7s.

22. Disposal of nightsoil?—Carted away to grass land, the cost of carting being as much as the nightsoil is worth.

23. Proportion of population using privies to water-closets? The greater part use privies.

26. Acreage under irrigation?—300 acres, Duke of Sutherland's land.

37. Crops grown?—Pasture, meadow, green crops for cattle, and human food.

Manchester.

2. Population, 380,000.

4. Collecting area of the town's sewage, 5,000 acres.

5. System of sewerage?—Principally the pail system. A projected new scheme has been commenced, sanctioned by Local Government Board. (See Chapter X. for full particulars of disposal.)

7. Single system of sewers.

15. About 9,000,000 gallons of sewage dealt with per 24 hours.

17. Settling tanks hold 9,000,000 gallons. The works are about half a mile from the nearest dwelling-house or public road.

20. Sludge will be pressed; system not working.

24. Precipitation intended, and the effluent to be run through land.

26. About 100 acres under irrigation,

Merton and Croyden Rural Sanitary Authority.

2. Population, 22,500.

3. Rateable value, £134,555.

5. Date of sewerage scheme, 1878.

6. Total cost of sewerage scheme?—£622 per acre sub-irrigation plots, artificial filters £750 per acre.

7. Single or duplicate system of sewers?—Duplicate.

11. Pumping of sewage?—Yes.

17. Settling tanks, number and size?—Two settling tanks.

20. Quantity of sludge formed, and how disposed of?—The sludge is pumped into filter presses.

28. Depth of drain pipes, six feet; distance apart, 400 feet.

29. Nature of subsoil under irrigation?—The land is of an open gravelly nature in some parts, whilst in others it is sandy and peaty, and in one portion clay was met with.

37. Crops grown?—The greater part of the land is cropped with rye-grass. A commencement with osiers has been made, and will probably become the future crop on this farm.

38. Annual sum realised by sale of crops, £463, in 1888.

New Malden.

2. Population, 3,000.

8. Cost of construction of laying storm-water system, £15,800.

9. Annual cost of maintenance of sewage works, £195.

11. Pumping of sewage?—Yes.

13. How pumped?—By means of gas-engines and three-throw plunger-pumps.

13. Capacity and cost of pump?—The normal duty of each engine and pump is the raising of 20,000 gallons of sewage per hour.

14. To what height is the sewage raised?—30 feet.

17. Settling-tanks, number and size?—Six, 20 feet by 20 feet, with average depth of 5 feet.

18. Settling-tanks, their total capacity?—75,000 gallons.

21. Cost of dealing with the sludge?—The sludge is air-dried in small filters, the quantity being as yet too small to necessitate other treatment; when dry it is either sold for a small sum per load or dug into the land.

24 and 25. Precipitation and Precipitant?—Experiments are being made with various reagents, and the chemical treatment of the sewage to be adopted remains under consideration.

Newcastle-under-Lyme.

The Leek Committee were received by Mr. James Pattison, Surveyor, and Mr. Alderman Heath, who took us to their sewage works, about one mile distant from the town.

From Mr. Alderman Heath we received very full replies to the question paper, so complete as to only require a very short report.

The crop of osiers was a very fine one; there was not so much sludge about them as at Fenton, yet the land and osier roots were quite covered either with sludge or a secondary precipitation of organic putrescent matter. The effluent water was flowing into the river Lyme in a perfectly limpid state, as clear as the water in Dovedale, no doubt from the quantity of bicarbonate of lime held in solution from the lime precipitant. This appearance is entirely a fallacious and misleading one, and is worse than a guide to indicate presence or absence of organic matter, and proves the absolute necessity for chemical analysis of all effluents. The deceptive nature of this effluent was easily shown when Mr. Alderman Heath showed us Dr. Reid's analysis, made only a few days previously, from which it appeared that this clear effluent contained in solution no less than 10·35 parts of free ammonia and 2·15 of albuminoid ammonia in a million. Free ammonia was being abundantly given off in the carriers round the osier beds, its characteristic and pungent odour being very perceptible and strong.

Secondary precipitation had very copiously occurred all along the carriers, and a whitish precipitate of several inches thick lined the entire length of the carriers.

The chief defect of this system appears to be the want of a safer precipitant and of drains under the osiers, the effluent being simply run on and off the surface, and what is not absorbed running into the river. I consider the results very unsatisfactory.

2. Population, 20,000.

3. Rateable value, £50,282 4s. 9d.

4. Collecting area of Board's sewage, 6,500 acres.

5. System of sewerage to date?—Surface irrigation, without drainage; the present system, 1882; no re-sewerage.

6. Total cost of sewage scheme, £4,000 for the sewerage works.

7. Duplicate system of sewers.

9. Annual cost of maintenance of sewage works, £1,100.

10. Annual loss in maintenance of works?—No material returns.

11. The sewage is pumped out of tanks into beds to settle.

12. How pumped?—Ram pump. There are two 9 inch rams and two engines.

13. Capacity of pump, 13 inches.

14. The sewage is raised to a height of about 12 feet.

15. Volume of sewage dealt with per 24 hours?—An average of 700,000 gallons in moderately dry weather, but the quantity is variable.

16. There is one outfall.

17. Settling tanks, how many and what size?—Two mixing tanks, small; four precipitating tanks, each 125 by 25 feet, and feet deep.

19. Analysis of effluent—Dr. Reid's analysis, October 3rd, 1890:

Free ammonia, per million parts	10·35
Albuminoid ammonia, ditto	2·15
Chlorine	6·00
Total solids, grains per gallon	2·22

20. The quantity of sludge formed is 2,000 to 2,500 tons per annum. It is heaped up to dry, and given away to farmers.

21. Cost of dealing with the sludge?—No cost. At first the farmers were predjudiced, and had to be paid to fetch it away, but now they fetch it, and are not paid.

22. Disposal of night soil?—Generally given away. It is sold in certain parts of the year. Farmers near to Newcastle fetch it, and pay 3s. per load; for farmers at a distance the Corporation fill the town carts during the night, and the farmers bring their horses, and pay 2s. 6d. per load, and cart it away; when a block occurs it is given away.

23. Proportion of population using privies to that using water closets, 3,500 to 4,000 privies, 400 water-closets.

24. Precipitation ?—Yes.

25. Precipitant ?—Lime and carbon refuse, costing 19s. per ton, from Mr. Kimon's chemical works, at Bradford, Manchester; through W. H. Cowburn, agent, Manchester. It is said to be the only place where this product can be obtained. It is a waste product in the manufacture of prussiate of potash.

26. Acreage under irrigation ?—Eight acres only now used; but there are 12 acres more in reserve.

29. Nature of land under irrigation ?—Red loamy subsoil, not very good.

30. Cost per acre of preparing irrigation land to receive the sewage ?—£66 5s. is for laying out the land and making carriers.

32. Land under irrigation is leasehold.

34. Rental per acre paid by Urban Authority to owner, £5.

35. The management is by Board.

36. Not sub-let.

37. Crops grown ?—Osiers and a little grass.

38. Annual sum realised by sale of crops ?—Last year £37, and £5 for grass. In previous years £50 and £60 have been realised.

39. Annual sum realised by sale of other products ?—None.

40. The system before Dr. Reid's analysis was considered satisfactory. In 1873 the borough was under a sequestration order at the instance of the Duke of Sutherland, and had to deposit £4,000 in the Court of Chancery, as a guarantee that it would proceed to purify the sewage so as to produce a satisfactory effluent. The Duke of Sutherland was satisfied, and consented for the guarantee to be returned, which has been done, along with £1,560 interest which had accrued.

Nottingham.

2. Population, 235,000.

4. Collecting area of Board's sewage, 10,000 acres.

5. Broad irrigation on land.

10. Annual loss in maintenance of sewage works ?—Loss on sewage farm £2,000 to £3,000 per annum; rent £5 per acre.

14. Volume of sewage dealt with per 24 hours?—About 7 million gallons.

17. Settling tanks?—Have none. There is a public road through the farm.

20. What quantity of sludge is formed, and how disposed of? None.

23. Proportion of population using privies to that using water-closets?—About one-fourth are water closets, and remainder pail closets.

24. Precipitation?—None.

26. Acreage under irrigation, 635 acres.

28. Depth of drain pipes, their size, and distance apart?—The effluent flows into the river Trent.

40. The system is quite satisfactory, and no complaints are made of the smell arising from the works.

OLDHAM.

2. Population estimated at 146,000.

3. Rateable value, £575,000.

4. Collecting area of the Board's sewage?—Area of borough, 4,729 acres.

5. System of sewerage and date?—We have not yet established any system of sewage in the borough. We have a main drainage system, with single sewers. The waters are discharged into rivers outside the borough, and carried forward untreated. A scheme for the treatment of the sewage is under consideration, prepared by Mr. G. Asling, C.E., of Sheffield.

22. Disposal of nightsoil?—Night collection on pail-system. Treatment at various depots, and ultimate sale as manure, after being mixed with shoddy.

23. Proportion of population using privies to that using water-closets?—There are very few water-closets in the borough.

OXFORD.

2. Population, 53,000.

4. Collecting area of the town's sewage, 4,650 acres.

5. Irrigation and intermittent filtration.

7. Single or duplicate system of sewers?—Duplicate.

9. Annual cost of maintenance of sewage-works?—Pumping, £800 per annum. Farm about pays its working expenses.

11. Any pumping of sewage?—Yes.

15. Volume of sewage dealt with per 24 hours?—One and a quarter million gallons.

17. Settling tanks, how many and what size?—None. Works about 600 yards from nearest dwelling-house or public road.

20. What quantity of sludge is formed, and how disposed of? We do not separate the sludge.

24. Precipitation?—None.

25. Precipitant?—None.

26. Acreage under irrigation, 369 acres.

28. Depth of drain-pipes, their size, and distance apart?—The effluent flows into the river Thames.

40. The system is quite satisfactory; no complaints are made of the smell arising from the works.

Peterborough.

2. Population, 26,000.

4. Collecting area of the town's sewage, 1,784 acres.

5. System of sewerage, broad irrigation.

9. Annual cost of the maintenance of sewage works, £200.

15. Volume of sewage dealt with per 24 hours, 750,000 gallons.

17. Settling-tanks, how many and what size?—Two tanks, each containing 2,800 cubic feet (much too small). There are two cottages and public road 200 yards from the works.

20. What quantity of sludge is formed, and how disposed of? 1,000 yards. It is dried by exposure to weather, and is occasionally given away. Now selling at 3d. per cart load.

26. Acreage under irrigation, 301 acres.

28. Depth of drain pipes, their size and distance apart?—The effluent flows into open watercourse.

40. Do you consider your system satisfactory?—The system is quite satisfactory. No complaints are made of the smell arising from the works.

PORTSMOUTH.

2. Population, 130,000.

3. Rateable value, £500,000.

6. Total cost of sewerage scheme?—The pumping-station and rising-main cost £25,000; tanks and outfall-works, £45,000; total cost, £70,000.

9. Annual cost of maintenance of sewage-works?—Pumping station, £2,450; tanks, £250; total, £2,700.

11. Pumping of sewage?—Yes.

12. How pumped?—A pair of 4 horse power Otto gas engines.

13. Capacity and cost of pump?—A pair of 150 horse power compound condensing rotary beam engines, each capable of delivering 500,000 gallons of sewage per hour.

17. Settling-tanks, number and size?—There are three storage-tanks, side by side; each tank is 160 feet by 150 feet.

18. Settling-tanks, their total capacity?—4,500,000 gallons.

REIGATE AND REDHILL.

2. Population, 25,000.

4. Collecting area of the town's sewage, 6,000 acres.

5. System of sewerage, broad irrigation.

7. Single or duplicate system of sewers? Duplicate.

9. Annual cost of maintenance of sewage works?—£180; includes bailiff's salary, £3 per week.

15. Volume of sewage dealt with per 24 hours?—720,000 gallons.

17. Settling-tanks, how many, and what size?—Straining-tanks only. The works are about 200 yards from the nearest cottages, and adjoin the main road.

20. What quantity of sludge is formed, and how disposed of? 200 tons yearly. It dries in heaps in the open air, and is sold to farmers and others at 5s. per load.

25. Precipitant ?—A little used.

26. Acreage under irrigation ?—Two outfall works; about 130 acres.

28. Depth of drain pipes, their size, and distance apart?—The effluent is run into a small river.

40. Do you consider your system satisfactory ?—Fairly satisfactory; have seldom complaints, never proceedings. Sometimes complaints are made of the smell arising from the works.

ROCHDALE.

REPLIES BY MR. S. S. PLATT, Borough Engineer.

1. Name of town ?—Rochdale.

2. Population ?—75,000, estimated.

3. Rateable value ?—£250,000.

4. Collection area of town's sewage ?—Whole borough, 4,180 acres.

5. Date of sewage scheme ?—1880. Farm completed 1888.

6. Total cost of sewage scheme ?—£76,000 under 1880 order, of which £40,000 is for works of sewage, and £36,000 for land for farm and works of disposal; in addition to that, about £38,000 had previously been spent as sewerage works (intercepting and main sewers).

7. Single or duplicate system of sewers ?—Single system, with automatic storm-overflows at every available point.

8. Cost of construction of laying storm-water system ?—Very little extra expense, as main intercepting sewers run along the valleys near the rivers and brooks.

9. Annual cost of maintenance of sewage works ?—£700 for working expenses. No rent.

10. Annual loss in the maintenance of sewage works ?—£500, disregarding rent, interest on capital, sinking fund, &c.

11. Any pumping of sewage ?—No.

12. How pumped ?—None.

13. Capacity and cost of pump ?—None.

14. To what height is the sewage raised?—None.

15. Volume of sewage dealt with per 24 hours?—At present about 1 million gallons, but expect 1¼ millions eventually.

16. How many outfalls?—Two.

17. Settling-tanks, how many and what size?—Three for secondary outfall, which will bring about one-fifth volume of sewage, all 6 feet deep; average capacity, each 125,000 gallons.

18. Cost of settling-tanks and their total capacity?—Not been in operation yet, as only just completed.

19. Analysis of effluent?—We have no analysis.

20. What quantity of sludge is formed and how disposed of? None.

21. Cost of dealing with the sludge?—None.

22. Disposal of night-soil?—Rochdale pail system of collection of excreta and house refuse. Excreta manufactured into a concentrated manure.

23. Proportion of population using privies to that using water-closets?—Practically the whole—only about 550 w.c.'s for good houses (which invariably have pail-closets for servants), and about 150 old privies and ashpits.

24. Precipitation?—Not yet in full operation.

25. Precipitant?—Not yet decided upon.

26. Acreage under irrigation?—50 acres.

27. System of irrigation?—Intermittent filtration.

28. Depth of drain pipes; their size; distance apart?—Average 5 feet; 4 and 6 diameter; 10 yards apart.

29. Nature of land under irrigation?—Alluvial loamy soil and gravel.

30. Geological formation? — Post-glacial over lower coal measures.

31. Cost per acre of preparing irrigation land to receive the sewage?—

	Per acre.
Under-drainage	£24
Preparation (*i.e.*, levelling, &c.) of filtration area and roads	80
Distribution pipes and valves	36
River and fence walls	20
Per acre	£160

32. Is the land under irrigation freehold or leasehold?—Freehold.

33. Cost per acre of the freehold?—£250 per acre, which includes river bed, which intersects farm very much.

34. Rental per acre paid by urban authority to owner if leasehold?—None.

35. Is the management by the Board or by sub-letting?—By Corporation.

37. Crops grown?—Rye grass, vetches, oats, cabbages, mangolds, turnips, and 4 acres of osiers (willows).

38. Annual sum realised by sale of crops?—£200.

39. Annual sum realised by sale of other products?—None.

40. Do you consider your system satisfactory?—Yes, but we could do with more land, as at present only two-thirds of sewage is connected to farm, but no more suitable or available land within reasonable distance except by pumping. Expect we shall eventually have to erect precipitation-tanks for principal outfall, to enable present lands to deal with increased quantity of sewage. The abstraction of the excreta from the sewage (by pail system) renders the liquid sewage weaker than the average, and contains less solids.

Sheffield.

2. Population, 328,337.
3. Rateable value, £1,088,294.
4. Collecting area of Board's sewage, 4,000 acres.
5. Date of sewerage scheme, 1883.
6. Total cost of sewerage scheme?—Cost of works only, £32,000.

7. Single or duplicate system of sewers ?—Single.

9. Annual cost of maintenance of sewage works, £6,000.

11. Any pumping of sewage ?—None.

15. Volume of sewage dealt with per 24 hours, 10 million gallons.

16. Outfalls ?—One.

17. Settling-tanks, number and size ?—30 tanks, 36 feet and 40 feet, 6 feet.

18. Analyses of effluent:

Free ammonia, per million parts..............	5·15
Albuminoid ammonia, ditto....................	0·50
Chlorine, ditto	68·0
Total solid grains per gallon	33,274

20. Quantity of sludge formed, and how disposed of ?—20,000 tons per annum, given away for agricultural purposes.

21. Cost of dealing with the sludge ?—Sixpence per ton.

22. Disposal of nightsoil ?—By cartage to a distance, and sale to farmers.

23. Proportion of population using privies to that using water-closets ?—About three-fourths use privies.

25. Precipitant ?—Lime.

40. Do you consider your system satisfactory ?—Yes.

SWANWICK.

2. Population, 4,000, draining to works.

4. Collecting area of the Board's sewage, 15 acres.

5. Date of sewerage scheme, 1886.

6. Total cost of sewerage scheme ?—The cost of land, including £400 paid for furnace ashes, was £1,123.

7. Single or duplicate system of sewers ?—Duplicate.

9. Annual cost of maintenance of sewage-works ?—The working expenses are not available, but probably amount to as much as the sales of produce.

17. One settling-tank.

23. Proportion of population using privies to that using

water-closets ?—Water-closets not general, but there is a public water supply.

28. Depth of drain-pipes ?—Four feet; size, 4 inches; distance apart, 5 yards.

29. Nature of subsoil under irrigation ?—Stiff clay.

32. Land under irrigation, freehold or leasehold ?—Leasehold.

34. Rental per acre paid by Urban Authority to owner, if leasehold ?—£2 per acre.

37. Crops grown ?—Osiers, which were found not to thrive in consequence of the small volume of sewage available, and the sets were removed.

Swinton and Pendlebury.

2. Population ?—Estimated 22,000 in the whole of the district.

3. Rateable value, £90,000.

5. System of sewerage and date ?—The International system, precipitation and filtration: Swinton, October, 1890; Pendlebury, July, 1890.

6. Total cost of sewerage scheme ?—Swinton, £3,800; Pendlebury, £10,000; but over £10,000 had previously been spent in making a sewage farm.

9. Annual cost of maintenance of sewage works ?—Swinton, £400; Pendlebury, £200.

10. Annual loss in the maintenance of sewage works ?—£600.

11. Pumping of sewage ?—No.

16. Outfalls ?—Two.

23. Proportion of population using privies to that using water-closets ?—We have very few water-closets compared with the population.

24. Precipitation ?—Yes.

25. Precipitant ?—Ferozone, with Polarite filtration.

33. Cost per acre of freehold ?—£250 at Pendlebury.

34. Rental per acre paid by Urban Authority to owner if leasehold ?—Swinton, £5 per acre.

35. Farming management by the Board or by sub-letting? By the Board.

37. Crops grown?—At Swinton, turnips and mangold-wurzel.

38. Annual sum realised by sale of crops?—Varies very considerably.

40. Do you consider your system satisfactory?—Yes, very.

Stretford.

2. Population, about 21,000; only about 6,300 contributing to sewage farm.

3. Rateable value, £126,211.

4. Collecting area of the Board's sewerage scheme?—About 175 acres built upon, consisting of villages and villa residences.

5. System of sewerage and date?—Broad irrigation, April, 1885.

6. Total cost of sewerage scheme, £20,000.

7. Single or duplicate system of sewers?—Duplicate.

9. Annual cost of maintenance of sewage works?—About £450; this is only the loss in disposing of the sewage, and does not include interest for outlay for farm land, &c.

11. Pumping of sewage?—Yes.

12. How pumped?—By two pulsometer pumps.

13. Capacity of pumps?—Capacity, 78,000 gallons per hour.

14. To what height is the sewage raised?—22 feet the highest lift.

15. Volume of sewage dealt with per 24 hours, 260,000 to 468,000 gallons.

16. Outfalls?—Two effluent outfalls from farm. Only one sewage outfall into a receiving tank.

17. Settling-tanks, number and size?—One.

18. Cost of settling-tanks, and their total capacity?—Cost, about £1,000; capacity, about 380,000 gallons.

19. Analyses of effluent?—Never been analysed yet. We rarely have anything wrong with effluent.

20. Quantity of sludge formed, and how disposed of?—About

18 inches of sludge are deposited per annum; this is lifted by hand-labour, and when dry is carted out to the land.

22. Disposal of night-soil?—By contract.

23. Proportion of population using privies to that using water-closets?—About one-fourth of population have water-closets.

26. Acreage under irrigation, 40 acres.

28. Depth of drain-pipes?—3 to 5 feet; size, 9 by 12, mains with 6 by 4; agricultural drain-pipes 20 yards apart.

29. Nature of subsoil under irrigation?—Alluvial deposit, excellent land.

30. Geological formation, ditto.

31. Cost per acre of preparing irrigation land to receive sewage?—Nearly £30 per acre for carriers and under-draining.

32. Land under irrigation, freehold or leasehold?—Freehold.

33. Cost per acre of freehold, £100.

35. Farming management by the Board, or by sub-letting.—By Board.

36. Crops grown?—Osiers, cabbages, mangolds, rye grass, oats, and potatoes.

38. Annual sum realised by sale of crops?—This is the first year of entire farming by ourselves; only osiers previously farmed by us.

40. Do you consider your system satisfactory?—Yes.

Taunton.

2. Population?—20,000.

4. Collecting area of the town's sewage?—1,200 acres.

5. Precipitation.

10. Annual loss in maintenance of sewage-works?—Recently £250 loss; more now, owing to grumblers.

15. Volume of sewage dealt with per 24 hours?—1,500,000 gallons.

17. Settling-tanks, how many and what size?—600,000 gallons. The works are about 200 yards from the nearest dwelling-houses or public road. Going to enlarge tanks by order of Local Government Board.

20. What quantity of sludge is formed, and how disposed of? About 1,000 yards per annum. It is dried in the air and sold at 50s. per pit.

24. Precipitant?—Lime, 10 cwt; Droitwich salt, 1¼ cwt.; carbolic acid, 3 quarts per half-million gallons of sewage.

26. Acreage under irrigation?—Area of works, three acres.

28. Depth of drain pipes, their size and distance apart?—The effluent is run into the river Tone.

40. Do you consider your system satisfactory?—Very satisfactory. No complaints are made of the smell arising from the works.

TOTTENHAM AND WOOD GREEN.

2. Population?—Estimated 85,000.
3. Rateable value?—£344,172.
4. Collecting area of the Board's sewage?—4,673 acres.
5. Lime precipitation.
7. Single or duplicate system of sewers?—Duplicate.
9. Annual cost of maintenance of sewage works?—Last completed year £13,668, including loan repayments.
10. Annual loss in the maintenance of sewage works?—No profit.
11. Pumping of sewage?—All pumped.
12. How pumped? One 100 H.P. compound beam engine, with bucket and plunger-pumps; one 45 H.P. horizontal pumping-engine.
13. Capacity of pumps? Can pump from 2 to 4 million gallons per day.
14. To what height is the sewage raised?—About 35 feet.
15. Volume of sewage dealt with per 24 hours?—Minimum 1½ million gallons.
16. Outfalls?—One.
17. Settling-tanks, number and size?—Series of five tanks.
18. Cost of settling-tanks and their total capacity?—Capacity, 4½ million gallons.
20. Quantity of sludge formed and how disposed of?—The

sludge is pressed into cakes, which finally retain about 50 per cent of moisture. 10,000 tons of cakes made per annum; making and disposal about £5 per ton.

23. Proportion of population using privies to that using water-closets?—All water-closets.

ANALYSIS OF EFFLUENT.

	Parts per million. Free Amm.		Alb. Amm.
September, 1877	36·4	......	2·40
November, 1878	40·0	......	3·50

WALTON-ON-THE-HILL.

2. Population?—40,000.

4. Collecting area of the town's sewage?—1,904 acres.

5. Irrigation.

10. Annual loss in maintenance of sewage works?—Loss, £10 to £20.

15. Volume of sewage dealt with per 24 hours?—770,000 gallons.

17. Settling-tanks, how many and what size?—700,000 gallons. The road adjoins the farm.

20. What quantity of sludge is formed and how disposed of? None.

24. Precipitation?—None.

26. Acreage under irrigation?—104 acres, enlarging to 179.

28. Depth of drain pipes, their size and distance apart?—The effluent flows into the river Alt, Little Brook, and Knowsley Brook.

29. Nature of land under irrigation?—The water, after filtering through land, which is particularly suitable, being very porous, is as clear as spring water. The cause of complaints was running off water direct to river Alt.

40. Do you consider your system satisfactory?—Are just completing new sewage scheme, which will relieve us from complaints. No complaints are made of the smell arising from the works.

Wednesbury.

2. Population ?—25,200.

4. Collecting area of the town's sewage ?—2,124 acres.

5. Precipitation by lime and alumina.

7. Single or duplicate system of sewers ?—Duplicate.

9. Annual cost of maintenance of sewage-works ?—Last year £312.

15. Volume of sewage dealt with per 24 hours ?—60,000 gallons, including infiltration water.

17. Settling-tanks, how many and what size ?—Eight tanks of 50,000 gallons each. The main road is about 300 yards from the works; a foot road passes through. The nearest house is 200 yards from the works.

20. What quantity of sludge is formed, and how disposed of? Practically none at present. Pressed into cakes by Johnson and Co.'s. filter presses. We fear there will be a difficulty in disposing of it.

22. Disposal of nightsoil ?—Only about 130 private connections made up to date. They comprise houses, water-closets, slaughter-houses, cellars, &c. Urinals to number of 279 inclusive.

24. Precipitation ?—Chemical precipitation with quiescence through tanks, and final application to land.

25. Precipitant ?—Five grains each of milk of lime and sulphate of alumina per gallon of sewage.

26. Acreage under irrigation ?—21 acres (12 acres sewage-beds).

28. Depth of drain-pipes, their size and distance apart ?—The effluent flows into the river Tame.

40. Do you consider your system satisfactory ?—Yes.

Wimbledon.

2. Population ?—26,154.

3. Rateable value ?—Poor rate, £187,310; general district rate, £169,551.

4. Collecting area of the town's sewage ?—3,220 acres.

5. System of sewage purification?—The Amines Process chiefly, and irrigation in part. The sewage from the high-level sewer can either be chemically treated or passed upwards through specially prepared filters, in which case the whole of the sewage is then run over the land before being discharged into the river.

7. Single or duplicate system of sewers?—Duplicate. The surface-water is conveyed to the natural watercourses by a series of surface water-sewers (for the most part formed of sewers originally constructed to take both foul and surface-water), and by the surface-water outfalls. The surface-water sewers have reduced the amount of sewage pumped as follows: In 1885, the year before the surface-water collecting commenced, 190,000,000 gallons were pumped from a population of 12,000; 15·8 million gallons were pumped per 1,000 persons per annum; rainfall, 21·5 inches; 12·3 million gallons were pumped per 1,000 persons, and per inch of rainfall; whilst in 1889, with a population of 15,324, only 186·6 million gallons were pumped, being 12·2 million gallons per 1,000 persons, and 9·6 million gallons pumped per 1,000 persons per one inch of rainfall.

9. Annual cost of maintenance of sewage works?—£700 per annum.

11. Any pumping of sewage?—Yes.

12. How pumped?—The high and middle-level sewers gravitate to the works, while the low-level sewer has to be pumped at Raynes Park, and again at the works. At the Raynes Park pumping station the sewage of Cottenham Park and part of Warple Road is raised to a height of 36 feet by two single-throw plunger pumps, 12 inches in diameter, plungers with 24-inch stroke, worked by two 31 H.P. gas-engines. At the farm pumping station the whole of the low-level sewage, and that from the Durnsford Road sewers, is lifted 22 feet by two-throw plunger pumps, 24 inches diameter, plungers with 4 feet 6 inch stroke, worked by two condensing engines, 14-inch diameter pistons, with 3 feet stroke.

13. Capacity and cost of pumps?—20,518,900 gallons of sewage were pumped at the Raynes pumping-station, at a cost of £142 8s. 10d. 180,762,500 gallons were pumped at the farm pumping station, at a cost of £486.

14. To what height is the sewage raised?—At Raynes pumping-station the sewage is raised to a height of 33 feet, and at the farm pumping-station the sewage is raised to a height of 22 feet.

15. Volume of sewage dealt with per 24 hours?—650,000 gallons.

17. Settling-tanks?—There are six settling-tanks, thus having a surplus capacity equal to 80 per cent of the day's flow as a provision for rainfall excess, but principally to allow for quiescence during the subsidence of the precipitate and matters in suspension.

18. Capacity of settling-tanks?—530,000 gallons.

20. What quantity of sludge is formed, and how disposed of? The sludge works comprise a building with two of Johnson's and one of Manlove's sludge-presses, an air-compressor, and other appliances. The sludge is compressed into cakes, for which 1s. a ton is obtained at the works when there is a demand. Five tons of sludge, as taken from the settling tanks, are reduced to one ton of cake, four tons of water being squeezed out. The total quantity produced in the year was 3,507 tons.

21. Cost of dealing with the sludge?—The cost of pressing the sludge was 2s. 3d. per ton, or 32s. per million gallons.

22. Disposal of nightsoil?—There appear to be no privies, but the scavenging of refuse amounted to 2·6 cubic yards per house, and cost 2s. 4d. and 1s. 2d. per yard. It was disposed of in part to a brickmaker, and the remainder dug-in over an area of three acres at the sewage farm. Total cost of collecting 11,000 cubic yards was £24 per week.

24. Precipitation?—Amines, and running over land.

25. Precipitant?—Two and a half tons of lime and one-fifth ton of herring-brine per 1,000,000 gallons. The chemical

treatment during the early part of the year consisted, as in former years, of the addition to the sewage of lime and alumino-ferric. Owing, however, to the successful experiments carried out by the Amines syndicate with their process during the months of July and August, that process has since been applied to all the sewage which receives chemical treatment. The average cost of chemical treatment during the year amounted to 38s. 6d. per million gallons.

26. Acreage under irrigation?—75 acres.

28. Depth of drain-pipes, their size and distance apart?—The effluent water is run into a tail-race of the Wandle.

37. Crops grown?—Rye-grass, mangold-wurzels, osiers, and garden produce.

38. Annual sum realised by sale of crops?—Receipts, £1,202 1s. 1d., including produce consumed by Board's horses, amounts annually to about £300 in value. Expenditure, £833 0s. 1d.

39. Annual sum realised by sale of other products?—Cows kept, but in decreasing number.

40. Do you consider your system satisfactory?—No complaints made. Only complaint last July, from a heap of sludge.

WOKINGHAM COLLEGE.

2. Population?—About 600.

3. Rateable value?—£1,860, not including about ten private houses on the sewage system.

4. Collecting area of the Board's sewage?—The College estate, about 300 acres.

5. System of sewerage scheme and date?—Irrigation, 1883. The effluent water is carried over adjacent meadow and arable land in troughs of galvanised iron.

6. Total cost of sewerage scheme?—£4,992.

7. Single or duplicate system of sewers?—Single.

8. Cost of construction of laying storm-water system?—No separate account.

9. Annual cost of maintenance of sewage works?—£250 paid to the Native Guano Co. to work it.

10. Annual loss of maintenance of sewage works?—About £220, the manure being worth about £30.

11. Pumping of sewage?—The sewage coming in at night is collected in a tank, and pumped up in the morning.

12. How pumped?—Small engine and centrifugal pump.

14. To what height is the average raised?—10 feet.

16. Outfalls?—One.

17. Settling-tanks, number and size?—Two small tanks used alternately.

23. Proportion of population using privies to that using water-closets?—No privies.

30. Geological formation?—Bagshot sand and clay.

38. Farming management by the Board, or by sub-letting? It is managed by the Native Guano Co.

40. Do you consider your system satisfactory?—Very satisfactory as far as disposal of sewage is concerned.

Upwards of sixty of the sewage-disposal methods of towns have been described in this chapter, and with more or less fulness; but as it has not been found possible in every case to obtain that most important information which I, at the outset, stated to be indispensable to enable sound inferences to be drawn, namely, the amount of ammonia and albuminoid ammonia present in the effluent after treatment, the answers given to the last question, "Do you consider your system satisfactory," must be taken only in a general way. Some, no doubt, are reliable; some I know to be more or less unreliable; in others the wish is naturally father to the thought; and when the County Councils can be empowered to fix and to insist on a standard of purification for the effluent, I am quite sure that many systems will have to be reorganised, and some certainly abolished. That such a standard will some day be obligatory, I have no doubt whatever. Sanitation and river-cleanliness alike demand it imperatively, and if the application of such a test of limitation can be rightly delayed, it should only, in my opinion, be on the score of local difficulties, the cost of too soon replacing an inefficient system by an effective one, or by a judicious postponement of action until there has been evolved a satisfactory solution of the difficulties which surround the subject. A lengthened study of bacteriology has quite convinced me that sufficient importance had not yet been conceded to this part of the sewage-nuisance difficulty. I have therefore placed it in the front rank, and have tried to show that it ought to form almost the most important consideration; for what system can be commended which can permit the rapid growth of these organisms, the generators of putrefaction, and in many cases of disease, in either pasture-lands, in precipitates, in effluents, outflows, or in sludge-accumulations. Therefore these thoughts, above all others, lead me to the conclusion I have stated above, that a standard of high effluent purity, and the utmost freedom attainable from putrefactive organisms, should be aimed at, if not at present demanded, as an axiom of the first importance.

CHAPTER XV.

On Crude Sewage.

In the foregoing chapter I have recorded a sufficient number of examples of sewage disposal to enable thoughtful inquirers to draw useful deductions and comparisons from them, without my passing each one in review. I will deal, therefore, with the general question of sewage and its disposal in its twofold aspect, the getting rid of this refuse product, and how far sewage can be utilised and rendered innocuous. Its chemical composition has been ascertained, its fertilising qualities are also now correctly known, and a consideration of these two points may be appropriately introduced here, to summarise and further illustrate what I have previously stated about them.

The Commissioners appointed to report on the Mersey and Ribble basin have given much valuable information respecting the composition of sewage, the main points of which have been succinctly tabulated by Mr. Santo Crimp, from analyses of sewage collected and averaged from upwards of thirty towns. I repeat his table,* but alter the proportion from parts per 100,000 to parts per million, for the sake of uniformity with preceding analyses.

Average Composition of Sewage.

In parts per 1,000,000.†

Description.	Total Solid Matters in Solution.	Organic Carbon.	Organic Nitrogen.	Ammonia.	Total Combined Nitrogen.	Chlorine.	Suspended Matters.		
							Mineral.	Organic	Total.
Midden Towns......	824·0	41·810	19·750	54·350	64·510	115·40	178·10	213·00	391·10
Water-closet Towns..	722·0	46·960	22·050	67·030	77·280	106·60	231·80	205·10	446·90

* Sewage Disposal Works, p. 10.

† To convert parts per million into grains per gallon, it is only necessary to multiply by 70 and remove the decimal point one place to the left; and, inversely, to convert grains per gallon into million parts, remove the decimal point one place to the right, and divide by 70.—T.W.

OR IN GRAINS PER GALLON.

Description.	Total Solid Matters in Solution.	Organic Carbon.	Organic Nitrogen.	Ammonia.	Total Combined Nitrogen.	Chlorine.	Suspended Matters.		
							Mineral	Organic	Total.
Midden Towns......	576·80	29·260	13·820	38·040	45·150	80·780	124·670	149·100	273·770
Water-closet Towns..	505·40	32·870	15·430	46·920	54·100	74·620	169·260	143·570	312·830

And in another table he very lucidly gives this information in a different form, adding to it money values.

COMPOSITION AND ESTIMATED VALUE OF THE SOLIDS OR URINE AND FÆCES AND OF THE MIXED EXCRETA OF A POPULATION.

Constituents per Ton.	Ammonia (=7d. per lb.)	Phosphoric Acid.		Potash (=8d. per lb.)	Estimated value per Ton.
		Soluble (=4d. per lb.)	Insoluble (=2d. per lb.)		
NATURAL STATE	lbs.	lbs.	lbs.	lbs.	£ s. d.
Urine	239·40	29·40	—	33·90	0 15 10
Fæces	354·50	—	266·20	94·60	1 7 6
Mixed Excreta of population	231·30	27·00	19·30	38·30	0 15 8
1,000 tons of average London Sewage ...	2193·70	276·10	242·00	506·50	0 0 1¾

And then he compares sewage with farmyard manures and guano, as follows:—

1lb. of human excrement=13 lbs. of horse-dung.

1lb. of human excrement=6 lbs. of cow-dung.

Excreta (solid and liquid) of one adult=droppings from one sheep.

Yearly excreta (solid and liquid) of one adult=75 lbs. of Peruvian guano.

The annual value of human excreta per person has been variously estimated. Lawes and Way value it at 8s. 5¾d., Voelcher at 9s., and Thudichum at 20s.; we may therefore safely estimate it at 10s.

Professor Tidy's valuation of the ammonia in urine per annum per person is 8s., and that of fæcal matter to be 1s. 3d.

Mr. Henry Robinson* states that the amount of ammonia present in sewage varies from two and a half grains to fifteen grains per gallon, and that the chemist's theoretical value is one farthing per ton for each grain of ammonia per gallon; thus the value of sewage per ton would vary from two and a half farthings to fifteen farthings, according to its dilution or strength. The average quantity per person per annum in a town population is 12·6 lbs. of ammonia.

His estimate of the average amount of solid matter per gallon of sewage is 90 grains, of which 28 grains are organic and 62 grains inorganic; and that the average weight of solid and fluid excreta per day per head of population is 2½ lbs.; of this the average weight of moist material is 39·1 oz., and the proportion insoluble in water is 291 grains.

Dr. Tidy estimated that London sewage contained two-fifths of a ton of nitrogen, one-seventh of a ton of phosphoric acid, and one-thirteenth of a ton of potash, in each million gallons.

Dr. Letheby has shown that

1 gallon of midday London sewage contains solid matter in solution	55·7 grs.
1 gallon of midday London sewage contains solid matter in suspension	38·2 grs.
Total	93·9 grs.
1 gallon of midday London sewage contains organic matter in solution	31·2 grs.
1 gallon of midday London sewage contains organic matter in suspension	16·1 grs.
Total	47·3 grs.

I extract the following table, prepared by Mr. Henry Robinson and Dr. Letheby, from an analysis by Prof. Way:—

* Sewage Disposal, p. 6.

		Grains per Gallon	
Rugby sewage, total solid matter		92·5	
Do.	proportion of organic matter		29·0
Do.	of mineral matter		63·5
	Total		92·5

	Grains.
Nitrogen per gallon	6·18
Phosphoric acid	1·68
Potash	2·81

Mean analyses of London sewage by Letheby, Hoffman, Witt, May, and Voelcker:

	Grains.	
Organic matter	27·72	per gallon.
Nitrogen	6·21	„
Phosphoric acid	1·57	„
Potash	2·03	„

Too much must not be expected from sewage in a money-value sense. It has not that high commercial value which some enthusiasts have accorded to it, and the same may be said of the various processes of sewage treatment. The way to look at the question of sewage disposal is how to deal with it in the most fertilising manner, as well as in the most permanent and economical way consistent with sanitary precautions.

Mr. Boulnois estimates the value of crude sewage at one half-penny to twopence per ton, or tenpence per head of the population per annum, and that one hundred tons of sewage, or 22,600 gallons, will cover an acre of land one foot deep.

The Rivers Pollution Commission estimated the value of 100 tons of average sewage at 17s., or about 2d. per ton. Other estimations, however, place its value as high as 3s. 4d. per ton.

Crude sewage is matter not to be trifled with, and all sewage disposals should never be allowed too near a town, but should be dealt with as far as practicable from it. M. Lefevre, president of the Paris Société of Geometry, found that land decreased in value in the neighbourhood of the Paris sewage disposal, and the health of the inhabitants in the vicinity was affected by it. He stated his opinion to be that it was a public m fortune to dispose of crude sewage on land immediately outside a town.

To give an idea of the quantity of sewage per population, I have selected the following towns, showing their sewage-flows each twenty-four hours, and in gallons per person.

Name of Town.	Population.	Sewage now per 24 hours. Gallons.	Gallons per inhabitant.
Birmingham	606,000	15,000,000	25
Bradford	224,507	8,000,000	35
Crossness	300,000	9,000,000	30
Chiswick	20,000	700,000	35
Derby	94,000	2,350,000	25
Edinburgh	262,733	10,750,000	41
Leeds	327,324	10,000,000	30
Leicester	143,153	8,000,000	42
Grantham	17,000	1,000,000	59
Norwich	93,000	4,000,000	43
Rugby	12,000	600,000	50

Thirty gallons of sewage per head per day may be taken to be about a fair average.

CHAPTER XVI.

THE PRECIPITATION, SETTLEMENT, AND FILTRATION OF SEWAGE.

IF chemical treatment for the precipitation of the soluble organic matter of sewage is adopted, we must aim at four desiderata, namely:—

1. The minimum evolution of ammonia, escape of which into the air is waste.
2. A harmless effluent.
3. A harmless liquor from the sludge-presses.
4. A minimum amount of sludge with a minimum amount of liability to putrefaction.

The effluent should be so purified as to be fit to turn into a river containing at least eight volumes of water to one of effluent without having been previously used for irrigation. It is but sorry economy to turn into the river an effluent containing much organic matter and the bacterial germs of typhoid fever, pneumonia, phthisis, and other disease-producers.

Drs. Frankland and Hoffman, in their report to the Metropolitan Board of Works on the deodorisation of sewage by chemical means, state: "We beg to express our opinion, based upon the experience acquired during our investigations, that the disinfection of vast volumes of sewage can be more easily accomplished chemically than is generally believed, and than we ourselves anticipated at the commencement of our inquiry."

This subject of precipitation is so bound up with the description of the many systems of sewage treatment described in other parts of my book, and with analytical results, particularly in Chapters IV., V., VI., VII., VIII., IX., XI., XII., XIII., XIV., XVI., XVIII., XIX., and XX., where it will be found largely entered into, that it is not at all necessary to treat it at great length in this place; neither is it necessary to repeat the various precipitants already so fully described, more than to state that it will be found that they may be chiefly grouped as lime, alumina, and iron in various states, and with or without admixture with other substances. The proportions used vary according to varying sewages and the nature of the precipitants. Generally speaking, about 1 ton of the precipitant to each million gallons, or 15·68 grains per gallon, is a necessary quantity.

PRECIPITATING AND SETTLING-TANKS.

Great care should be taken in the construction of precipitation and settling-tanks. They should have sufficient capacity to allow of a rest of from one to two hours to allow the precipitate to settle, and for rainfall.

Mr. Santo Crimp is of opinion that the precipitating and settling-tanks should have a capacity equal to 45 per cent of the day's flow, to allow for two and a half volumes of rainwater, and the room necessary to be provided for the sewage when it is in the quiescent settling-stage. Birmingham and Burnley have extra tank room equal to 56 per cent of each day's sewage flow.

He rightly condemns the tanks usually employed, and recommends a construction with segmented bottoms in cross section, with a fall longitudinally of about 1 in 100. By this form * the effluent is better clarified in settling and the sludge more economically removed. I have, by his kind permission, quoted *in extenso* his remarks on these points in Appendix C.

At Nuneaton two large circular tanks have been constructed. They are similar to the Dortmund system of Herr Kunbühler, and are concave at the bottom. Another form of tank is used at Essen, and is called the Rockner-Rothe system. It is shallow, and has in the centre a large vessel like a gasholder, from which the air is partially exhausted; the effluent after precipitation rises in the vessel until it reaches an outlet discharge-pipe.

---

* Sewage Treatment and Sludge Disposal, p. 14.

Filtration of Effluents.

The section of an efficient filter is given in my seventh chapter, and its mode of construction described; but as this "polarite" filter is protected by a patent, in consequence of the filter containing a stratum or layer of "polarite," it can only be used by arrangement with the International Co. But polarite, although an excellent oxidiser of organic matter, is not a necessary factor in ordinary filtration. Very efficient filters can be made of alternate layers of fine and coarse sand, with a bed of gravel underlying them. If cinders are used in a bed of six or eight inches, the layers of sand and gravel need not be more than six or eight inches each, but if without cinders thicker beds are necessary. I have also given sections of other filters. (See plate 7, page 77; plate 8, page 114; plate 10, page 147; fig. 46, page 159, and Appendix G.)

Mr. H. Robinson gives the sections of two kinds of filters,* one as constructed by Mr. Higgins for filtering the turbid water of the River Plate, the other an ordinary waterworks filter of the following construction:—

1st layer, 2 feet 6 inches of fine sand at top.

2nd layer, 6 inches of coarse sand.

3rd layer, 6 inches of shells.

4th layer, 2 feet 6 inches of gravel at bottom.

This filter allows about 700 gallons per square yard per day to pass through, and with excellent results.

Filtration through sand or soil is a process of oxidation, and is successful where not more than 5½ gallons of sewage per cubic yard pass through the filtering bed in 24 hours. The Rivers Pollution Report proved that the result is one of oxidation, from the organic matter being converted into carbonic acid and nitric acid. By such downward filtration through soil "the organic carbon and nitrogen were reduced to the proportions of London drinking water, namely, from 5·0 to 7·0 of carbon and

* Sewage Disposal p. 18.

·6 to ·7 of nitrogen in a million parts, the original amount in the sewage having been 43·8 of organic carbon and 24·8 of organic nitrogen."

On the other hand, the amount of nitrogen, as nitrates and nitrites, was increased from nothing to from 30 to 50 parts in a million.

Upward filtration does not cause oxidation.

The Rivers Pollution Commissioners found that, given a properly constituted soil, a properly drained and arranged for intermittent sewage flow, "the sewage of a water-closet town of 10,000 inhabitants could be cleansed upon five acres of land."

It appears from Mr. Warrington's researches that nitrification of sewage ceases in the presence of antiseptics, probably because the nitrifying bacterial organisms are destroyed. Thus it seems to be an advantage where sewage has to be treated and put into a river without the intervention of land filtration; but I have considerable doubt as to the accuracy of this remark. It is safe to say that it is an over-statement, or one open to many exceptions. For example, the successful monthly grass-crops grown at Wimbledon sufficiently prove that the "Amines" treatment, which I suppose would come under Mr. Warrington's definition of antiseptic, does not stop or prevent nitrification, but apparently the reverse.

In considering the question of filtration expenses, whether by artificial filters or by land irrigation, the most important factor is that of the local sewage flow, which varies very much and very curiously in different localities. The average daily flow of sewage is at the rate of about 30 to 40 gallons per inhabitant, but the quantity varies very much, probably on account of both local circumstances and the abundance or scarcity of water supplied to the town. Thus in Dudley, with a population of 47,000, the daily flow of sewage is only 13 gallons per inhabitant, whilst at Salisbury, with a population of 15,000, it is as high as 166 gallons; but those are the extremes, as the table on p. 235 shows.

CHAPTER XVII.

On Sewage Effluent Waters.—Deductions as to the most Effective Method of Treating Sewage.

As crude sewage is not universally applied to land, and as the treatment of sewage by precipitation is now rapidly being adopted, it will be necessary to consider the composition and values of the two products of sewage after chemical treatment, namely, the resulting effluent waters, which have either to flow upon land or directly into rivers, and the precipitate or sludge. The two following chapters are devoted to these portions of the subject.

On Organic Impurities in Crude Sewages, in Effluents, and in Drinking Water.

In order to emphasise one of the axioms I have insisted upon throughout, that of ascertaining the amount of organic matter in the effluent, I will give in a tabulated form the amounts of free and albuminoid ammonia which upon analysis have been found in the effluents of various systems of sewage treatment, and for comparison I place by their side the analytical results of the crude sewage wherever I have been able to obtain them. I also group with these the analyses of a series of good and bad potable waters in various parts of the country for purposes of comparison.

Here, therefore, is tabulated a series of most important facts which will help to guide those interested in hygiene and sanitation in a safe direction, both in aiming at purity of sewage effluents and also in water for drinking purposes.

In addition to showing the amounts of free and albuminoid ammonia per million parts, I have also converted the proportions into grains per gallon. I regret I was not able to obtain information on these points from many of the towns who were kind enough to answer many of the questions asked of them; they were not able to supply the desired data.

TABULATED STATEMENT OF FREE AMMONIA AND ALBUMINOID AMMONIA IN CRUDE SEWAGE AND IN EFFLUENT WATERS AFTER TREATMENT BY THE VARIOUS SYSTEMS OF SEWAGE DISPOSAL, AND IN SOME NATURAL POTABLE WATERS.

ELECTRICAL TREATMENT OF SEWAGE, OR THE WEBSTER ELECTROLYSIS.

	Name of Analyst.	Crude Sewage.				Effluent.			
		Parts per million.		Grains per gallon.		Parts per million.		Grains per gallon.	
		Free Amm.	Alb. Amm.	Free Amm.	Alb. Amm.	Free Amm.	Alb. Amm.	Free Amm.	Alb. Amm.
At Crossness (eight analyses)	Sir H. Roscoe	not given	12·430	not given	·87	not given	6·430	not given	·45
,,	,,	,,	16·710	,,	1·17	,,	5·570	,,	·46
,,	,,	,,	16·710	,,	1·17	,,	7·140	,,	·50
,,	,,	,,	16·00	,,	1·12	,,	4·00	,,	·28
,,	,,	,,	16·00	,,	1·12	,,	8·00	,,	·56
,,	,,	,,	19·00	,,	1·30	,,	8·570	,,	·60
,,	,,	,,	12·00	,,	·84	,,	1·500	,,	·105
,,	,,	,,	16·00	,,	1·12	,,	2·00	,,	·14
,,	Mr. A. E. Fletcher	19·0	6·50	1·33	0·455	17·0	6·0	1·190	·42

INTERNATIONAL WATER AND SEWAGE PURIFICATION CO. LIMITED.

	Name of Analyst.	Free Amm.	Alb. Amm.	Free Amm.	Alb. Amm.	Free Amm.	Alb. Amm.	Free Amm.	Alb. Amm.
At Acton	Sir H. Roscoe	not given	not given	not given	not given	4·06	·250	0·284	0·017
,,	Mr. Carter-Bell	,,	,,	,,	,,	0·160	0·150	0·0112	0·0105
At Swinton (5 analyses)	,,	8·80	1·60	·61	0·11	7·20	0·60	0·50	0·04
,,	,,	26·00	4·40	1·82	0·30	6·00	0·40	0·42	0·02
,,	,,	22·00	2·80	1·54	0·19	4·80	1·00	0·33	0·07
,,	,,	28·00	2·20	1·96	0·15	8·80	1·20	0·61	0·08
,,	,,	56·00	22·0	3·92	1·54	8·80	1·20	0·61	0·08
At Hendon	Mr. Arthur Angell	not given	not given	not given	not given	·31	·08	0·0218	0·0056

LIME PROCESS AND IRRIGATION.

	Name of Analyst	CRUDE SEWAGE.				EFFLUENT.			
		Parts per million.		Grains per gallon.		Parts per million.		Grains per gallon.	
		Free Amm.	Alb. Amm.	Free Amm.	Alb. Amm.	Free Amm.	Alb. Amm.	Free Amm.	Alb. Amm.
At Birmingham	Dr. Reid	not given	not given	not given	not given	·240	·220	0·016	0·015
,, the Tame	Mr. Wardle	,,	,,	,,	,,	1·660	·438	0·016	0·030
,, 11 miles below	Dr. Hill	14·00	2·80	0·77	0·19	not given	not given	not given	not given
At Tyburn outfall	,,	not given	not given	not given	not given	0·40	0·084	0·03	0·005
,,	,,	,,	,,	,,	,,	1·52	0·54	0·10	0·037
BROAD IRRIGATION.									
At Banbury	Not given	not given	not given	not given	not given	0·350	0·035	0·0245	0·0024
LIME PROCESS.									
At Salford	Mr. Carter-Bell	not given	not given	not given	not given	7·20	2·70	0·504	0·189
POROUS CARBON PROCESS.									
At Salford	Mr. Carter-Bell	not given	not given	not given	not given	·26	·21	0·118	0·014
INTERNATIONAL POLARITE PROCESS.									
At Salford	Mr. Carter-Bell	not given	not given	not given	not given	·40	·20	0·028	0·014
At Burnley, 1879	Mr. Angus Smith	,,	,,	,,	,,	not given	not given	not given	not given
Stream above Sewage Works	,,	,,	,,	,,	,,	0·25	0·24	·017	·017
Stream below Sewage Works	,,	,,	,,	,,	,,	2·36	0·70	·16	·49
Crude Sewage	,,	14·50	12·00	1·01	·84	not given	not given	not given	not given
Effluent	,,	not given	not given	not given	not given	13·30	3·05	·93	·21

LIME PROCESS AND IRRIGATION SYSTEM.

	Name of Analyst.	CRUDE SEWAGE.				EFFLUENT.			
		Parts per million.		Grains per gallon.		Parts per million.		Grains per gallon.	
		Free Amm.	Alb. Amm.	Free Amm.	Alb. Amm.	Free Amm.	Alb. Amm.	Free Amm.	Alb. Amm.
At Birmingham, 1879	Mr. Angus Smith.	not given	not given	not given	not given	not given	not given	not given	not given
Crude Sewage	„	27·00	10·50	1·89	·73	„	„	„	„
Effluent from Settling Tanks	„	not given	not given	not given	not given	32·20	6·60	2·35	·46
Effluent after Irrigation	„	„	„	„	„	12·75	2·10	·89	·14
SULPHATE OF ALUMINA AND LIME SYSTEM.									
At Coventry, 1879	Mr. Angus Smith.	not given	not given	not given	not given	not given	not given	not given	not given
Crude Sewage	„	25·80	17·00	1·80	1·19	„	„	„	„
Effluent from Settling Tanks (filtered)	„	not given	not given	not given	not given	24·00	2·45	1·68	·17
Effluent after Irrigation (drain out of order)	„	„	„	„	„	21·20	2·05	1·48	·14
Effluent after Irrigation	„	„	„	„	„	5·20	0·61	·36	·04
ALUMINO-FERRIC AND LIME PRECIPITATION, IRRIGATION, AND FILTRATION SYSTEM.									
At Aldershot, 1879	Mr. Angus Smith.	not given	not given	not given	not given	not given	not given	not given	not given
Crude Sewage	„	36·60	17·50	2·56	1·22	„	„	„	„
Effluent "A" from Sewage Farm	„	not given	not given	not given	not given	9·20	2·76	·64	·19
Effluent "B" from Sewage Farm	„	„	„	„	„	8·80	2·00	·61	·14
Effluent "C" from Sewage Farm	„	„	„	„	„	4·40	1·44	·380	·10
River Blackwater	„	„	„	„	„	0·108	0·3	·07	·02
Crude Sewage (camps)	„	120·00	52·50	8·40	3·67	not given	not given	not given	not given
Effluent from Sewage Farm	„	not given	not given	not given	not given	2·45	0·55	·17	·04
A. B. C. PROCESS, PRECIPITATION AND IRRIGATION.									
At Aylesbury, 1879	Mr. Angus Smith.	not given	not given	not given	not given	not given	not given	not given	not given
Crude Sewage	„	40·00	6·60	2·80	·462	„	„	„	„
Effluent from 3rd Settling Tanks	„	not given	not given	not given	not given	10·20	0·7	·7	·049

Urban authorities seem for the most part not to have attached sufficient importance to ascertaining the chemical efficiency or degree of purification, revealed only upon analyses, of the various methods of treatment. Not only do I urge that this should be done, but I think County Councils should insist upon periodical analyses of both effluents and potable water, nothing being more deceptive and illusive than a translucent or pellucid appearance.

The following analysis by Mr. Carter-Bell of a sample of Leek sewage of November of last year is a favourable illustration of the treatment of sewage by my basic persulphate of iron method. The sewage contained an abnormal quantity of dissolved organic matter, and was of the consistence of syrup. It is difficult to conceive of a fluid being more highly charged with decomposing matter. The sewer at the point from which this sewage was taken has not sufficient fall, and probably may have occasional accumulations of sewage.

Mr. Carter-Bell stated it to be the worst and most offensive sewage he had ever examined. There are occasional epidemic outbreaks at this part of the town.

SEWAGE COLLECTED AT 10 A.M., 31ST OCTOBER, 1891, FROM THE BROAD STREET MAIN, LEEK.

Parts in 100,000.	Equal parts of Salford and Pendleton Crude Sewage.	Effluent from Sewage, 20 grains to gal., Wardle's new method.	Leek Crude Sewage. No. 1.	Effluent from Leek Sewage, Wardle's new method, 20 grains to gal. No. 1.	Leek Crude Sewage. No. 2.	Effluent from Leek Sewage, Wardle's new method, No. 2.
Total Solids at 212°	156	114	182	...	154	...
Suspended Matter	34	nil	34	...	24	...
Ditto Mineral	14	"	15	...	14	...
Ditto Organic	20	"	19	...	10	...
Total Solids in Solution	122	114	148	...	130	...
Mineral ditto	82	85	112	...	90	...
Loss on Ignition	40	29	36	...	40	...
Chlorine	20	21	18	...	22	...
Oxygen required, 15 min.	4·2	2·3	2·106	1·5	2·1	1·46
Ditto 3 hours	7·1	3·9	3·80	2·1	3·1	1·92
Free Ammonia	2·6	1·10	12·0	10·0	16·0	12·20
Albuminoid ditto	·40	·18	1·2	·21	2·0	·34

Or 80 per cent of purification.

It is understood that the standard of purification which the Local Government Board would be willing to adopt is 2·50 parts of albuminoid ammonia per million parts.

POTABLE OR DRINKING WATERS.

Locality.	Name of Analyst.	Parts per Million. Free amm.	Parts per Million. Alb. amm.	Grains per Gallon. Free amm	Grains per Gallon. Alb. amm	Remarks.
Manchester, May, 1876 ...	Wanklyn.	0·03	0·08	·00	·00	Soft
„ August, 1867	„	0·01	0·06	·00	·00	„
Holmfirth, Flood-line	„	0·19	0·11	·01	·00	Contaminated
Edinburgh	„	0·00	0·07	·00	·00	Good
Glasgow, Loch Katrine ...	„	0·00	0·08	·00	·00	„
Sunderland	„	0·03	0·02	·00	·00	Very pure
Rhine at Bonn	„	0·01	0·06	·00	·00	„
Bonn Well Water	„	0·00	0·02	·00	·00	Organically pure
Nile, June	„	0·01	0·10	·00	·00	„
„ July	„	0·36	0·28	·02	·01	„
Thames at London Bridge..	„	1·02	0·59	·017	·04	Very bad
London Water Supply, Thames	„	0·01	0·06	·00	·00	
London Water, New River.	„	0·06	2·48	·00	·17	
London Water, Kent Co....	„	0·01	0·02	·00	·00	
Water of Great Purity......	„	0·00	0·05	·00	·00	
Safe Drinking Water	„	...	0·10	...	·00	
Dirty and Unsafe Waters are those which contain upwards of..................	„	...	0·10	...	·00	Unsafe
Dirty and Unsafe Waters to be absolutely condemned if over.......................	„	...	0·05	...	·00	„
Highwood	„	...	...	...	...	
Highwood Board School ...	J. C. Thresh.	·12	·08	·00	·00	Impure
Highwood, Chicken Row Pond	„	·08	·88	...	...	Much vegetable matter
Highwood Barrow's Cottage	„	1·25	·52	...	...	Much polluted
Boyton Road Stream	„	·04	·10	...	...	Usable
Chignal Pond.................	„	·25	·30	...	...	Bad
„ Pump	„	·04	·09	...	...	Suspicious
„ Sewell's Pump......	„	·16	·20	...	...	Bad
Great Baddow, Cottage Well	„	·00	·06	...	...	Much oxidised sewage
Galleywood, Francis Well ..	„	·40	·16	...	...	Bad
Runwell, Chalk Lane (rain water tank)	„	·80	·90	...	...	Very impure
Danbury, Bull Inn Well ...	„	·00	·20	...	...	Bad
Danbury, Baker's Pump ...	„	·94	·28	...	...	Polluted
Russell Green Well	„	·01	·10	...	...	Unsafe
Russell Green Well at Beer-house	„	·02	·05	...	...	Usable
Ingatestone Well	„	·01	·06	...	...	Much oxidised sewage
Great Waltham, Fanner's Farm Pump	„	·01	·09	...	...	Good

POTABLE OR DRINKING WATERS.—*Continued.*

	Name of Analyst.	Parts per Million.		Grains per Gallon.		Remarks.
		Free amm.	Alb. amm.	Free amm	Alb. amm	
Great Waltham Hill's Farm Pump	J. C. Thresh.	·01	·00	...	...	Very good
Great Waltham Public Supply	"	·03	·02	...	...	Good
Broomfield, Cottage Well...	"	·02	·20	...	...	Vegetable contamination
Broomfield Coffee Tavern Well	"	·04	·48	...	...	Very impure
Burnham	"	·01	·21	...	...	Suspicious, but clear
" Dilliway's Well..	"	·02	·10	...	...	Unsafe
" King's Arms Well	"	·32	·15	...	...	Bad
Burnham, White Hart	"	1·40	·10	...	...	"
" High Street	"	·01	·06	...	...	Good
Purleigh Rectory Cottage...	"	2·00	1·5	...	...	Bad
Writtle Cemetery Well ...	"	·04	·08	...	...	"
Oxney Green, Lawe's Cottage	"	·05	·24	...	...	"
Biddulph Moor, Leek	Wanklyn	0·03	0·14	...	...	Caused diarrhœa
Well, Leek Workhouse	"	0·02	0·34	...	...	"

If free ammonia is absent, Dr. Wanklyn considers that a potable water should not be condemned which contains albuminoid ammonia that reaches near 0·10 per million parts, but above this it is a suspicious sign, and that the presence of much albuminoid ammonia, little free ammonia, and almost entire absence of chlorides, are indicative of vegetable contamination.

Value of Sewage Effluents.

The fertilising powers of sewage effluents must naturally vary much, and are dependent upon the amount of organic nitrogen left in them after the sewage precipitation. It is scarcely necessary to tabulate the varying amounts, as they will be found in the foregoing tables of free and albuminoid ammonia. On reference to these tables, it will be seen that the Ferozone and Polarite process yields an effluent at Acton and Swinton with the minimum amount of organic or fertilising matter. At Swinton an average of five analyses gave 0·49 grains of free ammonia, and 0·05 grains of albuminoid ammonia per gallon; the average of two other precipitating methods, also lime and sulphate of alumina gave 1·77 grains of free ammonia and 0·33 grains of albuminoid ammonia; and between these extremes are shown a good many varying proportions resulting from the employment of the other processes.

Where an effluent has to be turned directly into a river or watercourse, no doubt the highest degree of elimination of organic impurity is essential; but where it can be used first for agricultural purposes in a proper manner, I hold that it is not so necessary to precipitate the sewage with such completeness, but to remove the solids and to run upon the land a clear effluent, and I think the more organic matter that may be left in the better, if the precipitation process is a sterling one.

The luxuriant growth of grass-crops at Wimbledon is, I think, a sufficient warrant for this reasoning; but what I have seen of sewage-farming where a crude unprecipitated sewage-flow has been used, convinces me that such a system is filthy, fraught with great risk both to cattle and to people living near. It is always more or less offensive, frequently polluting the air for considerable distances.

The agricultural value of an effluent also depends very much as to whether the storm-water is separated from the sewage or not. In rainy localities, where the sewage and the storm-water are united, the average fertilising value of the effluent is naturally very much reduced, but it must not be forgotten that even the addition of water to the soil is very beneficial, if administered intermittently and over sufficiently large areas. It dissolves the salts and ammonia in the soil, and causes a state of things highly beneficial to vegetation. I am inclined to urge for the most part that it is preferable to administer the effluent on to extended areas of land by pumping it, so that it can reach soils much more needing moisture than wet alluvial soils do. There are vast areas of land which might have their agricultural values much increased by the addition of very weak effluents, and I do not see why pumping should not be resorted to as largely for this purpose as it is for potable water supply to towns. In the neighbourhood of Leek there are admirable opportunities upon the poor, light Triassic sandy soils and banks for such a sewage-disposal, and so with many parts of the country. The suffusing and over-charging of wet alluvial or water-logged soils is wrong, arising from a mistaken notion of the economy of disposing of effluents or sewage by gravitation.

This distribution of effluents from precipitated sewage could be thus effected at reasonable cost even for long distances, and the method would have the advantage of being neither offensive by air-pollution nor injurious by overdosing to crops—in short, it would be cleanly, inodorous, and fertilising. It is thus, I think, the sewage disposal problem will be best met, namely, by a combination of a method of precipitation, an after intermittent land irrigation, and the ultimate discharge into a river of the twice cleansed effluent for further oxidation.

A moderate amount of organic matter in sewage effluent is rapidly oxidised by the presence of oxygen in river water.

The Rivers Pollution Commission found the quantities of oxygen in the following waters to be:

·637 cubic inches of oxygen in 100 cubic inches of rainwater.

·725 cubic inches of oxygen in 100 cubic inches of Cumberland lake water.

·588 cubic inches of oxygen in 100 cubic inches of Thames water.

·028 cubic inches of oxygen in 100 cubic inches of deep-well water.

The oxygen here found was absorbed from the atmosphere by the water in the greatest degree where the water had been longest exposed to the air, and in the least degree where it has been least in touch with the atmosphere.

Where a fairly purified effluent enters into a large river, the oxidation would be complete in a short distance.

It is stated that eight times the amount of river-flow to one of effluent produces satisfactory results, but it is easy to see that this depends upon the nature of the effluent.

Mr. Santo Crimp puts as a safe proportion that not more than fifty thousand gallons of effluent should be permitted to each million gallons of river-flow, or five per cent; but even then he does not suggest the degree of purity of the effluent, which, should it be bad, would be greatly better than crude sewage, but we may take an average effluent, showing, say, 3·00 of free ammonia and 0·60 albuminoid ammonia per million parts.

At Banbury the purified effluent finds its way into the river Cherwell, the river is stated to remain "good and pure, giving general satisfaction"; the analysis of the effluent being free ammonia 0·35, albuminoid ammonia 0·035 grains per gallon; or in parts per million, free ammonia 5·00, albuminoid ammonia 0·50. The standard of minimum impurity recommended by the Commissioners appointed in 1868 to inquire into the best means of preventing the pollution of rivers, was that no liquid should be permitted to go into a river if it contained more than two parts by weight of organic carbon or 0·3 parts by weight of organic nitrogen in 100,000 parts by weight. This would give an equivalent of albuminoid ammonia 0·36. These recommendations were not enacted.

The late Dr. Angus Smith, F.R.S., of Manchester, made some very important investigations upon the question of aeration of sewage, for the Local Government Board, in 1882, which he thus summarises: "In all cases putrefaction is delayed by aeration. The oxygen recovers itself in the aerated specimens better than in the non-aerated. Nitrates are formed more readily in the aerated than in the non-aerated specimens."

Deductions.

From all the information I have been able to obtain in this lengthened investigation and research, I am very strongly of opinion that the only safe method of sewage disposal is first by chemical precipitation, then land (or other) filtration, and afterwards river or other water-course oxidation. Precipitation by chemical reagents removes much dissolved organic matter from the sewage, yet it leaves some for further disposal; land-irrigation removes much of this, but not all, as has been often enough shown in these chapters. The diminished proportion, or what the land is unable to secrete or absorb in the course of fertilisation and filtration, is effectively dealt with by river oxidation; and it is this threefold action which best dequantitates sewage impurity, and in ratio of the most complete exhaustion, the action of the one beginning where the other ceases—results at once scientific, effective, and practical; and I believe the highest manurial value can be yielded by it—by which I mean that the greatest quantity of land can be brought under fertilisation.

I believe that this tripartite method would be advisable even if sewage-flows did not contain any excretal matter, for there is so much contamination passed into the sewers through sinks, chamber-slops and street rain-washings,* (which are exceedingly polluting) &c., that there would be a necessity for some treatment or other before they could safely be allowed to pass into a river.

I am thus led to the conviction that the best method of sewage disposal is that which precipitates a portion of the organic matter, and gives a clear effluent which can be used for irrigation, and after that river oxidation. With such a result, providing there is plenty of land and ample arrangements for intermittent irrigation, I repeat that it is not necessary or even advisable to remove a maximum amount of organic impurity from the crude sewage. It is another question when the effluent must be discharged into a river without previous land irrigation, or where there is only a small quantity of land available; in such a case the best precipitating process available, and the full equivalent proportion of the precipitant should be used. If in such circumstances filter-beds are employed, the effluent result is better, and in the case of very bad sewages, they are even necessary.

---

* See Appendix C on liquids from street surfaces.

CHAPTER XVIII.

On Sewage-sludge and its Disposal.

The chemical precipitation of sewage, and its subsequent treatment by irrigation and filtration where required, necessarily involve the disposal of the precipitate, which is composed of the dissolved organic matter thrown down, together with the precipitant and matters in suspension, all of which fall with it. This precipitate is termed sewage-sludge, and it is of this sludge that I propose to treat in this chapter.

Generally speaking, the value of sewage-sludge may be assessed at about as much as that of farmyard manure, but the following table gives some interesting figures both of actual sale values and calculated values, by which latter term is meant the fertilising value of the ammonia, calculated from the quantity shown by analysis in the sludge:—

Town.	Process.	Date.	Lime grains per gallon.	Lime per million gallons.	Fertilisat'n Value of Sludge per ton, calculated chemically.	Value of Sludge per ton, or price sold at.
Wimbledon	"Amines"	1891	23 to 30	28 to 37 cwt.	5s. to 6s.	1s.
Acton	International Water & Sewage Purifaction Co. Limited	1887	not stated	not stated	30s.	5s.
Aylesbury	A.B.C.	1879	,,	,,	33s.	not stated
Bradford	Lime	1876	10	12·88 cwt.	15s. 1d.	,,
Leeds	Modified A.B.C.	1876	16	1 ton	14s. 2d.	given away
Birmingham........	Lime	1879	not stated	not stated	10s. 9d.	dug into ground
Ealing.............	Lime & sulphate of alumina	1879	10¼	13 cwt.	not stated	not stated
Northampton	Lime	1874	—	12 bushels	,,	,,
London	—	—	12	15	—	—
Bolton	Lime & charcoal	1879	not stated	not stated	13s. 4d.	given away
Coventry	Sulphate of alumina	1877	,,	,,	20s.	1s. 6d.
Buxton	Iron water and milk of lime	1884-5	15	18 cwt.	not stated	given away
Peterborough	Broad irrigation	1891	not stated	not stated	,,	3d per cart load
Chester	Milk of lime	1875	,,	,,	,,	1s. 6d. per cart load
Kingston-on-Thames	Native Guano Co. A.B.C.	1891	,,	,,	£8 per ton	not stated
Reigate and Redhill.	Broad irrigation	1891	,,	,,	not stated	5s. per load
Leicester	Lime	1879	,,	,,	21s. 7d.	not stated
Newcastle-under-Lyme	Surface irrigation	1882	,,	,,	not stated	given away
Windsor............	Hillé's process	1877	,,	,,	11s. 5d.	not stated

The necessary difficulty connected with precipitation methods lies in the disposal of the precipitate, or, as it is usually termed, the sludge. In some places the sludge is manurially enriched by other fertilising matters; in others it is applied directly to the land in an unpressed state, or compressed into cakes. Sludge ought to be dealt with speedily, as it is prone to decomposition, especially where the precipitant is lime. The Amines process seems, however, to be an exception to this; the patentees claim that at Wimbledon the sludge does not putrify, but for a long time remains innocuous and inoffensive.

In the above table I have added the quantities of lime used, in order to help in calculating the cost of the various precipitation processes. The value of lime at the kiln may be taken at about 10s. per ton, to which must be added cartage or carriage by railway. I have also added a process-column, to enable reference to be made to previous papers, where the various systems are described and the various precipitating ingredients, with or without lime, are shown; 15 to 16 grains of lime per gallon are generally sufficient to precipitate ordinary sewage, say of a specific gravity of 1·00109. This quantity of lime reduces the specific gravity to 1·00104. (My own method reduces the specific gravity to 1·00087.) The amount of ammonia is not much reduced, but there is considerable reduction of albuminoid ammonia. About five-sixths of the phosphoric acid is precipitated.

As I have before shown, the use of lime alone as a precipitate is a failure, as both sludge and effluent are prone to secondary putrefaction; besides this, the escape of lime into a river poisons the fish; moreover, should the effluent into which any excess of lime has been used be discharged into any river, the organic matter in solution in that river-water is precipitated, and much danger and nuisance is occasioned by a speedy secondary precipitation and decomposition.

The sterilising action of lime upon bacteria is very small, unless entirely disproportionate and impracticable quantities are used, in which case the remedy is worse than the disease. The proportion required to sterilise would be 1,400 grains per gallon, or 87 tons per million gallons, 15·68 grains per gallon being equivalent to one ton per million gallons.

I will now give a series of analyses of dried sludge from eight towns, recorded in Mr. Robinson's work "On Sewage Disposal."

ANALYSES OF SEWAGE SLUDGE (AIR-DRIED).

Name of Town.	Aylesbury.	Birmingham.		Bolton.	Bradford.		Coventry.		Leeds.		Leicester	Windsor.
Process of Precipitation.	A.B.C.	Lime. 1	Lime. 2	Lime and Charcoal.	Lime.		Sulphate of Alumina.		Modified A.B.C.	Hanson's Process.	Lime.	Hillé's Process.
Date.	1879	1879	1879	1879	1876	1879	1877	1879	1876	1876	1879	1877
Water	12·60	12·70	13·16	14·34	8·90	6·92	14·04	10·04	9·56	16·40	11·93	11·76
Organic Matter, Carbon, &c.	35·60	19·19	20·04	26·18	33·75	34·53	20·58	23·09	20·82	27·92	22·18	12·06
Phosphoric Acid	2·11	·40	·72	·62	·80	·73	1·56	2·07	·64	·75	1·21	·87
Sulphuric Acid	2·70	1·45	·35	·61	·64	1·74	1·32	·56	2·15	1·02	·51	·49
Carbonic Acid	...	7·62	8·53	8·30	10·53	13·77	6·64	5·71	8·42	13·11	15·25	22·71
Lime	2·18	11·19	12·74	14·50	16·90	20·27	9·16	6·65	9·68	17·51	20·16	31·09
Magnesia	·18	·90	1·37	1·06	1·66	5·07	·86	·61	5·64	7·67	1·48	1·58
Oxide of Iron	6·20	2·70	3·20	1·98	2·11	2·01	4·14	2·66	4·61	2·32	2·66	1·68
Alumina	6·75	2·68	2·58	2·97	3·49	3·89	4·13	5·80	7·04	6·30	1·63	2·31
Sand &c.	33·50	41·13	37·93	29·50	21·80	10·23	37·83	42·00	31·60	7·36	22·30	14·16
	101·22	99·96	100·62	100·06	100·58	99·16	100·26	99·19	100·16	100·36	99·31	98·71
Phosphate of Lime	4·61	·87	1·57	1·35	1·74	1·59	3·40	4·52	1·39	1·64	2·64	1·90
Nitrogen	1·60	·52	·49	·61	·62	·66	·92	1·27	·66	·70	1·08	·52
Equal to Ammonia	1·94	·63	·60	·74	·76	·80	1·11	1·55	·80	·84	1·31	·63
Calculated Value per ton	33s.	10s. 9d.	11s. 5d.	13s. 4d.	15s. 1d.	15s. 4d.	20s.	27s. 2d.	14s. 2d.	17s. 2d.	21s. 7d.	11s. 5d.

(From H. Robinson's Book on Sewage Disposal.)

The percentage chemical composition of sludge is shown in the following table. The figures are derived from an average of the analyses of the air-dried sludge of eight towns made by Dr. Wallace, in 1879, for the Corporation of Glasgow, each town employing a different method of precipitation. The towns were Birmingham, Bradford, and Leicester, which used the Lime process; Aylesbury, the A. B. C.; Bolton, Lime and Charcoal; Coventry, Sulphate of Alumina; Leeds, modified A. B. C., and also Hanson's process; Windsor, Hillé's process. Two analyses at different times were made of the sludges of Birmingham, Coventry, Bradford, and Leeds, so that the following figures show an average of twelve analyses. The calculated fertilising values per ton of sludge were as follow: the A. B. C. process, 33s.; Lime, 11s. 5d.; Lime and Charcoal, 13s. 4d.; Sulphate of Alumina, 20s. to 27s. 2d.; Hanson's process, 17s. 2d.; Hillé's, 11s. 5d.

Composition of air-dried sewage sludge, by Dr. Wallace; averages of 12 analyses on the sludge of the above-mentioned eight towns:—

Water	11·86
Organic matter, carbon, &c.	24·61
Phosphoric acid	1·04
Sulphuric ,,	1·12
Carbonic ,,	10·96
Lime	14·33
Magnesia	2·34
Oxide of iron	3·02
Alumina	4·13
Sand, &c.	27·42
	100·83
Equivalents of phosphate of lime	2·26
Nitrogen	0·80
Equal to ammonia	0·97

The following analysis by Dr. Munro of the Wimbledon sludge, at the time when the sewage was precipitated with lime only, is interesting because of its being made on pressed sludge-cake, and not air-dried:—

WIMBLEDON.

Water	56·15
Organic matter	11·36
Insoluble silicate matters, such as sand, &c.	7·10
Phosphoric acid (P_2O_5) equivalent to tribasic phosphate of lime ($Ca_3P_2O_5$) 4·28	1·96
Carbonate of lime, oxide of iron, alumina, magnesia, and sulphuric acid, chloride of sodium, potash, &c.	23·43
	100·00

The organic matter contains nitrogen 0·41, or equivalent ammonia—0·50. The ordinary sludge filter-press takes away about half of the bulk of the moisture. If it were thought better to completely desiccate the sludge-cake, an ordinary destructor floor would reduce the sludge to a dry powder. There is, of course, a considerable amount of press-liquor driven from the sludge in the process of pressing, which should, if the precipitation process is good and thorough, be no less free from organic matter then the effluent.

It will be of interest, and of easier reference, if I tabulate a few of the answers to those questions which relate to the methods of sewage disposal, the quantity and disposal of sludge, the tank-system, and the average of land employed in some of those towns which were good enough to give the Leek Commissioners the information they asked, adding to them other towns described at pages 42-43 of Mr. Santo Crimp's book:—

Name of Town.	Mode of Disposal.	Chemicals used.	Sludge. Tons per year.	Tanks, Continuous or Intermittent.	Acreage of Land.	Years in operation.
Aldershot	Precipitation, irrigation, and filtration	Alumino ferric and lime	—	—	—	—
Aylesbury	Precipitation	—	—	Continuous	6	10
Banbury	Irrigation	—	—	—	136	—
Birmingham	Precipitation and irrigation	Lime	—	—	—	—
Blackburn	Precipitation and irrigation	Lime	640	Continuous	689	16
Brentford	Precipitation	Lime and alumino ferric	1,145	Intermittent	4	4
Burnley	Precipitation and filtration	Lime	22,000	—	65	—
Burton-on-Trent	Irrigation	—	—	—	572	—
Crewe	Irrigation	—	40	Intermittent	268	14
Ealing	Precipitation and filtration	Lime and sulp. alumina	11,000	Continuous	3½	20
Leeds	Precipitation	Lime	9,000	—	26	17
Maidstone	Precipitation	Lime	9,000	—	—	—
Northampton	Irrigation	—	—	Continuous	327	16
Oxford	Filtration and irrigation	—	—	—	330	7
Peterborough	Irrigation	—	500	Continuous	300	5
Reading	Irrigation	—	—	Continuous	770	13
Sheffield	Precipitation and filtration	Lime	20,000	Intermittent	22½	2

It is by chemical science that we learn that nitrogen is the main element we have to get rid of in sewage, and it is fortunate that we have in the roots of plants a ready receiver when in the form of nitric acid for it. Now, whether the urban authorities who want to get rid of the effluent and sludge, or the farmer who is willing to receive them, should supply additions to them, is a question which must be resolved by the incident of local circumstances; probably it might be worth while for either or both to do this. In the case of sludge, I think the urban authority might most economically add to it, and as effluent, perhaps, the farmer could more advantageously supply fertilising additions. That this consideration is one which ought to receive careful attention is clear from the experiments at Rothamstead, which proved that nitrification does not occur in the absence of phosphates.

Disposal of Sewage-sludge and Effluent at Southampton.

In 1885 the Corporation decided to adopt a more efficient system than the one then used, which was simply the discharge of the sewage into the tidal water.

The district of the town to which the new system has been applied has a population of 13,000, with a sewage-flow of 500,000 gallons each 24 hours. There are two efficient tanks for ferozone precipitation alternately acting. They are below the tidal level, and the sewage has to be lifted into them by two of Shone's pneumatic ejectors, one of 360 gallons capacity and the other of 700 gallons for this district. As soon as precipitation has taken place a valve is opened, and the ejectors force the effluent into the tide-way. As soon as the effluent has been drawn off, the action of the ejector is directed to force the sludge through four-inch iron pipes, nearly a mile in length, to one of Manlove, Alliott, and Co.'s destructors, which are used for burning the town garbage, the ejectors thus serve a threefold purpose. A pneumatic pressure of 40 lbs. per square inch is required for sludge-ejecting, and 10 lbs. for the effluent. The sludge is mixed with road-sweepings and sorted house-refuse in an incorporator, and it is then readily bought up as a good dry portable manure by agriculturists at 2s. 6d. per load, large quantities being also shipped to the Channel Islands, where it is in great demand for the cultivation of market produce for the London markets. A very important feature of these works is that the waste heat of the destructor is utilised for producing electricity, the engines driving a dynamo sufficiently powerful to feed 200 glow-lamps, which light up four neighbouring streets. But this is yet experimental, and the Corporation intend adding accumulators, so that the public buildings of the town can be so lighted.

This lighting, the Borough Surveyor, Mr. Bennett, A.M.I.C.E., to whom I am indebted for these particulars, says will be more economical than gas, as it will be seen no cost will be incurred for fuel, for he has ascertained that the house-refuse will be sufficient to maintain the steam whilst being burnt in the destructor.

The cost of the destructor and all buildings was £3,723; the sewage-disposal part of the works was about £3,000. The annual expense of burning the refuse is £221 per annum; the maximum quantity burnt per day of 24 hours is 50 tons, costing under 3½d. per ton for burning; the minimum quantity is about 25 tons.

This has maintained the steam for the purposes of the work for each 24 hours. The horse power is 31, or eight-tenths of a ton per horse power per 24 hours, and 75 lbs. of refuse, or 75 lbs. of refuse per horse power per hour.

The annual cost of the sewage works, is £308; the amount received from the sale of manure and compressed air which they sold last year was £800. The residues of the destructor sell annually for £300, in the shape of clinkers, concrete slabs, fine ashes for mortar and footwalks, to which saving should be added the coal which would have been required for working the engines.

CHAPTER XIX.

ON SLUDGE-PRESSING AND SLUDGE-CAKE.

THE most effective method at present of dealing with the wet sludge after precipitation and sufficient settling is by running it from the settling tanks to filter-presses, which I will proceed to describe, remarking, in passing, that there seems a possibility, which deserves some attention and study, of compressing the sludge in a larger and more expeditious manner. Is it not feasible by devising smaller settling tanks that the whole of the precipitate might be pressed in them at one operation?

The filter-press is an apparatus designed to filter the liquid through cloth or other filtering medium, and, as the cloth is firmly supported, the liquid may be forced through under considerable pressure. A large amount of filtering surface is arranged in a compact manner, and the mechanical details are simple, and not likely to get out of order.

The sludge entering the filter-press usually contains about ninety per cent of water and ten per cent of solid matter. The press deals with the sludge as fast as it comes from the settling tanks, and discharges it in the form of comparatively dry cakes, containing only about fifty per cent of moisture, and weighing only about one-fifth of the original weight of the wet sludge entering the filter-press.

The sludge-cakes, as discharged from the press, are about two inches thick, and in a convenient form for handling and transporting, and are practically without smell if a good precipitating process has been used. They are generally not in any way offensive, and, if stored, become dryer and harder.

The value of the sludge-cake as manure will, of course, depend directly on the quality of the sewage, and the nature of the chemicals used for precipitating. Several authorities state that sludge-cake gives about the same results as ordinary farmyard manure.

Sludge Filter-Presses.

The filter-presses I have seen at work are by two makers, Messrs. S. H. Robinson and Co., Carpenter's Road, Stratford, London, E., and Messrs. Manlove, Alliott, and Co., Bromsgrove Works, Nottingham.

Messrs. S. H. Johnson and Co.'s Filter-Presses for Sewage-sludge.

Their press consists of a number of narrow cells held in a suitable frame, the interior faces being provided with appropriate drainage surfaces, communicating with an outlet, and covered by a filtering medium, generally of jute, hemp, canvas, or other material. The interiors of the cells so built up are in communication directly with each other, or with a common channel for the introduction of the matter operated upon, and as nothing introduced into the cells can find an exit without passing through the cloth, the solid matter fills up their interior, the liquid leaving by the drainage surfaces.

To be of any utility in dealing with daily accumulations of sludge, the machines must be of considerable size, because even for a population of 30,000 about 30 tons have to be dealt with during the day. Difficulty of making large plates sufficiently rigid and tight at once appears, but as far as this is concerned it has been met by adopting the circular shape, a form which above all others for strength and resistance to the internal fluid pressure is at once apparent to the mechanical mind.

The results of the operation briefly described are: Within one hour every five tons of wet sludge, containing 90 per cent of water, can be deprived of 38 per cent of its water, giving a residue of one ton of hard pressed cake, containing 45 to 50 per cent of water. The cake so obtained is easily handled, is practically inodorous, air-dries very rapidly, and does not again enter into fermentation; it can be kept for any length of time without smell or nuisance if it has been properly precipitated.

The arrangement of plant for dealing with the sludge from a population of 30,000 comprises the following apparatus: Air-compressor; air-accumulator; two sludge filter-presses, 3 feet diameter; two sludge forcing-vessels, with their fittings, and the various distributing pipes for sludge and air; a tip, truck, and tramway for the removal of the pressed cake discharged from the machines. The cost of such a plant, with the requisite boiler power (about 10 horse-power actual), is about £1,000.

Thirty tons of wet sludge can be easily pressed into cakes, containing 50 per cent of moisture, equalling six tons, or one-fifth of the original bulk, consisting of five charges from each machine of 12 cwt., each in ten hours. The labour required is about two-thirds of the time of two men.

Messrs. Manlove, Alliott, and Co. also make an improved filter-press, which is worked by steam power. It abolishes most of the manual operations, which are very heavy. It is represented by Fig. 47.

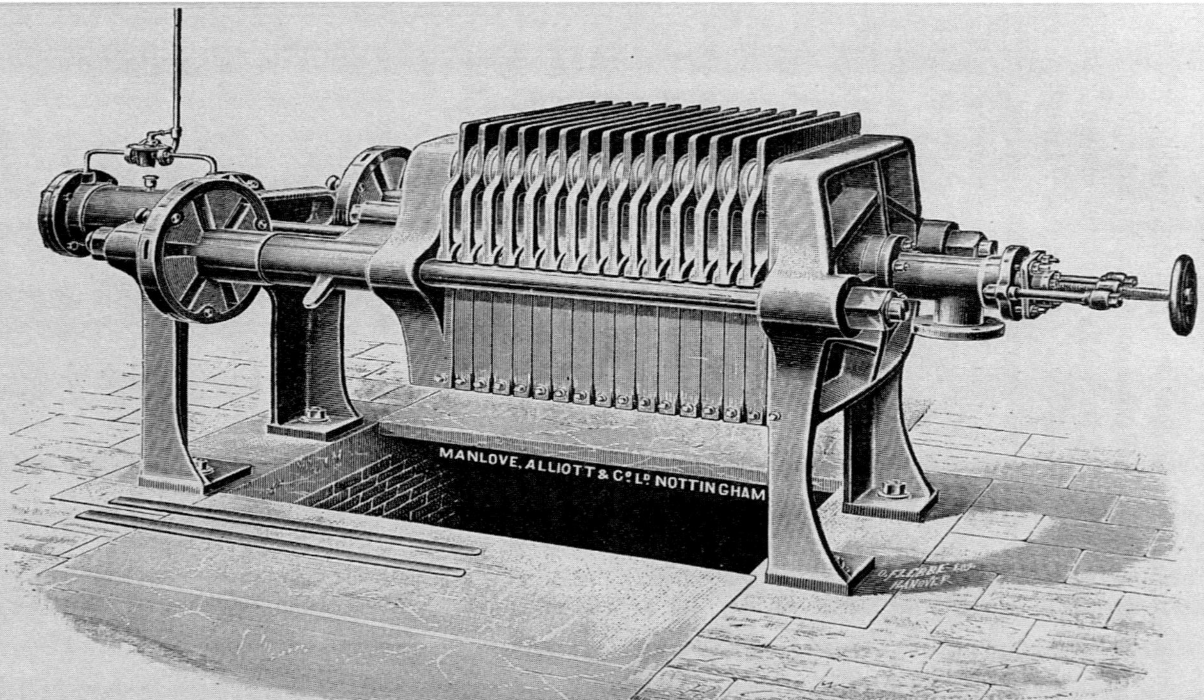

Fig. 47.—Steam Filter-Press for Sewage-Sludge.
Made by Messrs. Manlove, Alliott, and Co., Bromsgrove Works, Nottingham.

There is very little difference in appearance, principle, and working action between the machines of Messrs. Manlove, Alliott, and Co., and that of Messrs. S. H. Johnson and Co.

Sludge Cake.

The cost of thus pressing sludge-cake under ordinary sewage circumstances is from 2s. to 2s. 6d. per ton. The cost at Wimbledon varies from 1s. 10d. to 2s. 6d. per ton. Two per cent of lime is there added to the sludge before pressing, besides what is used for precipitation, to prevent stickiness and to allow of the sludge being quickly pressed. From five tons of wet sludge on the average one ton of sludge cake is obtained, where there is much adventitious matter from industrial operations the cost is higher; in a few instances it is as high as 3s. 6d. per ton, as at Chiswick.

CHAPTER XX.

On Irrigation, by the Pumping of Sewage and Sewage-Effluents by Steam, Air and Water-power.

The inexpensive method of disposing of sewage and effluents by gravitation has, in many places, led to great danger to health both to populations and to cattle. Such land is mostly moist alluvium or of an alluvial nature, and certainly not the best to receive a continuous application of sewage matter or effluents, and in such cases it would be much better for fertilisation to raise them to higher levels by pumping, and distribute them upon land better fitted to receive them, as I have before remarked. In most cases it would not be necessary for the lift to be great, for a few feet generally means the acquisition of a greatly extended area.

The three following methods are those most successfully adopted:—

1. By ordinary steam-pumping, as at Burslem and elsewhere.

2. By atmospheric-pressure, known as Shone's system of compressed air, as at Longton.

3. By the aid of natural water-power, as at Windsor and other places, by turbine pumping.

I have good authority for stating that raising sewage or effluents by means of a ram is not practicable.

The approximate estimates I shall give are not intended to supersede the necessity of calling in an engineer for more accurate ideas as to power, situation, and cost, but they may be roughly relied upon, having been very carefully worked out.

PUMPING OF SEWAGE BY STEAM-POWER.

No. 1. To deliver about 420,000 gallons of sewage in 12 hours to a vertical height of 100 feet, through half a mile of 10 inch pipes.

One pair of 12 inch by 24 inch Holman's double-action pumps, with cranks coupled at right angles, and driven through mortise spur-gearing by 20 nominal horse-power Tangye's horizontal condensing engine, with extra heavy flywheel, Meyer's variable expansion-gear [variable by hand], white-metal bearings, and continuous lubricators for oiling engines whilst running.

Engine steamed by one 25 horse-power steel Cornish boiler, fitted with four Galloway tubes 5 inch by 2½ inch, by 6 inch, Barnes ram-pump and boiler-feeder, 9 inch cast-iron retaining valve under each air vessel.

Necessary short suction-pipes, and steam-feed and exhaust-pipes for connecting the plant in engine-room.

Price in duplicate, £1,940 net.

No. 2. For duty as stated above, but delivering to a total vertical height of 53 feet.

Plant as above specified, but pumps driven by 12 horse-power Tangye engine, the engine steamed by 15 horse-power steel Cornish boiler, pumps, &c., as before.

Price in duplicate, £1,600 net.

These prices are approximate, and for plants in duplicate. They include delivery and fixing, but are exclusive of masonry and brickwork.

The figures have been kindly furnished to me by Messrs. Tangye, Limited, Engineers, Cornwall Works, Birmingham.

There are also other systems of steam-power pumping, particulars of some of which will be found in the various answers to questions 11, 12, 13, and 14, in Chapter XIV.

One useful point to be borne in mind in the consideration of steam-power pumps *versus* turbines is, that turbines, being self-acting, will work the whole of the twenty-four hours, whilst steam-power would only be applied during a day of twelve hours.

The Pumping of Sewage or Sewage-Effluent by Air-Pressure.

My information is derived from a descriptive pamphlet published by Messrs. Hughes and Lancaster, Engineering works, Chester, and 16, Great George Street, Westminster, London, who are the patentees and makers of Shone's pneumatic ejector, and from correspondence with them.

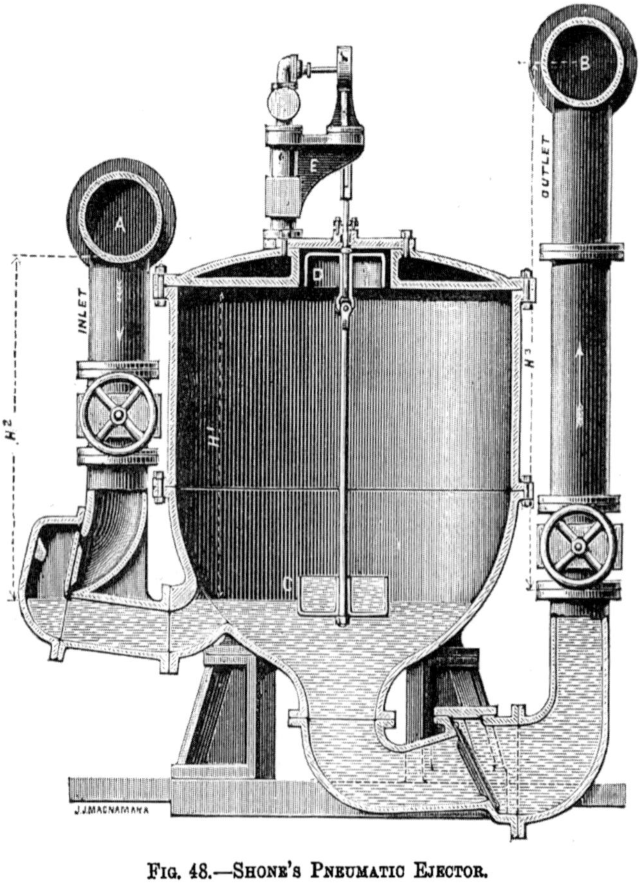

Fig. 48.—Shone's Pneumatic Ejector.

The above figure gives a sectional view of a Shone pneumatic ejector of ordinary construction, suitable for raising water, sewage, sludge, chemicals, and hot fluids of all kinds. Ejectors are made of any size or shape convenient for the special circumstances for which they are required. For sewage, sludge, pail-contents, preference is given to those having the lower portion of hemispherical shape.

The motive power employed is compressed air, and the action of the apparatus is as follows: The sewage gravitates from the sewers through the inlet pipe A into the ejector, and gradually rises therein, until it reaches the underside of the bell D. The air at atmospheric pressure inside this bell is then enclosed, and the sewage, continuing to rise outside and above the rim of the bell, compresses the enclosed air sufficiently to lift the bell, spindle, &c., which opens the compressed air admission valve E.

The compressed air thus automatically admitted into the ejector presses on the surface of the sewage, driving the whole of the contents before it through the bell-mouthed opening at the bottom and through the outlet pipe B into the iron sewage rising main, or high-level gravitating sewer, as the case may be. The sewage can only escape from the ejector by the outlet pipe, as the instant the air pressure is admitted on to the surface of the fluid the valve on the inlet pipe A falls on its seat and prevents the fluid escaping in that direction. The fluid passes out of the ejector until its level therein reaches the cup C, and, still continuing to lower, leaves the cup full until the weight of the liquid in the portion of cup thus exposed and unsupported by the surrounding water is sufficient to pull down the bell and spindle, thereby reversing the compressed air admission-valve, which first cuts off the supply of compressed air to the ejector, and then allows the air within the ejector to exhaust down to atmospheric pressure. The outlet valve then falls on its seat, retaining the liquid in the sewage rising-main; the sewage then flows into the ejector through the inlet, once more driving the free air before it through the air-valve as the sewage rises, and so the action goes on as long as there is sewage to flow. The compressed air for actuating the ejector is produced at some central station, and conveyed in cast or wrought iron pipes, laid under the streets, to the several ejector stations.

The outlet is from the bottom of the ejector, so that the whole of the sewage, including solids, sludge, grit, and everything brought down the sewer is discharged out of the ejector.

For these reasons, no screening or straining of the sewage is necessary, as is the case with pumps, and the great nuisance caused by the cleaning of pump-gratings and sump-wells is avoided.

The ejector forms an absolute severance of the house-drains of each district from the main sewer.

The ejectors are in successful use at Warrington for the transmission of pail-contents from central depôts in the town through 2½ miles of cast-iron main to the works at Longford, saving the Corporation over £1,200 per annum in cartage alone; and at Southampton for transmission of sludge through a length of 1,500 yards of four-inch cast-iron main. They are also in use for the same purpose at Plymouth, Shirley, Fenton, and at Freemantle.

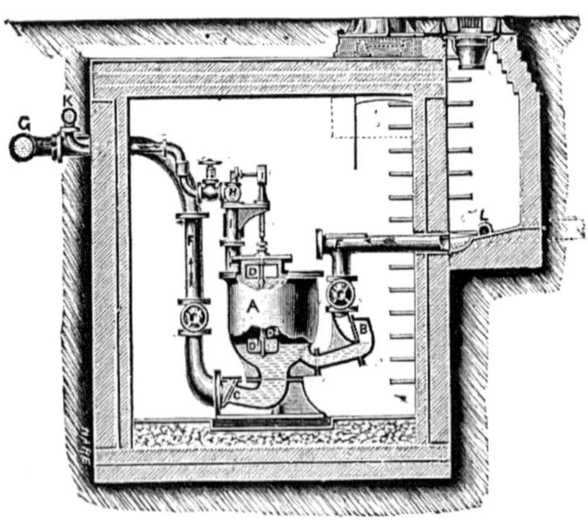

FIG. 49.—SHONE'S PNEUMATIC EJECTOR, IN BRICK CHAMBER. SIDE ELEVATION.

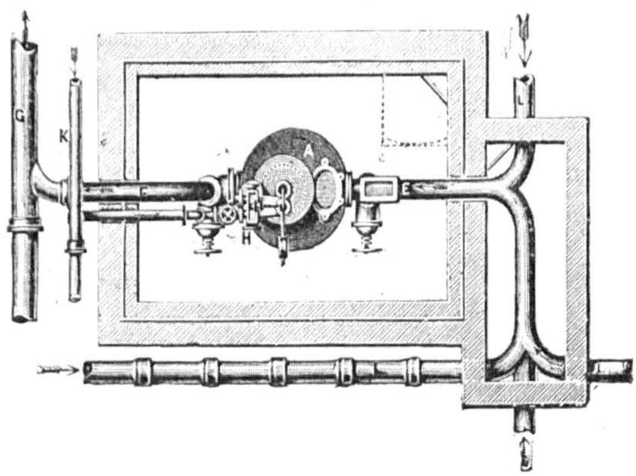

FIG. 50.—SHONE'S PNEUMATIC EJECTOR, IN BRICK CHAMBER. VIEW LOOKING DOWN.

Figures 49 and 50 show a pneumatic ejector placed in a brick chamber underneath the street, at the junction of four streets, and into which the sewage of the adjacent district gravitates by proper sized stoneware sewers, sufficiently well graded to render them permanently self-cleansing. The ejector station is practically the "outfall" of the district to which it is connected, and the district is as effectually separated from the other part of the town, so far as its sewage is concerned, as if it were delivered to an entirely separate destination. This will prove an important feature in localising and dealing with the causes of infectious diseases. Messrs. Hughes and Lancaster have given me an approximate estimate of the cost of dealing with either sewage or effluent by their method, for a population like that of Leek, of about 15,000, and suitable to the configurations of nature of land best fitted to receive the sewage; from which it will be seen that any number of ejectors can be placed at various distances from a central station, and there made to throw up the contents of the main pipe to any required height.

The following computation is based upon 25 gallons per head per day of sewage for towns drained on the water-carriage system—the bulk of the rainfall being precluded from the sewers.

This estimate is based on the assumption that four ejectors might be required: the first, three-quarters of a mile south of the town; the second, half a mile further down the valley; the third, half a mile still further away; and the fourth at the northern outfall, supposing all these were necessary. But with a 15,000 population they would not be. No doubt the first one would be amply sufficient for the present requirements. The coal for the 20 horse power engine would be about 3½ tons per week, say 185 tons per annum.

The quantity of liquid to be dealt with would be 375,000 gallons per day, and assuming that one half of this quantity will flow to the ejector in 400 minutes, the maximum quantity to be dealt with at the various ejector stations will be 492 gallons per minute. Assuming that the quantity would be equally divided between the stations B and C, then the quantity to be raised at each station would be 246 gallons per minute.

The size of air-mains would be 3 in. diameter, and the size of cast-iron rising sewage-main to the point of outfall would be 9 in. diameter. The rising-main from ejector-station would be also 9 in. diameter; but if it is wished to deliver sewage or effluent at points a mile further down the valley, it will be necessary to have branch-pipes and sluice-valves placed on the rising-main at the various farms where it is wished to draw off the supply of sewage. It will not, of course, be necessary to convey the sewage by gravitating sewers for the purpose of raising it at those points, because they are opposite the farms to be irrigated, but the sewage can be delivered direct to the various farm lands. Of course, if there should hereafter be houses to be drained at each extremity of the mains, then it would be necessary to put ejectors at these points.

The approximate cost of the works on the Shone system may be taken as follows:—

	£	s.	d.
2,640 yards 3 inch cast-iron air mains, from town to outfall, including trenching, laying, jointing, and testing	570	0	0
Four Shone's pneumatic ejectors, each of 250 gallons capacity, with all necessary fittings and connections, fixed.			
Two brick chambers, 11 ft. by 14 ft., XII. 6 ft. high, with manholes and covers	1,400	0	0
Two sets of steam-engine air compressors, each capable of dealing with the maximum flow of 492 gallons of sewage per minute to a height of 60 ft., the other compressor to be in reserve in case of accident.			
Two steam-boilers of best steel, 20 horse-power, each with all mountings and necessary fittings.			
One wrought-iron air receiver, with all necessary pipe connections, fixed complete	2,000	0	0
	3,970	0	0
Deduct the cost of three ejectors not required...............	1,050	0	0
Total ....£	2,920	0	0

The 9-inch cast-iron rising-main may be estimated at 12s. per running-yard laid complete.

This estimate assumes a lift, with friction, of 60 feet, which will require an air-pressure of 26 lbs. per square inch, but such a high lift would not be necessary.

Pumping the Sewage or Effluent by Water-Power.

Turbines.

Although this part of the chapter was written for the local press and for local purposes, the information contained in it is so suitable for general application, and may be found generally useful, that I have decided to retain it in its unaltered local form.

I have thought it well to go into the question of the practicability and cost of raising the sewage to suit such a level, or levels, as would dispose of it either by broad or effluent irrigation upon undrained lands on the soils and valley-sides which need its influence far more than the alluvia of the Churnet valley itself. If this plan is hereafter thought to be the most useful to adopt, there are several courses open.

I think the only objections that can be fairly raised against pumping by water-power at Leek are:—

1. When the flow of the river Churnet is inadequate, as it sometimes might be in exceptionally fine weather.

2. Inadequacy of water supply to yearly increasing quantity of sewage.

To both the first and second objection it may be urged that it would not be necessary to raise the whole of the sewage or effluent, as a large quantity can be disposed of by gravitation, and also that it would not be necessary to raise the surplus so high as 50 feet for a long time to come, as a contour-line of 20 feet would bring in a large area of undrained sandy land, available for broad or effluent irrigation. To a height not exceeding 20 feet the sewage could be lifted by inexpensive centrifugal pumps, as at the Hanley Sewage Works.

But in regard to this I may almost predict with safety that, if the Leek Commissioners are permitted to utilise and fertilise all the land upon which the sewage could be made to flow by gravitation, the amount of sewage remaining to be disposed of by pumping would be very small indeed, if any; the first effort in this direction having been recently accomplished by the Sewage Sub-Committee, who have taken a higher level for part of the south sewage-outfall, from which it is distributed over upwards of two additional acres of the ox-pasture of Big Birchall Farm, making altogether about 16 acres which can now be irrigated. This includes the swampy ground adjoining the railway, which is nearly 7½ acres; but with regard to this latter portion and other valley areas, there are important points to be well considered before the sewage is wholly or partially disposed of by gravitation on these areas, particularly as to the contamination of the Staffordshire Potteries Waterworks Springs.

Although this is but a very small irrigating addition, it points to what might be effected by tapping the sewage at much higher levels and conveying it by pipes either along contour lines, or—by what is very easily done—running the pipes on brick supports across the lower undulations if broad irrigation is preferred.

TURBINES—APPROXIMATE COST.

For raising sewage or effluent with a turbine of 6½ horse power.

No. 1. To be capable of delivering 8,000 gallons of sewage per hour to a total vertical height of 100 feet, through 7 inch pipes, half a mile long.

A pair of 5 inch by 12 inch Holman's double-action pumps on bases, coupled at right angles with slipper-guides. Approximate price in duplicate, £200.

No. 2. To be capable of delivering about 15,000 gallons of sewage per hour, to a total vertical height of 50 feet, through 9 inch pipes, half a mile long.

A pair of 7 inch by 12 inch Holman's double-action pumps on bases, coupled at right angles with slipper-guides. Approximate cost in duplicate, £265 net.

No. 3. To be capable of delivering about 30,000 gallons of sewage per hour to a total vertical height of 20 feet, through 12 inch pipes, half a mile long.

Three 8 inch by 12 inch Holman's double-action pumps on bases, coupled at right angles with slipper-guides. Approximate price in duplicate, £440 net.

No piping is included in the above prices.

At Windsor there is a fall of 3ft. 8in., which, with a flow of 14,500 gallons of water per hour, gives a turbine-power equal to eleven horses.

The Town Clerk of Walsall informs me that the turbine in use at the Walsall Sewage-Farm was made by McAdam Brothers and Co., Soho Foundry, Belfast, and its cost, including fixing, exclusive of the building, was £314. Its yearly maintenance has been very trifling. Its maximum power with 2,990 cubic feet of water per minute is 18 horse-power, but the turbine is so arranged that it can be worked with a varying flow of water. The cost of the turbine-house and the pumps was £650. (Pumps, £400; house, £250.)

We have sufficient instances of the possibility, value, and economy of this system quite near to Leek. The motive power of the flint-mills at Consal is the fall of the Churnet into a large turbine of 80 nominal horse-power, from a volume of water of 183,000 gallons per hour. In summer the water is sometimes reduced to 180,000 gallons per hour. These mills are the property of James Meakin, Esq., and are in the occupation of G. Goodwin, Esq., who has kindly supplied me with this information. The cotton-mills at Rocester are similarly turned. The turbine power is 300 horse; there are two turbines, 155 horse-power each. Fall, 19ft.; cost of turbines, £600 each.

The cotton mills at Mayfield and Tutbury, are also turned in this way. The following information about the Mayfield turbines has been given to me by J. Simpson, Esq., one of the proprietors. Their mills are turned by two turbines, 88 horse power each; a volume of water 8in. deep runs over a weir 140ft. wide, equalling 16,310 cubic feet per minute. The fall is 9ft. 6in. Cost of turbines, £440 each. The cost of fixing them was £440, and not 10s. has been spent on them since 1884, except in oiling, and that is very small.

During the dry weather of the middle of January of this year I carefully observed the flow of the river Churnet at Wall-Bridge. The weir at this place is 24ft. 3in. wide, with a direct fall of 5ft. 4in.; the river-bed speedily falling again to a further depth of about 3ft. below the bridge. The water in the driest winter time was passing over the weir in a stream 4½in. deep. This indicates a quantity of 9,460,751 gallons per 24 hours, or 6,570 gallons per minute. I think this may be considered a fair average for the year—in fact a low average, and it is probable that an average of about 11,000,000 may be nearer the mark.

Should the average waterflow of the river equal 11,000,000 gallons per 24 hours, or 7,632 gallons per minute, it would be equivalent to an additional 1½ horse power, and from 20 to 25 gallons more per minute would be raised.

The maximum amount of power available for pumping by this volume of water, say 6,570 gallons per minute, I am informed by Messrs Easton and Anderson, the eminent engineers of Erith, would be about 6½ horse-power, and this I have had further confirmed by Messrs. McAdam Brothers, turbine-makers of repute, of Belfast. This same water-power is available for repeated use wherever sufficient fall can be obtained; say, for example, at the Abbott's-meadows weir, the Wall-Bridge weir, and at Birchall-Dale, thus tripling it. At each of these places sufficient fall could be obtained for either a turbine or water-wheel, and if one pumping station should, in consequence of the hill-and-valley nature of the neighbourhood, be impracticable, two or three could be adopted. I have ascertained that, on the assumption that the sewage was divided between three pumping stations, there might be roughly about 100 gallons per minute raised at each.

The lift being taken for comparison at 20ft., 100ft., and 150ft., the cost would be approximately as follows:—

No. 1. Delivery to a height of 20ft., a turbine, with fan 20in. diameter driving a centrifugal pump, would cost about £160.

No. 2. Delivering to a height of 100ft., a turbine, with fan 42in. in diameter, driving a set of three-throw pumps, would cost about £350.

No. 3. Delivering to a height of 150 feet, a turbine, with fan 56 inches in diameter, driving a set of three-throw pumps, would cost about £450.

Should the fall be less, the sizes of the turbines would have to be increased. Before any definite information could be arrived at, it would be necessary to have the situation reported upon by a competent engineer, who would give an opinion as to the practicability of turbine or water-wheel raising. Should this idea be found to be practicable, the cost of pumping the whole or part of the sewage or effluent need not be considered a serious factor or bugbear. Estimating the amount of sewage for our population of 14,000, at the rate of 30 gallons per head per diem, we should have—

420,000 gallons per day,
17,000 gallons per hour,
300 gallons per minute.

The average of forty towns with populations varying from 606,000 to 7,795 is 39·56 gallons per head, but that of Leek is probably about 30. The Sanitary Inspector, Mr. Farrow, thinks that the Leek water-supply may be safely taken as indicating the amount of sewage per head of population. In the twenty-four hours Leek receives 440,000 gallons of water. This, reckoning the population as 14,000, gives 31 gallons per head.

One indicated horse-power is equal to 33,000 foot-pounds per minute; this, multiplied by 6½, the estimated horse-power of the water of the river, gives 214,500 foot-pounds per minute. To calculate this power to the raising of sewage 100 feet high, the two ciphers must be taken off, leaving 2,145. One gallon of water, weighing 10 lb., forms the divisor of 2,145, which gives 214½ gallons per minute. From this must be taken 50 per cent for friction; so in round figures we get a pumping power of 100 gallons per minute, but there will have to be taken off a further 25 per cent for friction for each half-mile of conveyance of sewage through a 6 inch pipe. This power would raise double the quantity 50 feet high. An additional 1½ horse power, if the average flow is eleven million gallons, would increase the quantity of sewage lifted by 20 to 25 gallons per minute. Duplicate pumps for a 6½ horse water-power would cost £200, and for 9 horse power £250, exclusive of piping or buildings, and to this must be added the cost of the turbines.

The pumps now made for this purpose are horizontal double-action, with leather buckets 7½ inch for 6½ horse and 8 inch for 6 horse-power, with special wrought-iron hinged valves for sewage pumping.

The following observations may incidentally be found to be of service in considering the question of the flow and pressure of water. In calculating the quantity of water flowing down a river or stream, it is found that 344,542 gallons per 24 hours are carried across a flow-over sill 1 foot wide, if the depth of water flowing over the sill be 4 inches. Thus, this rule applied to the Churnet flow-over sill at Wall-Bridge, which is 24ft. 3in. wide, with a depth of flow-over in dry weather of 4½ inches, gives 9,460,751 gallons per 24 hours.

The following Table is useful for showing the pressure of water in lbs. at various heights:—

Height of water in feet	72	96	120	145	192
Pressure in lbs	30	40	50	60	80

CHAPTER XXI.

On the Composition and Growth of Plants, their Food and Fertilisation.

I HAVE thought it necessary that a chapter on plant-life and growth should find a place in my book, in order to show that the important question of fertilisation by effluents and sewage sludge should receive further study.

It is not pretended that all the information contained in it shall be new to experts in agriculture or in agricultural chemistry, but I trust it may be conceded that the close relationship between fertilisation, and the application to land of such by-products as those resulting from sewage treatment, particularly in the sense of avoiding the errors of both surfeit-manuring, and the application to land of waste matter which does not contain a due proportion of fertilising power, will furnish a sufficient *raison d'être* for such a chapter as this to appear in a work on sewage treatment and disposal, and it is not unlikely that both the student and general reader may obtain from it some useful information.

I will, therefore, proceed to give the composition of a few plants of farm production, with a view of showing how far the substances contained in sewage are likely to be beneficial whether added to land in a simple state, or artificially enriched by the addition of certain chemical substances in which sewage may be found deficient.

The Composition of Plants.

The food-plants, whether for man or beast, in a dry state consist of the following elements and of about the following proportions by weight:—

Carbon, one half, nearly; oxygen, about one third; hydrogen, about five per cent; nitrogen, varying from half to four per cent, sulphur to five per cent, phosphorus about a thousandth part.

These proportions may be approximately accepted for potatoes, wheat, oats, hay, and clover, but they, of course, vary; oats give the highest proportion of carbon, and clover next; potatoes and wheat the highest of hydrogen and oxygen; wheat of nitrogen, being 2¼ per cent, as against 1·50 per cent of hay and potatoes, whilst in wheat-straw there is only 0·33 per cent of nitrogen. Oats contain 2·20 per cent of nitrogen.

This is shown in part by the following table, which exhibits the actual composition of 1,000 lbs. of some varieties of the more common crops, when made *perfectly dry*:—

	Carbon lbs.	Hydrogen lbs.	Oxygen lbs.	Nitrogen lbs.	Ash lbs.
Hay	458	50	387	15	90
Red clover hay	474	50	378	21	77
Potatoes	440	58	447	15	40
Wheat	461	58	434	23	24
Wheat-straw............	484	53	389½	3½	70
Oats	507	64	367	22	40
Oat-straw	501	54	390	4	51

The inorganic constituents of plants I will describe further on when I come to treat of vegetable ash.

Plants not being able to absorb these four organic elements in their uncombined state, chiefly derive them from the following compounds—carbonic acid, ammonia, nitric acid, and water. Soil and the atmosphere are the carriers and feeders of these elements, in which their presence is constant.

Professor Way found the composition of well-made hay from mean results of analyses of a number of the more important grasses to be as follows:—

Water	16·6
Albuminoids	15·81
Fats	3·18
Non-nitrogenous digestive matters	34·42
Woody fibre	22·47
Ash	7·59
	99·53

Fresh grass contains 68·8 per cent of water.

In weathered or late-cut hay the albuminoid or nitrogenous matter, as well as sugar, is very much less; fermentation and decomposition set in, when acetic acid, aldehyde and ammonia are formed, destroying the food-value of such hay.

The following is the percentage composition of several fodder-plants, seeds of cereals, roots and tubers, and the various parts of corn, from Johnson and Cameron's Agricultural Chemistry :—

FODDER PLANTS.

	Water.	Albuminoids.	Fats.	Sugar Starch.	Woody Fibre.	Ash.
Rape	87·06	3·13	0·37	4·00	3·56	1·61
Green rye	75·40	2·71	2·55	9·14	10·50	1·36
Sorgho (*Holcus sacharatus*)	81·80	2·19	0·56	10·97	4·03	0·99
Prickly comfrey	88·41	2·71	...	6·89	...	1·99
Yarrow (dried)	...	10·34	2·51	45·46	32·69	9·00
Cattle cabbage, Drumhead (Anderson)						
Outer leaves	91·08	1·63	...	5·06	...	2·23
Inner leaves (Voelcker)	89·42	1·50	0·08	7·01	1·14	0·85

Sorgho is stated by Sir Charles Cameron to be rich in sugar, and much relished by cattle.

PERCENTAGE COMPOSITION OF SEEDS OF CEREALS.

(Deducted chiefly from Wolff's and Cameron's Analyses.)

	Water.	Albuminoids.	Fats.	Starch, Sugar, &c.	Fibre.	Ash.
Wheat	15·0	12·0	1·5	67·0	2·8	1·7
Barley	16·0	9·5	2·5	63·2	6·3	2·5
Oats	14·0	11·5	6·0	56·5	9·0	3·0
Indian corn	14·5	10·0	7·0	61·4	5·2	1·9
Millet	14·0	14·0	3·0	58·5	8·0	2·5
Buckwheat	14·0	9·0	2·5	60·2	12·0	2·3
Rye (Irish)	16·0	9·0	2·0	64·0	8·0	1·0
Rice	14·0	5·3	1·0	76·7	2·5	0·7

PERCENTAGE COMPOSITION OF ROOTS AND TUBERS.

	Swedes.	White Turnips.	Aberdeen Yellows.	Mangels.	Beet Sugar.	Carrots.	Parsnip.	Potato.
Water	89·46	92·00	90·56	88·00	83·50	87·50	85·00	75·55
Albuminoids	1·45	1·08	1·40	1·40	1·30	1·20	1·36	2·06
Fat	0·20	0·15	0·20	0·20	0·12	0·20	0·34	0·75
Sugar	4·60	3·00	3·90	5·60	10·50	6·90	3·00	0·56
Carbo-hydrates	2·55	2·21	1·90	2·91	2·30	2·07	7·90	17·98
Insoluble cellulose	1·12	0·96	1·04	1·13	1·24	1·10	1·40	2·12
Ash	0·62	0·60	1·00	0·76	1·04	1·03	1·00	0·99
	100·00	100·00	100·00	100·00	100·00	100·00	100·00	100·00

PERCENTAGE COMPOSITION OF THE VARIOUS PARTS OF CORN.

	Wheat.		Maize Meal.	Oatm eal.	Irish Oatmeal.	Oat, Chaff and Hulls.	Barley, Chaff, and Hulls.	Rye Flour.
	Bran.	Flour.						
Water	13	12·6	9·00	8·7	9·26	14·3	14·3	14·0
Albuminoids	12	11·8	10·00	12·7	16·18	4·0	3·0	11·0
Fat	2	1·2	4·98	7·5	8·00	1·5	1·5	1·6
Carbo-hydrates	50	74·1	71·40	62·0	57·53	28·2	38·7	72·0
Woody fibre	15	0·7	3·40	7·6	6·99	34·0	30·0	1·5
Ash	5	0·7	1·22	1·5	2·04	18·0	13·0	1·6
	100	101·1	100·00	100·0	100·00	100·0	100·5	101·7

The chemical composition of the ashes of plants indicates their inorganic food requirements.

Johnson and Cameron state that—

1 ton of wheat straw when burnt yields from 120lbs. to 330lbs. of ash
1 ton of oat straw „ „ 120 „ 200 „
1 ton of wheat grain „ „ 45
1 ton of oat grain „ „ 80
1 ton of oak wood, only 4 or 5 lbs.

The following table shows the amount of mineral ingredients in the ashes of the more important crops. The figures given have been calculated by Messrs. Johnson and Cameron from the average results of a large number of analyses made by various chemists.

Minerals in the Ashes of Various Crops.

	Wheat Grain.	Barley Grain, with husks.	Oat Grain, with husks.	Rice Grain.	Indian Corn	Peas (pods).	Beans (pods).	Potatoes (tubers).	Yellow Globe Mangels.	Long Red Mangels.	Cow Cabbage Leaves.	Wheat Straw.	Oat Straw.	Meadow Hay.
Potash (K_2O)	31·54	21·28	37·48	23·70	37·95	44·50	45·40	61·60	35·09	25·18	34·85	12·16	20·60	26·80
Soda (Na_2O)	2·66	4·00	2·40	4·00	3·00	2·00	3·34	1·00	28·90	32·10	4·60	1·00	6·00	5·80
Magnesia (MgO)	12·10	9·10	7·80	12·10	7·50	8·00	7·30	5·00	2·50	2·26	4·40	4·00	3·66	4·32
Lime (CaO)	3·14	2·40	3·00	3·15	3·40	5·50	6·00	2·40	2·50	2·20	19·60	6·82	8·10	12·86
Ferric oxide (Fe_2O_3)	trace	0·15	0·60	trace	0·40	0·85	0·40	0·85	0·66	0·50	0·17	1·62	0·83	0·76
Phosphoric anhydride (P_2O_5)	48·50	33·17	20·80	55·00	44·80	32·55	34·66	17·67	6·16	2·16	16·63	3·20	4·16	6·95
Sulphuric anhydride (So_3)	0·08	2·10	1·80	0·05	1·50	4·73	3·50	6·25	3·19	40·0	17·31	5·78	13·32	4·45
Silica (SiO_2)	1·88	27·50	25·90	2·00	1·45	0·67	0·90	1·00	3·00	1·60	2·44	65·34	50·08	31·04
Chlorine (Cl)	0·10	0·30	0·22	trace	trace	0·30	1·40	2·23	18·00	30·00	trace	0·60	3·25	7·02
	100·00	100·00	100·00	100·00	100·00	100·00	100·00	100 00	100·00	100·00	100·00	100·00	100·00	100·00

The next table by Arendt gives an interesting comparison between the composition of healthy and unhealthy oats :—

	Potash and Soda.	Lime.	Magnesia.	Ferric Oxide.	Phosphoric Acid.	Sulphuric Acid.	Silica.	Chlorine.
Luxuriant oat plants	45·30	6·1	2·9	0·4	8·2	4·8	27·	6·7
Strong oat plants.....	34·3	5·4	2·3	0·5	8·5	4·1	39·9	5·8
Weakly oat plants	30·4	5·2	2·3	1·0	8·8	5·6	42·0	4·7

PLANT-FOOD.

It appears to me to be quite necessary in such an inquiry as this to show the relationship between plant-food and the composition of plants themselves, even at the risk of being tedious to the non-scientific reader, and the burdening of the pages with chemical tables; it must never be forgotten that the chemical facts in relation to plant-composition and to plant-food are as definitely ascertained as are the nature and proportions of metals in their ores, and any inquirer, be he farmer or sanitarian, cannot arrive at any true basis of reasoning on this subject unless he takes these facts into consideration; they are the skeleton upon which the whole body of the question is sustained. If a plant in its healthy state is composed of certain chemical substances, it must have obtained them from the soil and the atmosphere in which it has grown from a seed or cutting, and healthy growth must not be expected in any plant where its natural food is not available, or where certain substances are superabundant, lacking, or ill-applied.

Plants, as I have shown, assimilate the elements carbon, oxygen, hydrogen, and nitrogen, but not directly as elements; they decompose organic compounds, and in a good many instances they also split up inorganic compounds, and use up what they require of both acid and base, nitrate of potash, an excellent manure, for example; the organic compound sources are carbonic acid, water, and nitrogenous substances.

Of inorganic compounds, as has been shown from the composition of the ashes of plants which have been burnt to expel the organic matters, are the nitrates, sulphates, phosphates, carbonates, and chlorides of lime, magnesia, soda and potash, oxide of iron (Fe_2O_3), oxide of silicon or silica (SiO_2). These substances are necessary for plant-food and plant-substance; the more their uses and action are known, the better will the judgment be guided in suiting the manure to the land and to the crops required. It is waste to add to soil a salt it already contains in sufficient quantity; it is advantageous farming to add to soils those substances in which they are deficient. There is always a chemical reason why some soils give larger returns of certain crops than others; wheat, for example, which crops better in clayey than sandy soils; but if soils are enriched with those manurial substances in which the soil is poor, the beneficial effects as surely ensue as effect follows cause, if the substances are properly administered. As this is undoubtedly so, it is plain that sewage, whether applied in its crude state, as effluent, or as sludge, cannot be a specific fertiliser for any or all soils, or for any or every kind of crop. I believe many erroneous ideas have been formed and much prejudice created by the inadvertent and promiscuous application of sewage and sludge, and that when their composition is better understood it will be much more advantageously used, also that it may be beneficially treated with other substances added to them as varying soils and crops may need, which will greatly enhance their manurial values, and make them more applicable to agricultural and garden fertilisation.

The Atmosphere.

As the functions of the atmosphere form an important part in the operations of plant-life, a short account of its composition should not be omitted.

There is a remarkable structural difference between it and the other compounds I have been considering, inasmuch as it is composed of gases not in chemical combination or solution, but is merely a mechanical mixture of them.

Its normal composition is—

Nitrogen	77·98
Oxygen	20·61
Water	1·40
Carbonic acid	0·04

with traces of ozone, ammonia, nitric acid, and minute quantities of other probably adventitious gases and even solids.

The proportion of carbonic acid in the atmosphere is small, being not more than one gallon in 2,500, whilst the proportion of nitrogen in 100 gallons is 79 gallons to 21 gallons of oxygen. The proportion of ammonia is very much less, being less than one part in fifty millions.

Nitric acid is still less, being only found in traces except at the times of electrical meteorological disturbance; yet it is always found in rain-water. On an acre of land at Rothamstead, Johnson found that 7 lbs. of combined nitrogen fell annually, one-ninth as nitric acid and the rest as ammonia. This is absorbed by the roots of plants. Nitric acid is found in plants, especially in potatoes and similar juicy crops.

WATER.

Water is the result of a chemical union of oxygen and hydrogen. Its formula is H_2O, or aggregated molecules consisting of two atoms of hydrogen and one of oxygen in chemical combination. It is needless to state the universal prevalence of water. In the hottest climates the rising and setting sun disclose its presence in the air. We are familiar with it as rain, snow, ice, fog, and steam, but the sea and the earth are its great receptacles. Its great purpose agriculturally is as a solvent. It dissolves from both the air and soil the salts, compounds, and gases necessary to vegetable growth and structure, and carries them through the soil to the roots of plants. The chief function of water is to be the food-carrier to plant-life by its wonderful power of solution, now considered to be a chemical property. It often becomes laden, as the bee with honey, with dissolved plant-food of various kinds, the plants on being fed with it having the power of decomposing not only the food, but perhaps to some extent even the water itself. In whatever part of the plant or tree in which oxygen or hydrogen are separately required, they, like a galvanic battery, have force and power enough to rend the molecules asunder and to rearrange and assimilate them in that laboratory, the most wonderful of all—the plant cell.

Water being, as just stated, a dual chemical compound of the two gases, oxygen and hydrogen, 8 lbs. of oxygen and 1 lb. of hydrogen are required for every 9 lbs. of water, and this is owing to an atom of oxygen being eight times greater than the hydrogen atom. To its solvent power is chiefly owing the presence of our luxurious vegetation. It carries oxygen, hydrogen, nitrogen, and all the soluble chemical compounds which plants require, to their roots. It is absorbed by them in root, branch, and leaf, and they also possess the power, as I have just observed, of decomposing both water and its dissolved constituents into their elementary states, and selecting or refusing at will what is or what is not required for their growth. It is in this quiet flow through the vessels of plants that the mysterious chemical changes take place which give to woods their varied structure, and to vegetation those strange and multitudinous properties of taste, odour, food, or poison. How wonderful is this synthetic power of plant-energy, rending asunder with ease compounds which are most stable, and again drawing them into chemical unions of the most complex and subtle composition, having properties as numerous and as various as the species of plants themselves. This being so, it will not, I think, be denied that in any consideration of plant-food it is not out of place that ample consideration should be given to its composition and application, especially where it has to be successfully used in the production of crops for animal food.

Carbonic Acid.

Carbonic acid, or carbon dioxide (CO_2) or in aqueous solution $CO(OH)_2$, is a gas composed of carbon and oxygen in the proportions of 28 lbs. of carbon to every 72 lbs. of oxygen. It is present in the atmosphere when pure to the extent of ·035 per cent, or 1 gallon in 2,500. Water absorbs its own volume of this gas. Carbonic acid is absorbed from the air by plants in the light, and is also given off in the dark, oxygen being then absorbed. Light causes plants to deoxidise the carbonic acid, and so permit the carbon to enter upon its functions of structure-forming and to the creation of the multifarious carbon compounds which form the juices of plants.

The atmosphere derives its carbonic acid from the breath evolved by living animals, the decomposition of animal and vegetable matter, whether artificial or natural, quickly as in burning, or by slow combustion, as in natural decay, by fermentation of various kinds, especially in dung and excreta generally, which give off considerable quantities and which are absorbed by water, and are thus imparted to the soil.

Ammonia.

This is a compound of nitrogen and hydrogen. Its formula is NH_3, or in molecules of three atoms of hydrogen to one atom of nitrogen in chemical union. To every 14 parts by weight of nitrogen there are three parts of hydrogen. The sources by which ammonia is imparted to the atmosphere are various. I may mention at least two artificial means. It is an important by-product in the manufacture of coal gas. Its pungent odour is, for example, distinctly noticeable as it is being evolved from the sewage-effluent in the carriers at the Newcastle-under-Lyme osier-beds, and also when caustic alkali is added to a salt of ammonia. It is copiously given off when animal matter is distilled.

It is also variously communicated to the air in nature in

several ways, chiefly during the decay and decomposition of both animal and vegetable matter, especially in the putrefactive stages. It is also being formed in soils in which the atmosphere does not gain ready access, by the decay of root and other vegetation, and it is there held by other chemical agencies (which would be perhaps needlessly digressing too much to describe here), until imparted to the roots by the medium of water, or perhaps rather by what is more pertinent to my object now, it helps the soil to be a storehouse of ammoniacal—*i.e.*, nitrogenous—food, to be used in those proportions which best befit healthy and unforced growth.

Nitric Acid.

In all discussions about agriculture, sewage and water analysis, we necessarily hear a good deal about nitrates and nitrites, and of their good and bad qualities—good as manure, bad as existing in potable water. They are formed by the action of nitric acid, the composition and function of which it may be useful to explain. This corrosive acid is a compound of the second, third, and fourth of the elements I tabulated at the beginning of this chapter, namely, oxygen, hydrogen, and nitrogen, inert bodies when uncombined.

The formula of nitric acid is HNO_3, molecules of 3 atoms of oxygen chemically combined with one each of hydrogen and nitrogen. Pure nitric acid contains 14 parts by weight of nitrogen and 48 of oxygen, but it is always found associated with water. It is formed by the oxidation of ammonia, and is found in the juices of plants after at first existing in them as nitrates.

This acid forms salts. It combines with potash as saltpetre, soda as nitrate of soda, lime as nitrate of lime, and it is in these states that it becomes a fertilising agent, and it is from these salts that plants chiefly derive their nitrogen. Amongst other nitrogenous-giving matter is urea, whose main source is the urine of man and the lower animals.

CHAPTER XXII.

Soils and their Nitrification—Manures.

In a country like ours, with such a wonderfully diversified geology, there must, of necessity, be soils and subsoils equally varied, for soils are mainly the result of the disintegration of rocks, whose name is legion, although of only two groups, sedimentary and igneous. Limestone may be considered as scarcely coming within either of these, being mainly a secretion from living organisms. Besides these mineral or rock constituents, soils contain organic matter derived from the decay of vegetation, and also to a much less extent the remains of animal life.

The Inorganic Constituents of Soils

During ages of the decomposition of these slowly accumulating matters, a substance called humus has been derived, the composition of which is, to some extent, yet doubtful. It is said to consist of humic, and four other organic acids, in which carbon is the predominating element. These acids—or, let us say, this humus—has the useful property of absorbing and retaining ammonia. Its property is to hold it firmly; its function is to be the well from which the roots of plants draw their sustenance in the proportions necessary to their healthy growth. Some soils are rich in humus, and some poor. Is it not, therefore, apparent that in the distribution and application of manures, especially from refuse sources, that the properties of soil should be considered, and that the application of sewage matter should be in relation to the precise requirements of the soil? In peaty soils the organic part constitutes upwards of one half its weight; clay soils contain only 10 to 12 per cent; whilst on well-cultivated farms with rich soils about 25 per cent is generally found, but almost universally it is much less than this. Wheat soil

19

is considered good if it contains only about 5 per cent of organic matter.

Soils principally consist of rock detritus—*i.e.*, fine or coarse sedimentary matter, and of decomposed rock constituents. Clay is one of the finest sedimentary separations, and is composed chiefly of silica and alumina, both oxides of their respective metallic elements, but generally in chemical combination, mostly as aluminic silicates; silica also occurs uncombined. The insoluble quantity of the inorganic portion of soils is about 95 per cent, and consists chiefly of silicate of alumina, silicates and carbonates of lime and magnesia, phosphatic matter, iron oxide (which gives to soils their red colour), sulphates, and other compounds in less important proportions. A marly soil is that which contains more than 5 per cent of carbonate of lime; it is a calcareous soil if it contains more than 20 per cent of carbonate of lime, as in the chalk districts. Most soils are more or less sandy or arenaceous, such sand being simply the detrital grains of rock-masses, broken and worn down by denudation and weathering. To explain the various constituents of sand would be to enter more into the domain of geology than would be admissible here, but I may be permitted the observation that to farm with intelligence, a knowledge of geology is as indispensable as that of chemistry, for no other acquirement would correctly define the real and great difference which exists in the composition of the many kinds of soils and subsoils which carpet the various land-surfaces of our planet, such, for an example, as the alluvial deposits in the Churnet Valley. That on the north side of Leek being stiff, cold, and rather impermeable, rejecting the allurements of sewage fertilisation, whilst that in the south is lighter, more sandy, the subsoil less water-logged, and receives sewage with benefit. This great change in a very small area of valley sediment is due to easily explainable geological and mineralogical causes, and so with other districts. The skill in applying manure is that which overcomes the natural obstacles of soils, and which converts them into fertilised and cropping areas, whether corn or green-crops, according to the intellectual grasp in the application of the right manures at the right times; the clays to be lightened and not choked, the sandy soils to be fed with the requisite proportions of organic matter. It is, therefore, evident that an indiscriminate flooding of town sewage on all kinds of land alike is neither the best way of getting rid of it, nor the best way of using it for fertilisation.

The inorganic part of soils contains also soluble matter, but only sparingly. It is found to consist chiefly of the carbonates, sulphates, chlorides, nitrates, and phosphates of sodium, potassium, and magnesium.

These are the inorganic sources from which vegetation of all kinds derives those constituent parts which, after combustion, are found in its ashes as mineral salts.

Sprengel, a celebrated German agricultural chemist, analysed three soils, with the results stated in the following tables which are extremely instructive,* and clearly show the causes why some soils are fruitful and others barren.

No. 1 is fruitful without manure, because it contains a proper proportion of inorganic and organic matter, lime and saline ingredients, in beneficial quantities.

No. 2 is poor in saline matter, soluble and insoluble, and in lime, and therefore requires manuring according to the kind of crops wanted.

No. 3, the organic matter is too small, lime is almost absent, and there is too large a proportion of sand, or 40 per cent.

When washed with water they gave respectively, from 1,000 parts of soil—

	No. 1.	No. 2.	No. 3.
†Soluble saline matter	18	1	1
Fine clay and organic matter	937	839	599
Silicious sand	45	160	400
	1000	1000	1000

* Johnson and Cameron's "Agricultural Chemistry and Geology," pp. 139, 140.

† Common salt, chloride and sulphate of potassium, sulphate of calcium, with trace of sulphates of magnesium and iron and phosphate of sodium.

The finer portions, separated from the sand and soluble matter, consisted, in 1,000 parts, of—

	No. 1.	No. 2.	No. 3.
Organic matter	97	50	40
Silica	648	833	778
Alumina	57	51	91
Lime	59	18	4
Magnesia	8½	8	1
Oxide of iron	61	30	81
Oxide of manganese	1	3	½
Potash	2	trace	trace
Soda	4	"	"
Ammonia	trace	"	"
Chlorine	2	"	"
Sulphuric acid	2	¾	"
Phosphoric acid	4½	1¾	"
Carbonic acid	40	4½	"
Loss	14	—	4½
	1000	1000	1000

No. 1 is a very fertile alluvial soil from East Friesland, formerly overflowed by the sea, but for sixty years cultivated with corn and pulse crops, without manure.

No. 2 is a fertile soil near Göttingen, which produces excellent crops of clover, pulse, rape, potatoes, and turnips, the two last more especially when manured with gypsum.

No. 3 is a very barren soil from Luneberg.

These saline proportions of soil vary very greatly. In soils poor in saline matter there may not be more than a mere trace, whilst, on the contrary, in the Punjaub, I have seen hundreds of square miles of land white-over, as if covered with snow, with crystalline and amorphous incrustations of saline matter, called in Hindee "Reh," a deliquescent salt consisting of sulphate of soda, with a variable proportion of chloride of sodium. Some soils are so highly impregnated with it as to contain as much as 80 per cent, and even higher.

In the jungles at Pokhuria, in Manbhoom, I found it in nodular excrescences, with but a mere modicum of soil intermixed with it. Land so impregnated is incurable and valueless, and when wanted for garden purposes the Reh has to be dug out to a depth of 2 feet, and fresh soil added; nitrate of lime as a manure is recommended to be used in such cases.

Saline matters in right proportion are, however, necessities for plant-growth, wheat crops requiring more than grass. So, therefore, a knowledge of the quantity of the salts in sewage is useful, and still more how to judiciously apply that knowledge for cropping on lands best suited to receive it. In this way only must sewage be considered as of money value, not only by the farmer who is asked to use it, but by the urban authorities who want to get rid of it. Looked at with this knowledge, and in this way, sewage becomes a property as a manure much too valuable to be wastefully thrown into the sea or rivers, or indiscriminately flooded upon alluvial or other grass soils, or allowed contact with any kinds of crops whatever.

I feel certain the true value of sewage in its application to cropping has never yet been fully estimated—at least the exceptions are few—and that the majority of sewage-disposal systems are highly objectionable and wasteful to a degree, which will at no distant date be more fully seen.

The most extensive application of sewage must, I think, be to grass, in meadow and pasture land, as by far the greatest quantity of land in the country is cultivated in these forms.

Soils suitable for Sewage and Effluent Irrigation.

The light soils of the New red-sandstone banks, such as those at Leek, resting as they do on a very sandy subsoil and frequently on the porous rock itself, are especially suitable for copious effluent sewage-disposal, and, being extremely absorbent, cannot retain the fluid fertilising-matter like less porous or clayey soils; therefore they are less liable to become choked or poisoned by a surfeit of organic refuse-matter; clayey soils, on the contrary, not being permeable, the sewage remains on the surface, and cannot be absorbed; consequently sewage application is generally injurious instead of being beneficial to healthy cropping; the bacterial element, not being able to infiltrate itself into the subsoil, becomes an increasing source of zymotic danger and risky cattle-feeding, and a frightfully responsible method if crops for human food are grown in such serious conditions. In winter effluents could be beneficially run over our ploughed land with advantage, even in frosty weather. At Beddington mangolds and rye-grass lands are so treated.

The following well-penned thought by Mr. Santo Crimp will furnish ideas for reflection to all people who wish to lessen zymotic risk, and to see supplied to the lungs of urban populations uncontaminated and pure air: "The influence upon health of the improper disposal of sewage on the one hand, and of the unsanitary conservation of refuse matter on the other, is well known to be of the most pernicious nature." In this sentence lies the full kernel of the question, and no municipal notions of parsimony, miscalled economy, ought to come between the danger and an efficient and thorough treatment and disposal of sewage.

I am strongly of opinion that much more good would be done to sandy soils on the flanks of valleys by raising the effluent to them, than by pouring it upon the alluvial meadow-lands. Such an application would enhance the value by at least 10s. an acre.

Nitrification of Soils.

It is well known that the combining of manures promotes fertlisation, owing, no doubt, to the fact that no one manure provides all the kinds of food which vegetation requires. Let me repeat that plants require nitric acid, and that they have the power to obtain it by decomposing the nitrogen compounds which exist in the soil, both those which are in it naturally and those which are applied as manure. Here is Nature's process; it is a most important one, and it is equally important that it should be understood and kept well in mind both by those who prepare manures, whether sewage or other kinds, and by those who use them. It is what is called nitrification. Take ammonia, a compound of nitrogen and hydrogen; when it is formed in the soil or applied to it, it becomes converted into nitrites—*i.e.*, its nitrogen combines with oxygen and those earthy bases in the soil I have mentioned, and forms nitrites; then from these nitrites nitrates are formed, these conversions being the results of agencies I shall presently describe. It is from the nitrates that the roots of plants obtain their nitrogen, but they abstract it from the nitrates as nitric acid and as nitrogen. This is called the theory of nitrification. It is a part of agricultural chemistry which has in recent years been much investigated by such eminent chemists as Mr. R. Warrington, from 1884, and MM. Schlœsing and Muntz, in 1878. Mr. Santo Crimp describes the chemical action in the following interesting and succinct way:—

"When a liquid, such as sewage, is applied to a porous soil, the suspended matters are arrested at the surface; this is the first action. The water, freed of its insoluble matters, descends into the soil, and each particle of earth being surrounded with an extremely thin coating of liquid, an enormous surface of water is presented to the air contained in the soil. The second action of the soil now comes into operation, which is similar to slow combustion; the organic impurities are reduced to carbonic acid, water, and nitrogen, as in active combustion, but the organic nitrogen itself, which is more difficult to oxidise than either carbon or hydrogen, is changed into inorganic compounds. The means by which these changes are effected are now understood, and are doubtless due to organisms."

This is worth explaining, especially as subsequent researches have added to our knowledge of nitrifying organisms. Mr. Warrington stated in 1884 that it was apparently a micrococcus—*i.e.*, an extremely minute living thing, invisible without the aid of the microscope. The termination coccus is from the Latin *coccum* or *coccus*, the name given to the insects whose bodies yield a scarlet dye, such as coccus-quercus, the oak-feeding insect, also called kermes; coccus-cactus, the cactus-feeding one, called cochineal; but it is from a shape-resemblance only that the name coccus has been applied. These dye-yielding cocci are of the size and shape of grain—that is, slightly elongated in one direction. The micrococci are similar in shape; some are round, but immensely minute, and are revealed to sight only by extremely high magnifying power. They are isolated or detached elongated and round vegetable cells, monad-like in form.

Thus it is not a little singular and wonderful that this smaller and unseen world of plants act as chemical agents in altering the composition of organic and inorganic substances into such states as to enable the greater vegetable world to assimilate them as food from the soil. That such micro-organisms are endowed with such powers we have plenty of evidence near at hand, as in the yeast-plant, saccharomyces cerevisiæ, which turns the sugar of malt into an alcoholic beverage, producing what is called fermentation, which would not occur if the beer were kept from contact with it. Also in the putrefaction of excremental matter, and in the known power of certain species of bacteria to change nitrogenous matter into other specific compounds.

Soil is a storehouse of bacteria, both putrefactive and pathogenic. Michel computed that he found present in a single gramme of soil in one locality 750,000 germs, in a second 13,000,000, and in a third 1,100,000.

At first it was thought that the nitrification of plant-food in the soil was induced by one species of micrococcus; but now bacterial investigation has so far advanced as to cause the belief that two are required to complete the process; one for converting the ammonia or other nitrogenous organic matters in the soil into nitrites, and another species for changing the nitrites into nitrates; and so thus there are two nitrifications before the plant-roots can absorb their nitrogenous food as nitric acid.

This theory of nitrification is founded upon actual observation; its operations and action have passed the region of speculation into the domain of proved and well-established facts. These two-fold processes of nitrification have been experimentally demonstrated by Warrington and Winogradsky.

What is called the nitrous ferment has been isolated in 1890, by Dr. Percy Frankland and Mr. Warrington in England, almost simultaneously and independently of each other, and by M. Winogradsky, a month later, and announced in the "Annales de l'Institut Pasteur." The methods of isolation of the nitrous ferment are described in Dr. Percy Frankland's "Society of Arts" Cantor Lectures, at p. 952 in the Society's journal. He shows in these lectures that the nitrous ferment, or nitrifying organism of the first part of the two-fold process can be cultivated to the exclusion of other species of bacteria, by a nutrient solution having the following composition:—

COMPOSITION OF SOLUTION EMPLOYED FOR NITRIFICATION.

Ammonium chloride,	·5 grams.	in 1,000 cc. of distilled water.
Potassium phosphate,	·1 ,,	,, ,,
Magnesium sulphate,	·02 ,,	,, ,,
Calcium chloride,	·01 ,,	,, ,,
Calcium carbonate,	5·01 ,,	,, ,,

For four years, Dr. and Mrs. Percy Frankland carried on a series of experiments with this mineral solution, noticing its effect by inoculation with the nitrous ferment upon minute portions of garden soil, recording the dates when nitrification first appeared, the shortest time being about six days. The form of the bacteria which caused the nitrous ferment is shown in the following figure:

FIG 51. NITROUS FERMENT BACTERIA (Percy Frankland).

This was only the first stage of the process of nitrification, these organisms having the power only of converting ammonia into nitrous acid, or, in chemical equation,

$$NH_3 + 3O = H_2O + HNO_2$$

(Ammonia.) (Oxygen.) (Water.) (Nitrous acid.)

But another reaction waited explanation, namely, the agency which converted this nitrous acid into nitric acid; or,

$$HNO_2 + O = HNO_3$$

(Nitrous acid.) (Oxygen.) (Nitric acid.)

in which state alone soil becomes a fertilising medium.

The micro-organism which produces this nitric ferment has been isolated, and is shown in Fig. 52.

FIG. 52. NITRIC FERMENT (Winogradsky).

Its action has been demonstrated by Warrington and Winogradsky, by making cultivations of it in soil without ammonia, but containing nitrous acid. This microbe, unlike the nitrous ferment, does not possess the power of oxidising ammonia, but can convert nitrous into nitric acid. These two-fold processes of nitrification are constantly in operation in all fertile soil.

Dr. Percy Frankland shows that "however well drained, tilled, and supplied with the mineral ingredients of plant-food, such as potash, lime, and phosphoric acid," without that most important constituent, nitric acid, being present, the soil would be "incapable of growing the barest pretence of a crop, either of corn, roots, or grass." Notwithstanding that it occurs in the soil in almost unappreciable quantity, its presence may be shown by testing the soil with diphenylamine. He states that "it is in the form of nitric acid that the nitrogen of manures principally gains access as nutriment to the plant."*

In fertile soils there is as little as one part in a million parts by weight of soil of nitrogen as nitrates present, and under

* Cantor Lectures, *Journal of the Society of Arts*, p. 950, 1892.

some conditions as much as 100 parts of nitrogen in the form of nitrates per million parts.

The great accumulations of nitrate of soda in the rainless districts of Chili and Peru, are due to these nitrifying processes.

But there is another fertilising operation of recent discovery which ought not to be omitted here, namely, the power of plants to assimilate the free nitrogen of the air. Although it has been long well known that the soil can be made more fertile by the growth of leguminous crops, such as peas, beans, vetches, and other kindred albuminoid plants, and that they contain more nitrogen than can be accounted for by the addition of manures and rain-water, the reason has only been divined in later years.

Hellriegel and Wilfarth have conclusively proved that this excess of nitrogen is derived from the atmosphere, and is effected by micro-organisms "flourishing in and around the roots of these plants," forming, when present, swellings or tuberosities, which never occur in these plants when growing in sterile soil.

The microscopic appearance in a cut section of these tubercles is shown in Fig. 53.

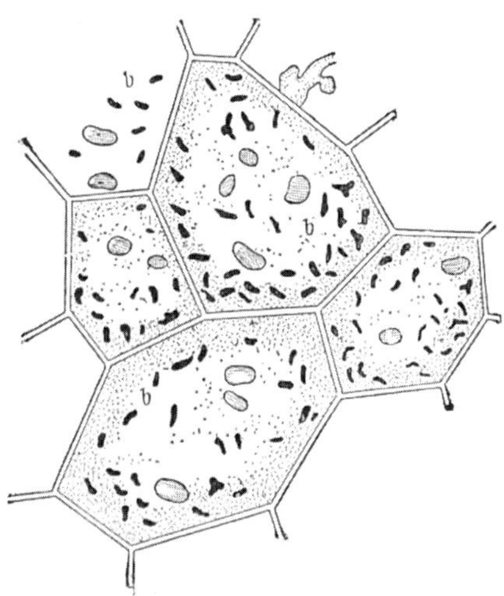

Fig. 53. Section through Tubercle of Leguminous Plant (after Laurent).
b bacteroids present in considerable numbers in each of the polygonal plant-cells.

The individual bacteroids, under greatly increased magnifying power, have the appearance shown in Fig. 54.

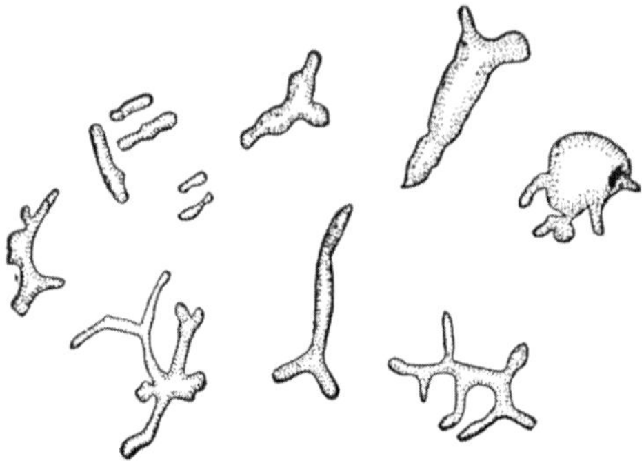

FIG. 54. INDIVIDUAL BACTEROIDS HIGHLY MAGNIFIED (after Laurent).

The term bacteroids is only provisional until morphologists have defined to which group of organisms they belong.

They can be cultivated artificially.

Their power is most remarkable. If these bacteroids are introduced by inoculation into the roots of leguminous plants, "a more abundant growth and fixation of nitrogen takes place, it is supposed by the vital activity of these micro-organisms present in the tubercles taking up the atmospheric nitrogen and elaborating it in a form which can be assimilated by the plant."*

The two following figures illustrate this action:—

1 2 3

FIG. 55. ROBINIA (after Nobbe).

1. A Robinia plant uninoculated. 2. Inoculated with pure cultures from pea tubercles. 3. Inoculated with pure cultures from robinia tubercles.

The inoculations were made on June 27, 1890; the plants were photographed on August 5, 1890.

* Dr. Percy Frankland.

They were again photographed on October 3, 1890, with the following appearances:

FIG. 56. THE SAME AS FIG. 55, BUT PHOTOGRAPHED ON OCTOBER 3, 1890.

So much for the extraordinary power which these micro-organisms possess of assimilating otherwise inert atmospheric nitrogen for food uses.

So wonderful is this discovery of the power of bacterially-inoculated plants to absorb free nitrogen from the air, that it is scarcely extravagant for the imagination to conceive a progress of scientific evolution which should make it unnecessary for plants to depend upon their nitrogenous requirements from any other source, and in so doing to be under less obligation to the present necessities of manuring.

MANURES.

The following is a list of the principal and best-known manures, but it is not an exhaustive one.

I will put them in three groups, suggestive of the three principal requirements of plant structure: Phosphatic, Mineral, and Nitrogenous.

There are two classes of the phosphatic series—first, raw phosphatic products; and secondly, dissolved or chemically-prepared phosphatic manures. Of the first, there are bones, coprolites, guano, basic-cinder or phosphatic slag; and of the second there are dissolved guano, dissolved bones, and superphosphate—*i.e.*, superphosphate of lime.

Of the second, or mineral group, we have lime, gypsum,

which is sulphate of lime; kainit, a mineral composed of chloride and sulphate of magnesia and chloride of sodium; sulphate of potash; chloride of sodium, or common salt; sulphate of ammonia; chloride and sulphate of soda; kelp, for its iodine and potash; protosulphate and phosphate of iron; sulphate of magnesia, &c.

The nitrogenous group consists of the various kinds of vegetable and animal manures and excreta, such as farmyard manure, vegetable mould and refuse, blood, fish, seaweed, crude sewage, sewage-effluent, sewage-sludge, &c., chiefly useful for the nitrogen they yield. Farmyard manure is a fertiliser both from the nitrogen and the phosphoric acid it yields. Nitrate of soda and nitrate of potash, although mineral salts, must be classed in this group.

There are two kinds of sewage-sludge—ordinary sludge, consisting of matters chiefly mineral, which were in suspension in the sewage, and the precipitated organic matter, resulting from the process employed, together with the precipitant, often chiefly lime. The other kind is prepared sewage-sludge, or ordinary sewage-sludge to which other fertilising matters have been purposely added, such as in the A B C process, the Manchester and other analogous methods, and that known as Forbes' process, by which phosphatic and antiseptic matters are added, the one for increased fertilising power, the other to deodorise, to disinfect, and to decolourise.

A knowledge of the constituents and action of these very differing manures is as essential to their right use and application, as is also the knowledge of the chemical and physical nature of the soils to which they severally have to be applied.

On the Nature and Fertilising Properties of Manures.

This most important consideration brings us nearer to the point aimed at—the effects of manures and their greater usefulness when mixed, on which subject I must necessarily treat in order to prove what I laid down in a previous chapter, that it is neither good farming nor good science to administer only one kind of manurial feeding to plants whose substance consists of so many elements, all to be derived from their food.

I select nitrate of soda, now known to be one of the most useful fertilisers, when properly administered, as an example. I take this substance because it is in the state nearest for conversion by plant-roots into the nitric acid they require for their nitrogen. But nitrogen is not the only food, and this manure without others is not safe to use; *i.e.*, generally speaking, plants must have phosphorus, potash, lime, and other kinds of food, and no healthy growth or good cropping can occur without these additions. The intention and relevancy of this apparent digression will be fully seen further on, when I come to the disposal and best fertilising application of sewage.

The following is the result of one of the Essex Agricultural Society's experiments conducted in 1891 under the supervision of Mr. Bernard Dyer, F.C.S., and Mr. Edward Rosling, of Chelmsford, on the cropping of oats and mangolds on one plot. Twelve tons per acre of farmyard manure alone were used. The value of the increased crop, as compared with no manuring, was £2 1s. 10d. per acre.

Another plot with ten tons of dung and three cwt. of superphosphate gave £3 7s. of increased value. Another plot with four cwt. of nitrate of soda, gave £3 3s. 6d.; whilst another plot, to which was given twelve tons of dung, four cwt. of phosphate, in the shape of basic-cinder (better for this purpose than superphosphate), and four cwt. of nitrate of soda, yielded £7 7s. 3d. in value of increased cropping.

This shows that when nitrogen, in its nearest form of assimilation by plants—*i.e.*, as nitrate of soda—is administered in conjunction with other manures, the cropping reaches its maximum.

In some experiments the nitrate of soda was mixed with salt with manifest advantage, and in other combinations with muriate of potash.

In 1887 this Society made experiments in growing mangold-wurzel, with the following results: A plot with no manure yielded 388 bushels of 56 lbs. per acre; on another, with 4 cwt. per acre of nitrate of soda alone, 533 bushels; on another, with 12 tons of dung per acre and 3 cwt. superphosphate, 419 bushels; on another, with 2 cwt. nitrate of soda added to the last, 626 bushels; on another, with 3 cwt. superphosphate and 4 cwt. nitrate of soda, 680 bushels.

Nitrate of soda is a quickly acting manure, and it is thus seen that mangolds grow best under its influence, but it must be added judiciously, and in one or two dressings, when the roots are ready to act upon the nitrate.

I would suggest nitrate of ammonia as a manure, if it could be obtained cheaply enough, as probably being the most readily assimilated by crops. Nitrate of potash is also a valuable manure.

Phosphatic manures include bones, both dissolved and undissolved, but they are also nitrogenous; guano is also in this category.

Superphosphate is a dissolved manure, but the basic-slag or cinder I mentioned above is an undissolved manure, like raw bones; it is a by-product, or refuse, from the manufacture of the phosphorus used in iron smelting. It is ground into powder. For the proper application of these phosphatic manures, whether in the raw or dissolved state, a knowledge of the nature of the soil is required. For a limey soil Mr. Dyer recommends for all crops a dissolved phosphatic treatment, whilst on soils deficient in lime he advocates the raw materials—bone-meal, raw guano, or basic-cinder. The pertinency of these remarks will be evident from what I have written in reference to dealing with sewage sludge and also with effluents in Chapters XVII. and XVIII.

I am not leaving out of sight the fact that there are many manufactured or compound manures which furnish both phosphates and nitrogenous fertilisers, but I am treating of principles in relation to the application and disposal of sewage, and not alone cataloguing manures and their action.

Manuring for potatoes is a very important subject. Mr. Dyer recommends the use of potash more than to any other crop, applied in the shape of Kainit, mixed with superphosphate, but dung is also required, and a top dressing of 1 cwt. of nitrate of soda or sulphate of ammonia. If the soil is limey or not, these precautions must be observed.

In the foregoing remarks it will be seen that the substances required for fertilisation are chiefly nitric, phosphoric, and sulphuric acids of the bases lime, soda, magnesia, and potash, and that the success in fertilising with the best results depends upon these substances being absorbed by the plants, according to their being treated with them in mixed proportions and according to the nature of the soils and the crops required.

These facts ought to be helpful in the consideration of the several states in which sewage occurs, such as crude sewage, effluents, or sludges. The phosphatic part in sewage precipitation is left in the sludge; the effluents contain scarcely any.

The state of dilution of sewage-effluent is important. It would be too costly to add any ingredients to its nitrogenous qualities if it is put on the land in a very diluted state; therefore it might be desirable to separate the storm-water from the ordinary sewage, but in doing so regard must be had to other methods of properly flushing the sewers, especially in low-lying and level parts, where putrefactive gases are soon formed by rapid bacterial growth and action, or the death-rate would soon be raised. To such a less diluted effluent the farmer could add phosphates, lime, &c.

Could not the fertilising qualities of sewage-sludge be greatly heightened by the addition of nitrate of soda, and proportionate quantities of the salts previously mentioned, and could not effluents be made more fertilising by the addition of raw or dissolved phosphates? I think it is most likely. I have mentioned that one patent process, that of Forbes, has a phosphatic addition, but I do not think it has been well enough thought out, or sufficiently experimented with. It would be well if much experimental work were regularly carried on at all sewage farms by adding to sewage other fertilising constituents and trying their effects, separately and conjointly.

I mentioned this idea to the farm bailiff of the Wimbledon Sewage Farm, in my visit there in September, and he thought it would be too expensive to add phosphates to the effluent there, inasmuch as, the storm-water not being separated, the effluent was in a too dilute state. I gave an account of the methods in operation at Wimbledon when describing the Amines process, but, having derived more information in a recent visit, that record would be incomplete without a fuller account of this interesting system, and also of the excellent farming in operation there.

I went down with Dr. Bahardhurjé, an eminent Bombay physician, who had come over to the Hygiene Congress, and to examine the best English methods of sewage disposal and air purification, both of which Bombay stands in pressing need.

We were met by Mr. Hugo Wollheim, the discoverer of the application of the trimethylamine plan in destroying bacteria in sewage. Mr. Snooks, the farm bailiff of the Wimbledon Sewage Farm, was also present. The whole arrangements for the disposal of the sewage here are so complete and interesting that any sewage-disposal investigators omitting to see these operations carried on would be rightly charged with culpable omission. The farming is carried out on a most perfect system, and the process, apparently one of the best, worth fully examining.

The following is a resumé of what we saw: There were a series of precipitating tanks, six in number, of a capacity of 530,000 gallons, worked in a series or separately, into which the treated sewage flowed, after being raised from the sewage level by plunge-pumps to a height of 18 or 20 feet. Into the unscreened and comminuted sewage of 15,000 persons was poured a mixture composed of lime to the extent of 30 grains per gallon and three grains of herring-brine, the salt liquid in which herrings had been cured, and in which the characteristic odour of trimethylamine was present. That is the whole process. On the brine coming in contact with the lime, aminol gas is formed. This gas, like ammoniacal gas, is extremely soluble, and at once pervades the sewage; precipitation takes place immediately, and all organisms are said to be destroyed and carried down in the precipitate, whilst the effluent runs away to the meadows from the tanks, perfectly clear and sterilised.

In parenthesis, it is interesting to know that it was Professor Tyndall who first discovered that herring-brine possessed antiseptic properties. At pp. 166-7 of his "Floating Matter of the Air," published several years ago, he states: "On the 5th December, the herring and sole infusions were both clear, but this was only a respite, for on the 6th white spots appeared on the latter, which extended until they covered the whole surface. The herring infusion remained clear for a week, after which small specks began to appear on its surface. They never reached the development of the scum which formed in the other infusions. It sometimes occurred to me that the oil of this fish exercises a certain antiseptic action. Last year I preserved infusion of herring perfectly pellucid for months, even in a chamber so leaky that the light could be seen through its chinks. I had, moreover, no failure—*i.e.*, failure to obtain growth—with any of the other animal infusions here enumerated."

The sludge contains about 2 per cent of phosphates, principally phosphate of lime, 1 to 1½ per cent of nitrogen, as nitrogen compounds.

Mr. Wollheim estimates the value of the sludge to be 5s. to 6s. per ton, but it is sold at 1s. per ton. It grows good crops of mangolds and cabbage. The grass is fertilised with the clear effluent.

In 1890, 175·7 millions of gallons of sewage were pumped from a population of 15,500 persons.

The following crops are cultivated:—

Rye grass, the most important; the meadows yield four to

five good crops a year, or about one every month during the middle part of the year.

Mangolds, cabbage, and various other vegetable produce, such as cauliflowers and rhubarb, are grown; peas and beans are not cultivated, on account of expense of labour. Osiers are grown of one sort, known as the new kind, but they are not cultivated to a large extent.

The cabbage crop	produces	£20	per acre.
The mangold	„	30	„
The grass	„	20	„
The osiers	„	12	„

A profit of £400 a year is made, but cost of precipitation and of works are not charged to farming account, nor any expenses before delivery of sewage outside the works after treatment.

The effluent is run on the grass-land for three weeks at a time, and is then moved from plot to plot. Here I thought I detected a flaw in the administration of the sewage. It is not natural to flood soil for three weeks at a time with any liquid, or for any crop except rice; the paddy fields of India, of which I saw many hundreds of square miles, were in many places permanently ankle deep in water; but the fertile and splendidly cultivated alluvium of the middle Ganges between Benares and Delhi, the garden of India, had magnificent crops of corn, and many varieties of farinaceous and nitrogenous food plants on extensive areas, in plots upon which water was but sparingly administered, the soil being dry rather than wet. But in Wimbledon the meadow-soil was saturated, the long grass growing in a more isolated manner than it does in a field naturally, more apart, like wheat stems, than as in sods.

It seemed to me that much shorter application than three weeks at a time would be sufficient. I suggested two or three days at intervals, intermittently, and that a much larger extent of land could be fertilised, with the result that very much more effluent would be utilised, and less, if any, discharged into the

Wandle. The farm bailiff told me the grass was in such demand that they could not grow enough of it. I saw it being carted away as it was cut. The precipitant, lime, is administered in proportions larger than in other lime systems; the herring-brine action comes into play, and is said to sterilise the effluent by destroying its bacteria, causing it to bear exposure to atmospheric and soil-germs without subsequent bacterial putrefaction. Only about one half of the organic matter in the sewage is precipitated as sludge; consequently there is a much smaller quantity of sludge to be pressed and carted away.

My purpose in the foregoing narration is to bring the argument nearest to the main question, that of sewage-disposal, and to examine the relation of plant-life to those organic and inorganic substances which are necessary for its sustenance, and especially how they can best be evolved from sewage and healthily applied. Vegetation is neither gluttonous nor drunken naturally, and it is surely a pertinent question how plants shall be wholesomely fed with a refuse matter, which at best is filthy, and liable to promote fermentation and structural decomposition in plant-life if injudiciously employed.

I have stated in a previous chapter the advantage of putting weak effluents on land, or even water alone—they absorb oxygen, nitrogen, and carbonic acid from the air, besides dissolving saline matters in the soil, all of which are absorbed by the roots of plants, which thereby become more fruitful.

I have shown that the best fertilising results are given by manuring with mixed nitrogenous and mineral manures; that nitrogenous manures give better crops than mineral, and mixed mineral and nitrogen better than either; that nitrate of soda is wholly absorbed by one corn crop, and leaves nothing for the next, and that phosphoric acid for root crops is especially good.

A full inquiry into this most important part of my subject would involve a consideration into the domain of agriculture much too wide and comprehensive to be attempted in this discussion, even if I were competent to deal with it. I have necessarily confined myself to outline-treatment, and have endeavoured to lay down principles for the guidance of sewage-disposal which are based upon the natural requirements of food-plants and the considerations concerning its economical utilisation.

My friend, Monsieur Marnas, of Thurins, near Lyons, informs me that he finds sulphate of ammonia particularly beneficial to the crops growing in that region, the principal ones being vines.

These observations and investigations also have close bearing upon sewage-sludge disposal, and lead up to the idea of the important necessity of studying how best to utilise it by such fertilisation additions as will best admit of its being recognised and welcomed by agriculturalists as matter of such manurial value as to cause it to be sought for without prejudice and without apprehension of its not being a useless commodity. In fact, now it is coming to be admitted that the future of sewage-disposal must be connected with some form or other of the precipitation of the organic matter, more or less complete, contained in the sewage, the question of the utilisation and disposal of sewage-sludge must rank as a study equal to that of the treatment of sewage, and upon which I have treated in a chapter devoted to the subject.

CHAPTER XXIII.

Etiology in Relation to Sewage Disposal.

"Salus populi suprema est lex."

There is, fortunately, an identity of idea if not of interest in the production of healthy food by the agriculturalist for man and beast, with the necessary watchfulness on the part of urban authorities to prevent the food supply of the population being introduced under other than healthy conditions, so that the energies of health may not be lowered or impaired. Disease, zymotic and other, may be kept in check by watchful oversight, by the agriculturalists to give us healthy food in exchange for our money and our waste products, and by an equally watchful determination that such waste products are supplied to them in a state the least harmful to cattle and to crops. The question of the transmission of pathogenic bacteria from cattle to man is a very difficult one to settle in the present imperfect state of our knowledge. That positive evidence will one day be produced I have no doubt whatever. There is, in fact, justification even now for more than inference; there is ample circumstantial evidence, as I shall proceed to show.

From a paper read by Mr. Ernest Hart, Chairman of the National Health Society, in 1883, at Huddersfield, I gather some important particulars.

It is not twenty years ago since infection by means of milk was acknowledged. In 1873, there was a great epidemic at Marylebone, in 1870 a typhoid outbreak at Islington, in 1872 outbreaks at Armley and Moseley, Leeds and Parkhead, all traced to infected milk. Up to and during the decade ending

1883 there were registered the following epidemics attributed to polluted milk:—

Typhoid fever 53	Number of cases	3,500	...	400	deaths.
Scarlatina ... 17	„	800	...	120	„
Diphtheria and throat illness 12	„	700	...	60	„
Total 82		5,000		580	„

About half the epidemics of typhoid fever were attributable to polluted water supply, but the most common way in which the poison reached the milk was by water infiltration and contamination of typhoid excrementitious matter, but in many cases epidemics began without such contact, or could not be traced to it; they were attributed to insanitariness of the surroundings, bad drainage, and contact with sewage. The following list will suffice to show the danger arising from these causes:

Date of Outbreak.	Locality.	Reporter.	Total No. of cases.	Sanitation of farm or dairy from which suspected milk was derived.	Exciting cause of outbreak.
1873 July to 1877, Nov.	Ascot	E. Ballard, M.D.	69 in 40 families	Business of dairy conducted in a slovenly manner. Drainage bad and offensive; cess-pit privy on the premises. Water supply derived from well, the pump case of which was in atmospheric connection with the drains. Empty milk cans inverted over sink and over opening of drain pipe in the kitchen. Water used for rinsing dangerous, being exposed to soakage of foul matters from privies and manure heap, and from absorption of foul emanations from drains which communicated with pump case.	No case of fever occurred at farm till August, 1876, when epidemic had lasted for three years. From this time onwards, water used for washing and rinsing milk cans had an opportunity of specific pollution.
1873 November	Brighouse, Yorks.	T. Britton, M.D. M.O.H.	68	Cows healthy, but cesspool, 9ft. by 6ft., on premises in connection with a drain through fold yard, in which was a large manure heap. Cows drank from this cesspool.	
1874 June	Taunton	H. J. Alford, M.D., M.O.H.	5	Well water of dairy "fearfully contaminated" with sewage, and smell very offensive. Well saturated with sewage, which was clinging to the bricks where it had trickled down. Privy choked, with no flushing apparatus; milk cans kept in a small back kitchen, in which were a pump, a small lighted furnace, and a large barrel full of pig's wash.	
1875 August	Jarrow	John Spear, M.O.H.	34	Water used at dairy derived from a well situated 18 yards from a privy and cess-pit, the latter merely a hole in the ground; well, bricked loosely round, and unquestionable soakage into it from privy and cess-pit. A quantity of water taken from the well found to be putrid two days afterwards.	Six of the farmer's family (including himself) found ill with typhoid.
1875 October	Eythorn. East Kent	M. K. Robinson, M.D., M.O.H.	21	Large collections of filth in neighbourhood of farmhouse and well. Latter found polluted on analysis. Duty of attending on sick and milking performed by same person. Lodgers (probably fever convalescents) received at this farmhouse.	Four cases of enteric fever at farm between October 3rd and end of November, 1875. Outbreak probably caused through infection of milk by person who was both nurse and milker. Fever had been in farmhouse three weeks before cases occurred in the village, but its existence was kept secret.
1876 Jan. and Feb.	Eagley and Bolton	W. H. Power (to Local Government Board), J. Robinson, M.O.H., Eagley.	195	Farmhouse depended for its water supply upon a brook which had been very much befouled with the excrement of men engaged in building a mill 200 yards off. All the families usually using the brook water for drinking purposes had, on this account, discontinued its use, except the farmer and one other family.	Large quantities of fæcal matters found on course of brook to the water, in which excrement must have soaked. Towards end of January, it was generally noted that something was wrong with the milk. It turned sour almost at once; it had a peculiar colour, tasted unpleasantly, and even smelt offensively. In many instances, after standing, there was left at the bottom of the vessel a sediment variously described as grit, sand, or dirt.

EPIDEMICS OF MILK-TYPHOID.—*continued.*

Date of Outbreak.	Locality.	Reporter.	Total No. of cases.	Sanitation of farm or dairy from which suspected milk was derived.	Exciting cause of outbreak.
1877 Oct. and Nov.	Tunbridge Wells	W. H. Rix, M.O.H.	68 in borough during two months in 49 houses	Dairyman got his milk from six sources, and there was a considerable mixture of milk. One source was from a small farm outside town, where sewage of village flowed through cowyard. In the village there had been cases of typhoid fever. Water that cans were washed with came from a hill on opposite side of stream. Farmer's wife and children were continually ailing.	
1878 Jan. to March.	Moss Side, near Manchester	E. Suctliffe, M.O.H.	32	Water of farm, whence came part of supply of dealer, derived from well in close contiguity to ashpits. This water found on analysis to be polluted from sewage contamination.	Two deaths from typhoid occurred at farm in February.
1880 July to Sept.	Mill Brook, Cornwall	E. Ballard, M.D.	19	Part of a slaughter-house used as a dairy, on shelf of which milk was habitually kept in pans. In one corner of the slaughter-house, nearest the dairy, a badly trapped and very offensive drain inlet. Drain was in communication with another which had received infected excreta, so that infected sewage air from inlet had free access to dairy and milk which stood in it.	Six cases of enteric fever occurred in the butcher's family in less than three weeks.
1880 October	Southport	H. H. Vernon, M.D., M.O.H.	32	Well in close proximity to premises, and much exposed to excremental pollution. Water declared on analysis to be nothing but liquid sewage.	
1882 August	Stone Chair, Halifax.	T. Britton, M.D. M.O.H.	11	Back kitchen of farmhouse used as a dairy. In a place adjoining was an open grate to a drain, which communicated with manure heap and pigsties. The untrapped grate was within two yards of the table on which the milk bowls were placed, and in the same place in which the milk cans were kept; an open door divided the places. Water supply contaminated; offensive accumulation of sewage from an overflowing cesspool in front of farmhouse.	An old man came on a visit to this farmhouse, and was laid up for some time. His wife came to nurse him, and was taken ill with billiousness, and died. The farmer's wife and child were also somewhat unwell.
1882	Rugby.	Wilson, M.D. M.O.H.	12	Well water of house of dairyman, who had died from disease, highly polluted. This water was used to wash cans, if not to dilute the milk.	Contaminated water was the cause of the dairyman's illness and death. The disease being once originated on the premises, the well water became specifically contaminated with typhoid poison.

Besides these cases there are numbers of others directly traceable in origin to typhoid patients or convalescents at the dairy farms. It is not for me to dispute the opinion of a Medical Officer of Health that a recent outbreak of typhoid in a Midland town was not connected with milk supply. Possibly not, but the cause has not been traced or ascertained, and we may be left still to wonder whether amongst the hidden causes there may not have been outbursts of sewer gas from insufficient sewer ventilation, atmospheric bacterial suffusion from the fields irrigated by crude sewage, or from drinking the milk of cattle feeding on these lands. The way in which the causes of infection reach the milk is mostly too obscure and mysterious for any inquirer to be dogmatic about, and it seems hardly possible of ascertainment in our present state of knowledge, but the deaths and lowering of the health of cattle and sheep near Lichfield, described in the next chapter, go far to prove that milk must suffer deterioration, just as muscular tissue and life-functions deteriorate by unhealthy food.

Nothing will induce me to think that the drinking of crude sewage by milk-giving cattle is other than dangerous, and it will not be conceded by any thoughtful inquirer, medical, scientific, or lay, that the direct conversion of liquid sewage into food for cattle, without previous purifying treatment before converting this refuse matter into vegetation, is not fraught with great danger. Will anyone venture to say there is no risk to the human species in the neglect of such a condition?

I have it upon unquestionable authority that one farmer had to choose between ceasing to send his milk to Manchester or removing his cows from sewage-irrigated fields; and I do think it would be wise for farmers to refuse to allow sewage to flow upon their land until it had undergone some treatment to deprive it, not only of its solid matter, but also of some of its organic matter in solution. Mr. Carter-Bell, the public analyst for Cheshire, adds the weight of his name to this opinion, and strongly advises this. It is true that the digestive organs of cattle have wonderful assimilating powers, even to the all but absolute destruction of the micro-organisms they consume with their food, and I believe it is owing to these powers that there is not more infection; but it cannot be so with their lungs, into which there must be ready and suffusive access of pathogenic bacteria ever arising from land irrigated with crude sewage. At the same time, we know the subtle influence on the flavour of milk and butter of so simple a vegetable as turnips; and it has been proved lately by Mr. F. Jean, the Director of the Laboratory of the Bourse of Commerce, Paris, who recently gave an address upon the influence of food in the composition of butter, in which he said, after relating certain experiments, that the oil in the food composing the rations passed into the milk, and subsequently into the butter, with its special properties retained, and in sufficient quantities to distinctly modify both the physical and chemical properties of the butter, and to give it all the characteristics of a butter to which a large proportion of foreign fat had been added. This non-assimilation was first observed by Kildan, a German chemist, and was also fully proved by M. Gerard, the eminent Parisian chemist.

It is well known that bacillæ, and especially their spores, pass into and out of the system unchanged; that some obtain baneful lodgment, and not unfrequently scourge humanity. Even seeds of plants and germs of the lower orders of the animal world can defy assimilation. I have previously mentioned the extraordinary growth of tomatoes on the sewage-fed ground at Manchester, arising from seed which had passed unharmed through the human economy. The growth of trichinæ in the muscular tissue by eating pork infected with this dangerous parasite is well known, and is not an unfrequent occurrence.

In the best regulated systems of sewage-farming the crude sewage is not allowed to come into contact with the leaves of the growing vegetation, but, "spreading laterally through the soil by means of furrows, reaches their roots," as at Dewsbury. In this there is, of course, much less objection than flooding the crops themselves, whether grass, mangolds, or other vegetables, especially so at the spring season of the year, when even sparse and short-blade grass becomes almost embedded in crude solid sewage. Those who are apologists for crude sewage application may find consolation in the refinement of cropping at Hitchen, where roses and strawberries are fertilised by sewage, and gooseberries at Doncaster!

I now turn to the reports of the Local Government Board, for which I applied to Dr. Buchanan, the Medical Officer of Health to the Board. In his absence, I received a letter from Mr. R. Thorne-Thorne, his colleague, saying that the Board had not yet investigated the subject of milk typhoid, but sending me the report of 1885-6 and of 1888, with pathogenic and bacterial illustrations of milk-scarlatina and diphtheria. These reports are sufficiently exhaustive in my opinion to inform any reader that it is not necessary to believe that zymotic infectious disease can only be imparted to milk by human beings suffering from such diseases, either directly or indirectly, but that the germs of such diseases can be and are imparted to the milk by cows by their having contracted such disease without such human influence. This being incontrovertibly so, is it not fair and logical to assume that the presence of disease-germs in food or in the air is often the cause of disease, and that the healthy feeding of cattle in this respect is of the first importance?

One of my chief objects in these chapters has been to demonstrate the importance of food as free as possible from pathogenic bacterial germs. In the first paragraph of the report of the Local Government Board of 1885-6, it is stated that the Medical Department has been called on to investigate the outbreaks of disease related to local milk-supplies, and that "sometimes the disease has been enteric fever, sometimes scarlatina, and sometimes diphtheria; and within the experience of the Board there have been fifteen instances where one or other of these diseases has been shown upon sufficient evidence to have been distributed with the milk-service of the families invaded.

"In the case of the scarlatina outbreak it was inevitable that infection of the milk by human agency should present itself as the readiest explanation of the facts; but as successive epidemics have occurred, and have been found capable of more exact study, distrust of this explanation has arisen, and the means by which the milk receives its infective properties has come to be regarded as unknown—as possibly being related to the milk as a secretion of the cow."

Mr. Power's report of the Hendon outbreak of scarlatina precludes pre-existent human disease, and he demonstrated beyond reasonable doubt that the infection was conveyed by newly-arrived cows from Derbyshire.

Dr. Klein, the eminent bacteriologist to the Local Government Board, was consulted, and in the course of his investigations succeeded in producing both local and general evidence of scarlatina in a calf inoculated from a cow suffering from a disease having affinities to scarlatina in the human subject. He found in ulcerated udders micrococci and streptococci, and succeeded in isolating and cultivating them separately, and with these bacteria he communicated the disease from animal to animal. Dr. Buchanan states in his preface to this report to the Local Government Board that this is a starting point for fresh observation and experiment, not only by the Medical Department, but by all who have opportunity of investigating the new and promising fields of research opened up by the Hendon experiences.

Into this domain of etiology, I, although but a humble observer, have been irresistibly drawn by the ever-increasing interest and importance of the subject, and if I begin to see the close relationship of pathogenic bacteria to vitiated cattle life, I shall not rightly be accused of being other than a contributor to the agriculturist's best interests, and the health of urban populations. I think no thoughtful person can infer that there is no risk in the food of cattle which comes directly into their animal economies from crude sewage, without purification, either by land-filtration or chemical means, or by both. An opinion prevails that infection of milk cannot take place until "after it has left the cow;" and I have met with more than one sanitary official who shares such an opinion. I think I have sufficiently proved the erroneousness of this idea.

Since this report was published much new ground has been opened up in bacteriology. We know more now about the rapid growth of bacteria in sewage; that fresh sewage in which not more than a few bacteria could be observed, even in so small a quantity as a cubic centimetre—that is, about as much as the 1-6th of a cubic half-inch—has been proved to contain, after a few days' exposure to air and sun, no less a number than a million and a half forms. Absolute immunity from danger of infection cannot be expected under such circumstances, although we are yet in ignorance as to the specific form of inoculation of this or that disease.

The official report of the Local Government Board, 1886-7, deals with diphtheria, especially in its relation to the lower animals; but as its conclusions follow the bearings of milk-scarlatina cases, there is no need to quote from the report, deeply interesting as it is. All point to this issue, that these diseases creep into the cattle somehow apart from human contact, and from the cattle into their milk, and if the present state of etiology is at fault in not enabling us to prove their cause and origin, it does but announce the beginning of another exploration field, by showing that our knowledge is but yet incomplete.

One of the most instructive occurrences of milk-typhoid is described in considerable detail by Dr. Shirley Murphy in his report to the Local Government Board of enteric fever in St. Albans, in 1884. I am indebted to Mr. R. Thorne-Thorne for a copy of it.

The report is too long to be wholly transcribed, but I am anxious to give its salient points, with a twofold intention, namely, that milk, at times, does receive infective properties from other sources than human, and also to show how mysterious and difficult it is to trace the origin of the danger. It would seem, in fact, that it is easier to show from what the infection does not come rather than to define what is the real inceptive cause. I will freely quote from the report which relates to an outbreak of enteric fever during the months of May and June in the city of St. Albans, and the area immediately surrounding it. The latter area contains about 60 houses, in eight of which cases of fever have occurred.

The outbreak has been investigated by Dr. Saunders, Medical Officer of Health to the Middlesex and Herts combined districts, who traced its origin to an infected milk-supply, received from a farm situated without the city boundary.

St. Albans has a population of 11,000, and had, in 1884, 2,500 inhabited houses. The bulk of the population are supplied with water from the deep wells of the Waterworks Co., and about 610 houses receive their water from private wells. The city is sewered; the sewers convey the sewage to a sewage-farm about two miles away. Enteric fever is not usually present at St. Albans. There have been only 14 deaths in ten years from typhoid, but in May, June, and July of 1884, 23 deaths occurred from typhoid out of a total of 131 cases. Of the houses attached to the sewers, 34, or 2·7 per cent, were attacked, and of those not so connected 48, or 3·7 per cent, were attacked by the fever. 3·9 per cent of the houses using the Company's water were attacked, and 1·1 of those using water from private wells. It was not thought, however, that the fever was in any way connected with water-supply, but from evidence described in the report in considerable and conclusive detail, Dr· Saunders connected the outbreak with the distribution of a particular milk-supply. Part of the infected milk was sent to London, and there caused enteric fever, fully corroborating the evidence at St. Albans. At the farm itself, both servants and visitors drinking the milk were attacked, but, strange to say, neither the farmer, his wife, nor family, were attacked.

The evidence given of the London distribution is very

interesting, traced as it was into the particular houses and streets where it had attacked the people.

It is curious that the milk from this same farm had caused a considerable outbreak of enteric fever both in St. Albans and London in the preceding year. This caused Mr. Murphy to give greater attention in examining the farm on the second outbreak. He failed to discover any signs of ill-health in the cows. They drank of water from two ponds, and from what was believed to be a spring; their food was mainly the grass of the farm, with mangold, chaff, meal, and cake during those parts of the winter when they were tied up.

There was no reason to believe that any persons employed on the farm, or their families, had previously suffered from enteric fever, and that therefore there was no reason to suppose that infective properties could have been obtained by the milk through association of the cows with any persons suffering from enteric fever.

They could not trace its origin to any want of cleanliness or precaution in washing the milk pails.

The well, about 100 feet deep, from which water was used for washing the vessels, was about 20 feet from the cess-pit. In 1883, the roots of a sycamore tree grew between both, and its roots were found to have penetrated the well-wall, and pollution in that year might have occurred. In 1884, a space was found beneath the floor of the privy containing water with a fæcal odour, but no indication could be found that any percolation into the well had occurred.

Dr. Shirley T. Murphy's most interesting report concludes with an excellent suggestion, and in a paragraph so important that I quote it *verbatim et literatim.*

"I have no data on which to base an opinion as to the nature of the conditions which gave to the milk distributed from this farm its power of producing enteric fever among its consumers. All that I was able to make out during my inquiry is stated above. But, for practical purposes, and especially seeing that an infective quality can reside in milk produced under circumstances of reasonable cleanliness and healthiness, it comes to be of importance that English people should adopt, as their invariable rule in the use of milk, the custom among many continental nations, of boiling all milk as soon as they receive it into their houses."

I think that this foregoing evidence is of considerable value in helping to prove the soundness of my thesis, that of the possibility, and I must now add the probability, of zymotic danger in fæcal and other decomposing matter being in constant contact with the surface of the ground, and upon crops which cattle have to eat, and of open sewage-carriers in fields to which they can drink, particularly the latter, for it is well known that cattle prefer to drink sewage-water rather than pure water. I attribute this to the abnormal quantity of chloride of sodium (common salt) it contains, no doubt from urinal addition.

This admitted difficulty of proving causation is surely an argument for far more extended and independent research.

I have received two letters, from which I will quote. One is from Dr. Thresh, Medical Officer of Health for the Chelmsford and Maldon Rural Sanitary Districts. He says: "I should strongly object to drinking milk from a cow drinking any kind of sewage; neither would I let a child of mine drink it. In my own district, although we have several sewage-farms, large and small, so far as I know no milch cows are kept thereon."

The other letter is from Dr. Sidney Barwise, Medical Officer of Health to the Derbyshire County Council:—

"My dear sir,—I strongly hold the opinion that the solids should be precipitated from sewage before it is poured upon the land. Dr. Tidy collected some cases where crude sewage was instrumental in spreading disease. There is a *prima facie* case in favour of raw sewage spreading entozoic diseases, and this alone is sufficient to necessitate precipitation. I also believe that, unless the land is exceptionally suitable, a sewage-farm will soon cease to do its work, and get choked, unless the solids are first precipitated; *e.g.*, the Burton farm.—Yours faithfully, SIDNEY BARWISE."

In a letter from one of the Government Medical Officers to Mr. Ernest Hart, editor of the *British Medical Journal*, the following remarks occur, and although they may at first sight be taken as somewhat modifying probabilities of danger, it is only fair on both sides that they should be quoted. They show at any rate by their confession of incomplete knowledge a datum line for research, and a fresh starting point for inquiry, and they in no way invalidate the possibilities I have advanced.

"Dear Mr. Hart,—I have been endeavouring to trace some such cases as Mr. Wardle refers to in his letter to you, and I have been making inquiries among friends who are likely to know of any such, but without much result. The position seems to be that whilst it cannot be dogmatically asserted that cows grazing on sewaged grass cannot in that manner contract illness, or produce milk which would convey disease, there is little or no reliable evidence on the point.

"The enclosed extracts, which I have made from some works where the matter is referred to, and also the accompanying copies of reports made for the Local Government Board by Dr. Simpson (the present Health Officer of Calcutta), in which there is some very indirect reference to the point, may be useful to your correspondent."

In the extracts alluded to in this letter one is on the spread of entozoic disease by the application of sewage to the land, in which, it is stated, that whilst there is as yet no positive evidence of the derival of tapeworm, round-worms, trichina, bilharzia, and distoma hepaticum from sheep fed on sewage irrigated land, there is an authenicated occurrence recorded of an epidemic.

An epidemic of "enterocolitis," due apparently to the presence of trichocephalus, occurred at Pierrefen (Var), amongst the inmates of an asylum. Between January, 1888, and March, 1889, there were 137 cases amongst the inmates (more than half), together with 17 employés. There was no epidemic outside the asylum. It was attributed to the watering of the gardens with sewage-water; the use of the vegetables was stopped, and the "illness ceased."

The following extract from "Sewage Treatment, Purification, and Utilisation," by J. W. Slater, F.R.S. (1888), will strongly support my syllogism:—

"Are the vegetables grown on sewage farms fit for human consumption, or for the food of cattle? On this subject very conflicting opinions prevail. Suspended excrement, healthy or diseased, as the case may be, comes in direct and constantly renewed contact with the roots and the stems of the grasses and other plants, and no matter how close upon perfection the drainage of the fields may have been carried, there it will adhere. Of this any candid person may convince himself by observation in such fields, or even by direct experiment. Let him, for instance, fit up a funnel loosely with a grass root, and pour sewage upon it. However well the filter acts, he will soon find the stems and the roots coated with an adhesive matter, concerning the nature of which no doubt can be entertained.

"If, then, such grass is cut from time to time, without any intermission of irrigation, and is given to cows, the fæcal matters in question are conveyed into the stomachs of these animals, and their tissues and secretions may become poisoned. That disease germs may pass into milk, and thus reach fresh victims, is an established fact.

"But we have to do not merely with sewage matters clinging to the outer surface of plants; the impurities penetrate also into the interior. Herr Lefeldt, in his report on the various systems of sewage treatment as pursued in this country, notices stems of grass from irrigation meadows full of unassimilated sewage matters (*Kloaken Stoffe*). Were the irrigation suspended for a sufficient time, these matters would doubtless be assimilated by the plants, which would then be perfectly harmless; but if the sewage flows on day by day, fresh excrementitious matter is absorbed as fast as—perhaps faster than—the former doses can be assimilated. Surely such grasses, or other plants placed in similar conditions, must be of very doubtful value as food, whether for man or beast."

I now come to the experiments carried out by Mr. Smee, jun., and published by him in a work entitled "Milk in Health and Disease." These experiments and their results have been met, I am compelled to say, with something very like the "conspiracy of silence." Certainly, so far as I can learn, no attempt has been made to prove them inaccurate, or to refute the author's conclusions. Two cows were set aside for experiment. The one, which we may call A, was fed on sewage irrigation grass; the other, B, on grass from an ordinary meadow. The milk obtained from each cow was kept separate and examined. It was found that the milk of A became not merely sour, but it putrified and stank much sooner than that of B. It was noticed that a favourite cat, exceedingly dainty in its tastes, entirely refused to lap the milk of A. The butter from A's milk became rapidly rancid, as compared with butter obtained from cows fed on ordinary pasturage. Cream from the milk of A required, in three successive lots, 1½ hours and 2½ hours to churn, and the butter was soft and smeary. Check-samples of cream from the cows fed on normal food required only 35 minutes, 1½ hour, and ¾ hour to churn, and the butter was firm.

So far, of course, this experiment is open to the objection that the bad quality of the milk and butter from A was due to some morbid condition in herself, rather than in her food. To meet this doubt Mr. Smee reversed the experiment, feeding B on sewage-grass and A on normal herbage. He also tried other cows. Still, the results reached were practically the same; the milk from every cow fed on sewage-grass was notably more prone to putrescence than that from cows fed on common meadow grass.

Mr. Smee made further experiments on the grass itself. He found that the juice of sewage-grass became more quickly and more offensively putrid than that of ordinary grass. Hay made from sewage-grass, if kept in a vessel of water in a warm place, quickly set up a putrid fermentation, whilst hay from ordinary grass, treated in the same manner, behaved quite differently.

I have not the slightest doubt of both the possibility and the probability of danger to cattle feeding on pastures irrigated by crude sewage, urban or suburban. I think it would be most interesting for experiments to be instituted to see if the bacteria of zymotic disease, and their germs, particularly of typhoid fever, could not be traced, say from human typhoid dejecta, and from cultures from this source, through the cow's stomach, and also, by inoculation of the cow, into the milk, and from there to such animals as the guinea-pig, rat, mouse, cat, or dog, and so producing and perpetuating typhoid disease. I will conclude my remarks on this important and abstruse subject by referring to the reports of Dr. Simpson, mentioned in the above letter.

The Wellington report on the prevalence of diphtheria:—

"A sewer running from south to north discharges the main drainage of Wellington into a stream that flows into the river Tone. As, however, this main sewer is insufficient in size, an intercepting sewer takes the overflow into an open ditch which lies some little distance east of the houses in the lower part of North Street, and into which part of the sewage from the work-house is also emptied. The sewage of the west portion of the town is discharged into the Westford stream; while that of the east portion, where the houses lie too low to be connected with the main, is discharged into an open ditch, which empties tself into a pool, called Mitchell's Pool. Here the solids are collected, and the pool is emptied once or twice a year, as occasion requires, and then the sediment is heaped up on the banks, along with other manure from the cowsheds close by. The pool is a most offensive nuisance; cases of disease of various kinds, but spoken of by the local doctors as resulting from 'blood poisoning,' are constantly occurring in the neighbourhood.

"There are two dairy-farms, supplying milk to the inhabitants of Wellington, that require special mention. One is at Mitchell's Pool, where the dairy, which adjoins the house, has running alongside of it the ditch that receives, a few yards further on, a part of the sewage of the town, and, in addition to this, the window of the dairy opens on the ditch at a place where the contents of an offensive privy run into it. The water-supply of the farm is an open spring, exposed to the surface drainage, that runs into it in rainy weather. The cows are kept in sheds close to the sewage pool. Altogether the surroundings are extremely filthy.

"Some facts in regard to disease in the lower animals may be noted. In 1883, at a farm where there have been several outbreaks of throat-disease in the family, a heifer, drinking the same water as that used by the family, became affected with a throat-affection and died; also a fine sheep-dog. In another farm a dog, a horse, and a cow died of a throat-affection, which was attributed to drinking water from a dirty pond. In 1881 there was a prevalence of throat-disease among horses, causing death in some cases."

"Along with Dr. Prideaux, I examined with the microscope, using a high power, several specimens of these vegetable growths, and in some cases discerned, amongst the various moulds, numerous bacteria and micrococci. In a special instance, in a house where a death from diphtheria had recently occurred, we found a cut potato, on which various moulds were growing. Examining these microscopically, micrococci were discovered, growing in regular form on the potato, which appeared to be a suitable soil for their cultivation. They were tested by various reagents, and coloured to ensure that we were not dealing with inorganic bodies, which might have been mistaken for ordinary fungi." I mention these interesting facts only because they may be worth working out.

Dr. Simpson's report on the prevalence of diphtheria in the urban and rural districts of Shaftesbury:—

"In the centre of the town the principal streets are drained by sewers, partly of brick and partly of pipe, which take the sewage into the lowlands on the south side of the hill. Formerly the sewage was used for purposes of irrigation on a large pasture field, called Holyrood-Mead, but owing to fears being entertained that the milk from the cows feeding on this land was injurious, the irrigation was discontinued, and the sewage was carried across Holyrood-Mead in a pipe-drain, and discharged into an open receptacle some distance from the town; it then flowed along an open rivulet that debouched into a brook, called Whitings-Water. At present, from want of repair, the brickwork that formed the sides of the outfall is completely swept away, and some of the bricks and pipes are found a considerable distance down the stream. The land close by is sunk and fissured by the action of the outflow, so that the sewage now finds its way by two or three courses into the rivulet, whose bed for some distance is choked up with solid sewage matter and drift. This condition is very offensive.

"Diphtheria, introduced from Shaftesbury in January, 1885, caused, in that month, three deaths in Enmore Green. A village in a more filthy state can scarcely be met with, and certainly not one where the soil-conditions are better adapted for the spread of such specific disease as can spread by means of excrements. The milk of the cows that drink the sewage water is sent to towns on the south coast. At the time of my visit one of the cows was suffering from a large abscess in the udder. The farmer says his cows are often ill, which he attributes to the foul water they have to drink. The farm and dairy were extremely filthy. Here, as in the other parts of the country I have visited, the Milk Shops Order of 1879 is a dead letter, and will be so until some systematic inspection of the sanitary condition of the farms is instituted.

"Whitehouse Farm, in the parish of Motcombe, has a well

with a stagnant pond within 20 feet of it; both well and pond have the drainage of the farmyard going into them. The milk goes to London."

In the parish of Cann there is a dairy farm, of the arrangements of which a plan is given. Dr. Simpson states that "its extremely insanitary condition is not only a danger to the neighbourhood, but also to the towns that are supplied with its produce."

I think these extracts show pretty conclusively that absolute danger and damage to the health of cattle fed under such conditions does occur, and that to suggest that milk contamination in all cases occurs after the milk has left the cow, is to narrow the conclusions far beyond reasonable conjecture, the balance of evidence being very much the other way, notwithstanding the difficulty of proof absolute.

The following remarks will sufficiently substantiate my thesis:—

I have received a letter from Dr. Thorne-Thorne, of the Local Government Board, informing me that the subject has been for some time, and is still, a matter of inquiry in the Medical Department of the Local Government Board, and he further states, "You will doubtless have gathered, from the reports which I have already sent you, that in the case of a disease such as scarlatina the infection of the milk is not exclusively brought about after it has left the cow."

This is surely sufficient to prove even more than I have stated, and I will add, that if it be so with scarlatina, is it not possible that it may occur also in any, or even all, the zymotic diseases; and that if there is at present absence of proof positive, there is the strongest circumstantial evidence, and we need not wait for its actual demonstration before applying to the utmost preventive measures against the use of polluted food and drink for cattle, of which, in my opinion, crude sewage is the most filthy and the most dangerous form.

Tuberculosis in Cattle.

The danger of the contraction of tuberculosis by persons eating the flesh and by drinking milk of tuberculous cows has received important confirmation, as shown by a meeting of the North-Western Branch of the Society of Medical Officers of Health in Manchester, in May, 1892. I quote the following report of the meeting from the *Manchester Guardian*:—

Dr. Jasper Anderson (Blackpool) read a paper on "Tuberculosis in Meat and its Exclusion from the Meat Market." He said the question was whether the use of meat from tuberculous animals was dangerous. It had been said that the danger had been greatly exaggerated. In the first place, they must find out whether such a danger existed, because if it did it could not be exaggerated. If it was proved that such a danger existed, they, as guardians of the public health, were pledged to protect the public from such danger by making use of the powers they at present possessed. Dr. Anderson quoted the results of a number of experiments made by competent observers, and said those results were sufficient to lead to the conclusion that meat from tuberculous animals should not be allowed to be sold in its raw condition. With regard to the effect of cooking, it was said that cooking removed the source of danger, but experiments showed that that was not the case. The disease had been communicated to animals through the medium of the juice of an underdone steak, and by meat which had been boiled for a quarter of an hour. The question of cooking, however, was beside the point. Meat was exposed for sale in its raw state, and there was no power to ensure that it should be thoroughly cooked before consumption. The majority of people liked roast-meat best when it was underdone, and when all fear of infection would not have been destroyed. It was also not unknown for people to eat portions of meat raw, and this was sometimes ordered by physicians in the cases of people—and especially of children—who were in such a state of health as to predispose them to contract tuberculosis. It had been objected that there was no evidence that man had contracted tuberculosis from eating the flesh of tuberculous animals. In the case of milk, however, there was abundant evidence. He urged that action should be taken to prevent the development of tuberculosis amongst cattle, and said it seemed more essential to him, from a sanitary point of view, and as opposed to the strictly economic side of the question, that tuberculosis should be included in the Contagious Diseases (Animals) Act than pleuro-pneumonia.

In the course of discussion the Chairman said they were all agreed that the milk of tuberculous cows was highly dangerous. He did not altogether agree in the inference that because such milk might be infectious, therefore meat must also be infectious. It became a matter of educating the public to see the great danger of using such meat. At present public opinion was not sufficiently strong to support a medical officer in seizing the carcase of a cow whose organs alone were infected. The public must be educated to feel that this was a highly infectious disease, but that, deadly as it was, it could be prevented by stopping it at its source.

Dr. Paget (Salford), Dr. Vacher (Birkenhead) and other members of the society, took part in the discussion, and on the motion of Dr. Anderson, seconded by Dr. Paget, the following resolution was unanimously agreed to:—

"That the flesh of any animal affected with tuberculosis, to however slight an extent, is, in the opinion of this branch, unfit to be sold for the food of man."

In February last an important deputation from the London County Council waited upon the President of the Board of Agriculture, to ask for powers to slaughter cows affected with tuberculosis, and to give compensation to the owners out of imperial funds.

Dr. Bott, the Chairman of the Public Control Committee of the Council, in introducing the subject, said that their inquiry as to the London cowsheds had brought out the fact that there were many animals actually diseased, and that the milk of these animals during their life, and their flesh after they were slaughtered, were sold as human food. Recently twelve cows had been sent up to a London slaughter-house to be killed, and it was found that their lungs were one mass of tubercle. It was well known that there were certain slaughter-houses in London where animals of this description formed a large proportion of those slaughtered. It was impossible to dress the carcases of these twelve animals in the ordinary way, and portions of the flesh were either salted to disguise its condition, or stripped from the bones and made up into what was technically known as "block ornaments" for sale in poor districts. To what extent the flesh of tuberculous animals was inimical to human life remained perhaps to be determined by the Royal Commission on Tuberculosis now sitting, but the question appeared to the Council to be of such importance and of considerable danger to human life, that they trusted the report would not be much longer delayed. The Council had done everything to improve the sanitary surroundings of London cowsheds, but still the number of cows affected with tuberculosis was very great. Some recent experiments made by the Board of Agriculture had shown that *this disease was transmitted to human beings through the flesh and the milk of infected animals*, that ordinary methods of cooking were insufficient to destroy the bacilli, and that the virus of the disease was communicated to the human being through milk. They believed that if the Board of Agriculture applied to tuberculosis preventive measures similar to those applied to pleuro-pneumonia a remedy would to a great extent be found. The compensation to the owner ought to be paid out of money voted by Parliament.

Mr. Gardner, in reply, said "he was glad to receive the deputation on this important and interesting subject. Any measure to safeguard the public health the Board of Agriculture was in perfect sympathy with, and they were anxious to render any assistance in their power; but all matters concerning the public health properly belonged to the Local Government Board. That was the central authority to which local authorities must look when they desired to prevent the sale of food, whether meat or milk, which they considered unfit for human consumption. Professor Brown was a member of the Royal Commission on Tuberculosis, who were to report on what was the effect, if any, of food derived from tuberculous animals on human health, and, if prejudicial, what were the circumstances and conditions with regard to tuberculosis in the animals which produce that effect on man. They hoped that the report of that Commission would soon be in their hands. Their researches and experiments had been of a very interesting character, and had been carried out in the most scientific and accurate manner. They had every reason to believe that the conclusions to which the Commission would arrive would prove of material assistance to various sanitary authorities in carrying out the provisions of the Local Government Act."

It is impossible to read this very serious account without recurring to the thoughts which have been the cause of this chapter on the Etiology of sewage being written, namely, the origin of bacterial diseases in the lower animals used for food, particularly in cows whose milk is so largely used for children's food. No doubt the occurrence and prevalence of tuberculosis in cows may have much to do with heredity, but may it not be reasonably assumed that the bacilli of tuberculosis occurring in such numbers in the sewage of towns, especially those towns whose trades have a tendency to induce and promote phthisis, may be—nay, must be—introduced into the bodies of cattle feeding upon land irrigated with crude sewage, both into the stomach and the lungs; and if the disease can be communicated to man through milk, it is clear that this secretion is not free from the danger of the passage of living tuberculous bacilli from the affected parts, and if so, the inference is clear enough that the great assimilating powers of the cow's stomach may not be always sufficient to destroy the bacilli eaten with the grass coated with sewage sediment, and soil suffused with bacteria of this and other pathogenic kinds. This, I think, is answer sufficient to those who content themselves with self-satisfying confidence that milk only receives disease powers "after it has left the cow," or that it is "impossible for a cow to be a conduit pipe in the transmission of disease."

The very fact of these lower fungi increasing in heated media in such astounding numbers, and in such a short time, makes it impossible to combat the thought that food charged with bacterial germs may and do pervade the flesh and milk of cows even for some time before the cows have fallen from a state which from all outward appearance could not be diagnosed as unhealthy.

It is to be hoped that the Royal Commission on Tuberculosis now sitting may have entered into this further field of research, and enable us to know with certainty the sources of this disease.

My friend, Lieut.-Colonel Surgeon Hendley, C.I.E., the chief Medical Officer of the Maharajah of Jeypore, who has the entire medical direction of this native state, has informed me that cows feed on the street-garbage and refuse in the native parts of the city of Jeypore, and the mortality from typhoid fever there is great, and is consequent on this unhealthy food, which first infects the cows and then naturally the milk, which, in his opinion, causes the disease amongst the people; and quite as strong testimony as this is the opinion of Dr. Klein, bacteriologist to the Local Government Board, in the following letter to me, dated May 12, 1892:—

"Dear Mr. Wardle,—With regard to your letter of April 30th, which has reached me this morning, I beg to say that I am distinctly of opinion that a good many cases of milk epidemic have been caused by milk which has *not been* contaminated after it had left the cow. If any Medical Officer of Health or Sanitary Inspector will take the trouble to inform themselves on the subject by looking at the various reports of the Medical Officer

of the Local Government Board, they will hold a less distinct opinion. That certain milk which has caused infection had been infected after it had left the cow does not, of course, prove that this is always so, and, as I said before, I am of the very decided opinion that in a good many of these epidemics no such contamination of the milk—after it had left the cow—had been traced. With reference to the other kindred subjects mentioned in your letter, I beg to state that the work which you have sketched out, is, of course, of the greatest importance, and I need hardly say has been for some years in my mind, but it would require many years' work to satisfactorily solve it, even supposing that the funds for such work would be forthcoming.—
Yours faithfully, E. KLEIN."

CHAPTER XXIV.

ON SEWAGE-FARMING. — SEWAGE POISON AND BACTERIAL DANGERS: THE LICHFIELD AND BURTON SYSTEMS OF SEWAGE-FARMING AND IRRIGATION.

IN support of the position which I have assigned to the extreme danger arising from the application to crops above ground of crude and unpurified sewage, and of the high importance of this branch of the subject of sewage treatment, I shall now proceed to adduce facts which will prove not only the risk, but also the loss of ratepayers' money which ensues from a too short-sighted initial parsimonious view of sewage-disposal, as well as from the adoption of imperfect methods.

In 1878, Lord Lichfield obtained a judgment against the Lichfield Corporation on account of defective sewage disposal, and an injunction was granted restraining the Mayor and Coporation from passing into the Curborough Brook (a stream running through two of Lord Lichfield's farms northward from Lichfield to the Trent) any sewage or foul water from the sewage farm of the city without first sufficiently purifying or deodorising it. Despite the injunction, the stream continued to be fouled, and cattle were injured by the pollution. Damages were awarded to the extent of £534 19s. 6d. to Lord Lichfield for loss of rent, and to Mr. Kean £245 for loss of stock. Mr. Stubbs, another tenant, received £100 compensation. Subsequently, in about the year 1890, the brook again became foul, and the tenant of one of the farms complained of losses in his stock in consequence of their drinking the water.

In December, 1890, some of Mr. Coxon's cattle on the Alrewas Hays' Farm, had access to the brook, and became ill; he called in Mr. Carless, an eminent veterinary surgeon, who found them suffering from enteric fever. One, in a hopeless state, he had killed, and on a *post-mortem* examination found the viscera perfectly healthy, except the mucous membrane, which was considerably thickened and congested. He was clearly of opinion that it was the water they drank from the brook that was the cause. Mr. Sperring, another veterinary surgeon, and five farmers (one of them a former tenant of the Alrewas Hays' Farm, Mr. Stubbs, the tenant of the Curborough Farm, and three others) reported to the same effect.

Lord Lichfield, by advice of counsel, applied for an order of sequestration against the property of the Corporation of Lichfield in that year.

The trial took place in 1891, before Mr. Justice Kekewich, sitting in the Chancery Division.

The evidence of various experts was taken, showing that cows and sheep died from drinking the water impregnated with sewage.

One tenant stated in evidence that he had lost ten times as many sheep as he had ever lost before on any previous holding.

Another tenant stated that he had lost by blood-poisoning a number of sheep and young beasts, some of which had access to the brook, and others had been fed on the hay grown on land irrigated with this water.

Dr. Hill stated he had analysed the mud, water, and fungi taken from the brook, and had found all three impregnated with sewage.

Several eminent veterinary surgeons and professors from the veterinary colleges at London and Edinburgh gave evidence to the effect that, in their opinion, the animals had died from sewage poison.

On the other side evidence was given of a contrary nature, contending that it was not proved that the animals had died of sewage-poison.

In a report given by Mr. Nevill, of Rugeley, land agent and surveyor to Lord Lichfield, after showing the deleterious nature of the polluted stream and the sewage-mud (no doubt the result of secondary decomposition), he states that the Lichfield Corporation subsequently diverted the point of discharge of the sewage from the Curborough Brook, sending it into the Elmhurst Brook and into a catch-carrier alongside the Curborough Brook, so that it came into the Curborough Brook at a point near the end of the Curborough Farm; and after the tenant had fenced off the intervening portion of the brook, when grazing with stock, he made no more complaint, but after a time the lower part, or that below where the Elmhurst Brook joins the Curborough Brook, to which point the sewage had been diverted, gradually became foul—the sticks and stones in the brook were covered with sewage-fungus, there was no clear bottom to be seen, and no vegetable life of any kind except the sewage-fungus. There was on the bottom and sides of the brook-course a deposit of black sewage mud, which on being disturbed, says Mr. Nevill, smelt as bad as a privy. Above the junction with the Elmhurst Brook the case was quite different; the water there was clear, the bottom of the brook bright, and aquatic weeds and plants were growing in it; but this ceased immediately below the junction, where the water became foully polluted with sewage, and was quite unfit for cattle of any description to drink, and that it would be specially injurious to young stock, and to beasts and horses which put their feet in the water when drinking, and so stir up the mud and drink it with the water. He also considered it dangerous for the purposes of irrigation, as the hay from land irrigated by such impure water is especially injurious to young stock.

In April, 1891, Mr. Justice Kekewich delivered judgment in the motion for sequestration, and said that under the terms of the injunction of 1878 he understood the Corporation to be held responsible for polluting the Curborough Brook with sewage, and they were to take care to free it from excrementitious matter. This he was satisfied they had not done. Even the defendants' witnesses were unable to say that the water was not fouled, polluted, or deteriorated.

He was not satisfied that Mr. Coxon's sheep had in any way suffered from the polluted water, nor was he satisfied that any of the animals which died from time to time, or were killed when on the point of death, died from drinking it. But he was quite satisfied that the cattle had suffered in their general health, and that Mr. Coxon had been deprived of pasturing them in the meadows abutting on the Curborough Brook. All healthy vegetation in the brook had disappeared, whilst it continued in the neighbouring brooks, and the evidence showed that a line was drawn where the water destroyed vegetation. He unhesitatingly came to the conclusion that the stream had been fouled, polluted, or deteriorated in the quality of its water by reason of sewage matter. He did not agree with the view put forward for the defendants that so many years had elapsed since the injunction was granted that it should not be enforced. A more plausible argument was, that Mr. Stubbs, the successor of the original complaining tenant, Mr. Keen, had, with Lord Lichfield's consent, agreed to refer the matter to arbitration, and compensation had been paid by the Corporation.

He assented to the proposition that a sequestration was an inconvenient remedy, but he could not hear the Corporation say that, having assented to an injunction for what they found difficult, the Court should be satisfied with their having done their best to provide a remedy. His duty was to see the order of the Court obeyed, and he must grant the sequestration, with costs. The order of the Court necessarily partook of a punitive character; but he held that was not Lord Lichfield's object; it was rather to preserve his property from deterioration, and if the Corporation could by any improved scientific means accomplish that, he had no doubt his lordship's object would be attained.

I am not sure what steps the Corporation have since taken to comply with Lord Lichfield's requirements, but I am informed by Mr. Heaton that the order for sequestration is still held in suspense, and meanwhile the Corporation have recently received, and he believes have settled, further claims on the part of the tenants.

Now, what is the moral to be naturally and according to common sense, derived from this trial? Is it that Lord Lichfield was intolerant, tyrannical, or selfish? is it that the Corporation of Lichfield were indifferent to healthy cattle-feeding, or to effective sanitary arrangements and sewage disposal? Not at all, for Lord Lichfield's character was as blameless and charitable as mortal man's could be; the Corporation of Lichfield had proved their desire to deal with the sewage of their city by having, perhaps too early in sewage-disposal days, gone to the expense of a sewage-farm, and naturally were desirous not to change their bad system for another problematical or costly one. No doubt their fear was not to burden the rates unduly, and so they stayed their hand, and by so doing burdened the rates with legal costs, and costs of compensation, with sequestration hanging over them. I believe that this healthy desire to keep down rates has in most places in England been cultivated to excess, and the effect has been in inverse ratio to the approach to parsimony and narrowness of spirit, as well as to lack of information on the subject.

Sewage-Farming at Burton.

I gather the following particulars from the Burton press reports of a Town Council meeting held at Burton, on the 6th of January last, and more particularly from the speech of Mr. Councillor C. Graham on that occasion. Councillor Harrison proposed that the great and exhaustive scheme of the Highways and Sewers Committee, to re-construct a large portion of the borough sewers, be adopted.

Councillor Graham, in a long and very interesting speech, showed that the sewage treatment and disposal hitherto adopted, that of sewage farming, had been on wrong lines, and that it had cost the Corporation £170,000 already; that there had been an increased cost during the last nine years of £7,000 a year, and this for a population of only 50,000. It is now found that the sewage scheme is not a complete one. Mr. Councillor Graham said the sewage-farm was a failure, and that it was no use to proceed on the old lines, but that, acting on the experience and the lessons taught them, they should endeavour to find out some new method by which the whole of the borough could be relieved of present difficulty. There was the refuse from the breweries to be dealt with in one way or another. The condition of the western portion of the borough was unsatisfactory and unhealthy, as shown by the recent epidemic of typhoid fever. The sewage lay stagnant in the pipes, two epidemics of typhoid fever had at intervals occurred, the cause of which was attributed by some to bad sewage, and by others to milk supply.

The rates were rapidly increasing, 10d. in the pound having been added to them during the past twelve months. The balances are in as low a condition as they ever had been since the town had been incorporated, and the slightest attempt to increase the expenses for anything of a permanent description will involve an increase in the rate to 5d. in the pound. The elaborate proposition (said Councillor Graham) to remove from the sewers the sewage of a particular portion of the borough, and then to pump it upon a swamp, would be a failure. Their whole system of farming the sewage was a failure, and was the task of Sisyphus rolling a stone uphill that was too strong for him to move. They were liable to injunctions and complaints from all quarters from the farm nuisance which was a menace to the whole locality, and that it was time to give up temporary measures and resort to something practical, instead of dealing with the refuse on a wrong principle. He said they were fast going to ruin, and were fast approaching bankruptcy, and with such rapidly increasing rates, three things would happen: Higher prices would have to be asked for the commodity for which Burton exists; higher cottage rents, and higher wages to pay for the advanced rents, thus making competition with outside breweries, now very keen and successful, quite impossible.

The Highways and Sewers Committee now asked for £25,270

more, which they proposed to spend on the reconstruction of sewers for 453 acres, leaving 1,568 acres of sewers untouched, and which must later on be dealt with. Thus no less a sum than £25,750 was asked for to be spent on improving only one-third of the borough.

He prophesied some years ago that the sewage works would eventually cost £200,000, and that when that sum had been expended the sewage farm would not be a success; and so it had happened, and he moved and carried the following amendment to the proposition of the Highways and Sewers Committee:— "That in the opinion of this Council no further expense should be incurred in providing sewage-works until a sub-committee has been appointed, and has conferred with the brewers of the town, and laid before them the failure of our present system; that they may devise, if possible, some scheme whereby brewers' refuse may be disposed of, either at the place where it is produced, or by means of separate services if continued to be sent on to the sewage-farm, and that the sub-committee report the result of their interview to the Council."

The nuisances arising from the Burton Sewage-Farm have at length become too great to be borne, and two actions against the Burton Corporation have been recently successfully contested, as the following interesting account, taken from the *Burton Chronicle* of March 23rd of the present year, will show:—

At the Derbyshire Assizes, on Thursday, the first cases down for hearing were two actions brought against the Mayor and Corporation of Burton-on Trent, in respect to alleged nuisances arising from their sewage farm at Eggington. The actions were tried by a special jury. The plaintiff in the first case was Mr. R. L. Homer Mole, solicitor, of Derby. The cause of action was the alleged improper working of the sewage farm, whereby the noxious vapours and foul smells had issued, and done damage to the property of the plaintiff.

Mr. Stanger, in opening the case, said the Corporation of

Burton-on-Trent in 1880, acting under statutory powers, purchased some land, to the extent of 384 acres, mostly in the parish of Eggington, a portion being in the parish of Etwall, and from that time to this they had carried the sewage of Burton on to the land.

In order to establish a nuisance it was necessary to show that there had been a sensible interference with a person's enjoyment. So far as the exercise of the statutory powers was concerned, if in carrying out those powers it would be impossible to avoid a nuisance—really impossible—it would be a defence. If it was possible to carry them out without creating a nuisance then the statute was no defence.

The plaintiffs said that the statute gave them no permission to carry on their farm in such a manner as to be a nuisance to the surrounding neighbourhood. Mr. Mole was a solicitor, practising in Derby, and in March, 1891, he took a house at Etwall, about six miles from Derby, where he had since resided with his family. Almost continuously during that time they had been annoyed by the abominable stench which had come from the farm, the nearest point of which was within three hundred yards of Mr. Mole's house. The sewage of Burton was made very much worse by a practice which plaintiff said the Corporation ought to have stopped, the practice of permitting the refuse from the breweries to mix with the sewage, and also permitting liquids of too great a heat to enter into the sewage, and be carried on to the farm. The consequence was that very noxious gases were produced, and a very large quantity of sulphuretted hydrogen was caused in the air. There were many ways by which the sewage could be treated, and the nuisance reduced to a minimum. One very important matter was to separate the solids from the liquids, and another was to treat the sewage by chemical process, and thus deodorise it; but the Burton Corporation had not adopted any proper method of treating the sewage, which was poured upon the land to the amount of six million gallons a day. They allowed the sewage to flow upon the land continuously, and the land was thus given no rest, there being no system of intermittent filtration. The effect was that for miles—certainly for a considerable distance past Mr. Mole's house—the stench from the farm was perfectly sickening and abominable. Mr. Mole, when he went to live at Etwall, was assured that methods were being adopted to reduce the nuisance to a minimum, and that the Corporation would perform their promises; but this was not done, and with the exception of a few days, or sometimes a few weeks, the nuisance went on continuously, annoying himself and his family. The house was sometimes full of the horrible stench, and they were even awakened in the night by it. They could not get rid of it, as, of course, opening the windows only made it worse. It did not cause any serious illness, but they were affected in health, and plaintiff's daughters had to go away.

They consulted Dr. Alfred Hill, who reported that the cause of the stench was the solids remaining in the sewage, and the high temperature from the steam and hot liquids. He recommended as a remedy the construction of tanks, and the proper chemical treatment of the sewage. In October, 1889, the Rev. W. M. Furneaux, head master of Repton School, wrote complaining of the "intolerable nuisance" caused by the farm. The Town Clerk replied that every precaution was being taken, and experiments were being made. In 1889 the Corporation employed Professor Church to report on the farm. His report referred to the prevalence of sulphuretted hydrogen, "that most pestilential of gases," and albuminoid matter. He recommended means for deodorising and precipitating the sewage by the erection of tanks, &c. A number of other complaints were read by the learned counsel, the reply almost invariably being that the Corporation were making every effort to avoid any nuisance, and that they were making experiments. A report was made by Mr. Giles, the manager of the sewage farm, recommending the precipitation tanks, &c., and his recommendations had not been adopted. Mr. Swindlehurst, the borough surveyor, also reported on the sewerage of Burton, strongly condemning it. The Corporation had also obtained recommendations from Mr. Mansergh and Sir Frederick Bramwell, but had not carried them out. In October, 1891, the Derbyshire County Council took the matter up, and made a complaint; in reply to which they had an assurance that the Corporation were making experiments, and that they had already expended £145,000 on the farm.

Mr. Mole, the plaintift, was called, and said the smell was not very perceptible at first, but in the summer and autumn of 1891 it became so bad as to cause him great inconvenience and sometimes illness. It prevented him sleeping at night, and caused sickness, headache, and sore throat. His daughters' health had also been affected, and he had to send them away. In July, 1892, in company with Dr. Barwise, the medical officer of health for the county, and Dr. Thresh, he visited the farm, the action having been then commenced. In one place there was a large pool of sewage, an acre or more in extent, and a foot to eighteen inches deep. The land in other parts seemed to be thoroughly saturated with sewage. Over a considerable portion of the area there was no sign of cultivation, and a lot of refuse, including hops, was lying about.

Mrs. Mole, wife of the plaintiff, gave evidence to the same effect, and said they had frequently to close the windows on the hottest summer days. A clergyman who had come to stay with them had to go away on account of the smell. The copper utensils were discoloured, and Dr. Barwise said it was caused by the smell.

Two daughters of the plaintiff also spoke to the badness of the smell, and its injurious effect upon their health.

Dr. Thresh, of Chelmsford, medical officer of health to the County Council of Essex, and an examiner at South Kensington, describing a visit to the farm, said the jury might imagine what the smell was like if they would take a bottle of sour beer, put a rotten egg into it, shake it up, and pour it on to a dish. (Laughter.)

His Lordship: Prescription for a stink. (Laughter.)

Witness, proceeding, said the sewage smelt strongly of sulphuretted hydrogen, which was very unhealthy. Witness went on to describe in detail the condition of the farm. If proper means had been taken, all the evils complained of might have been prevented. The main causes of the nuisance were the high temperature of the sewage, and the existence in it of brewery refuse, particularly yeast. Unless the solid matter was separated from the sewage, it would be impossible to avoid the nuisance. It was merely a question of expense.

Mr. Barwise, medical officer for Derbyshire, gave similar evidence, and said he had always thought the smell could not be described until he heard Dr. Thresh. (Laughter.) He described how the brewers at Stratford-on-Avon had adopted a system of dealing with hop-leaves on their premises. If this were done at Burton, it would considerably diminish the nuisance. He had seen large areas of the farm at Eggington covered with hop-leaves, which killed the vegetation. Last year there was very little vegetation on the land, but now there were signs of a good deal, because no sewage had been coming up. (Laughter.) He believed it had gone direct into the Trent at Stretton.

The Rev. Edward Price, Vicar of Braceby, Leicestershire, said he went in September, 1891, to stay with Mr. Mole, and on the second day he suddenly felt a wave of smell, which entered his mouth, and was so abominable that he turned aside and was sick for twenty minutes. He never smelt anything like it before. He wrote a letter to say he was going home next day.

Dr. Trinder, of Tutbury, son-in-law of plaintiff, said the smell was that of sulphuretted hydrogen, and he had noticed the ill-effects of it upon Mr. Mole and his family.

H. F. Longden a former neighbour of the plaintiff, said he

left Etwall in September, 1892, because he could not stand the smell.

T. H. F. Measham, of Etwall, landlord of the house occupied by the last witness, said he had not since been able to let it.

Thos. Cooper, signalman on the North Staffordshire Railway, living in a cottage close to the farm, said in 1890 he was in bed six months ill in consequence of the stench. On one occasion he struck a match over one of the manholes, and the gas went up like a flash of lightning. The stink had turned his watch black, and coloured a copper warming pan in his house like a rainbow.

Mr. Dugdale, for the defence, pointed out that the defendants were a public body, who were under statutory obligations to deal with the sewage. They were doing so under their statutory powers, and no person could proceed against them for a nuisance caused in the exercise of those powers unless it could be proved that they used those powers negligently.

They had done the best they could, and taken the best advice. The sewage-farm was laid out by Mr. Mansergh, the eminent engineer, and it was expected that the same results would follow that were found in other similar cases. It got to work in 1885, and then this filming up on the soil was perceived, and they found themselves in a difficulty. They had tried various methods of mitigating the nuisance. In 1889 they consulted Professor Church, and on their instructions Messrs. Matthews and Lott, analytical chemists, tried several experiments suggested by him, but they were a failure, as were some experiments made by Messrs. Matthews and Lott on their own account. The same year they tried the system of the Rivers Purification Company, in operation at Halifax, but that was a failure, and the compression system used at Wolverhampton was also tried in vain. Other experiments had been attempted without success. A precipitation scheme was under consideration, and the Corporation were also about to spend £42,000 on re-constructing the sewers in the lower part of the town. Could the jury, under these circumstances, come to the conclusion that the Corporation had been negligent in carrying out their powers. On the contrary, the evidence would show that they had done everything possible to carry out the duties cast upon them.

His Lordship, in summing up, said there could be no doubt about the difference which the planting of this farm must have made in the neighbourhood. The frightful stench which was caused had not been denied. It could not be said that the action was premature; on the contrary, he thought the inhabitants of the neighbourhood had been remarkably patient. He could hardly understand how the claim could be resisted—possibly it was in the hope of gaining time. The Corporation had been for years considering the matter, and it was evidently all they did, instead of taking effective measures, and going to reasonable expense for the purpose of getting rid of the sewage. There was no defence to the action in point of law, and he should be surprised if the jury found one in point of fact.

The jury returned a verdict for plaintiff for £300.

His Lordship gave judgment.

Mr. Stanger: Your lordship will grant an injunction?

His Lordship: Oh, certainly.

Mr. Dugdale: But suspend the injunction for a reasonable time?

His Lordship: A very limited time.

Mr. Dugdale: We are dependent upon the sanction of the Local Government Board.

His Lordship: I cannot make it depend upon the sanction of the Local Government Board. There has been too much delay already. I will suspend it till the 1st of July.

Mr. Dugdale: With liberty to apply?

His Lordship: Liberty to apply as much as you like (laughter), but not to me. I consider it a very long time indeed.

At the Derbyshire Winter Assizes, on the following day, the Court proceeded with the hearing of the second claim against the Mayor and Corporation of Burton-on-Trent for nuisance caused by the "foul smells and noxious vapours" proceeding from their sewage farm at Eggington. The plaintiff in the second case was the Rev. Rowland German Buckston, vicar of Sutton-on-the-Hill. Counsel engaged were the same as on Thursday.

Mr. Stanger said he had hoped that after the exhaustive trial of the previous day it would not have been necessary to litigate on the present action, and that the Burton Corporation would see their way to recognise the claim of Mr. Buckston and come to a settlement. It was, therefore, necessary to prove the case, but he should be able to do it much more shortly, as he understood that the Corporation were not now going to dispute that the sewage farm was a nuisance, and therefore they would be liable to all persons who would be affected by it.

Plaintiff's house was four miles from the farm, but the smell was of such a peculiarly deadly and penetrating character that it was perceived as much as nine miles away. Ever since the sewage farm came into operation plaintiff had been annoyed by the fumes, sometimes for several days consecutively. It was not claimed that it made him ill, but it caused discomfort in a high degree.

The plaintiff, being called, said the vicarage and ground at Sutton belonged to him, and he owned other property in the neighbourhood. He had lived there over fifty years. His house was about four miles from the farm in a straight line, but the exceedingly offensive smell was so strong that they often had to shut the doors and windows, and then could not keep it out. It also penetrated the church. They had not perceived the smell so much during the last winter.

Miss Buckston, daughter of the plaintiff, corroborated his evidence. Last year they had to put a stop to a tea party in the garden on account of the smell.

Francis Perret, plaintiff's butler, said the smells gave him headaches; and Arthur Shambrock, head gardener, also spoke to their offensiveness.

Henry Bridges, manager of a cheese factory at Sutton, said the smells made his children sickly.

Frederick Newham, farmer, Sutton, said the smells were very bad at his house, a quarter of a mile further than the vicarage from the farm.

His Lordship said that if a man's servants were ill or inconvenienced by the smell—whereby they might not stay, or it might be difficult to get servants—it was a loss to the master. If it interfered with his receiving visitors, he was also entitled to damages on those grounds. He put it to the jury whether plaintiff had sustained substantial injury—or he preferred to put it, substantial inconvenience—by the smells from the sewage. If so, to what damage was he entitled?

The jury found for the plaintiff, damages £200.

His Lordship certified for a special jury, and granted an injunction similar to that in the previous case, suspended till July 1st.

Paper mills are the greatest polluters of sewers and of rivers; woollen scouring the next, as evidenced at Nuneaton (Chap. XVIII.), and silk manufactories the least. The total nitrogenous or organic impurity is shown by Mr. Santo Crimp to be as follows:—

	Parts per million.	
	Organic Nitrogen.	Organic Carbon.
Paper mills	780·00	93990·00
Dye, print, and bleach works	3·99	42·26
Tanneries	96·40	700·47
Wool scouring	548·59	1324·80
Silk works	1·74	14·89

Brewers' refuse wash is very contaminating from its easily decomposable albuminoid matters. I have found my basic persulphate of iron or ozonine process the best purifier and preventive of such decomposition.

I am informed that the Stoke-on-Trent Sewage Farm is very unsatisfactory; that it is a nuisance, and cannot be farmed at a profit.

At Burslem, where the crude sewage is pumped to a plateau with a soil of almost brick-clay, the crops are rank and the surface suffused with putrescible matter, and support the objections to and danger of the application of crude sewage to land, which the foregoing instances so eloquently affirm.

Great as I believe the danger to be from cows feeding upon herbage coated with fæcal matter, I believe the risk to be much greater from the germs of pathogenic bacteria being carried about with every wind from such decomposing matter as it dries, and inhaled into the system of both man and beast.

What says the great advocate of sewage farming, the late Dr. Carpenter, in a paper written so long ago as 1877, on sewage application to cropping? He said: "Such a system of irrigation which caused sewage to cling to the stalks of grass and other crops out of the reach of the rootlets themselves, could not belong to properly managed sewage-farms."

"Some portions of a field improperly laid out may so suffer, but a field that is badly laid out ought not to be used for sewage-farming at all; and sewage ought not to be allowed to go on the ground so abundantly as to flood the crop above the rootlets, unless the crop is young, and some time is to elapse before it is taken from the land, in which case no harm will arise.

"Crops nearly ready for gathering ought not to be, and never are, sewaged on proper sewage-farms for some days previous to cutting. Thus the agriculturist who applies sewage to his crop of rye-grass, or anything else, shortly before it is ready for cutting, is acting wrongly, and disregarding a well-known axiom of sewage-farming, and is not only setting up the chance of a possible danger, but is also losing some of his financial return."

And leading up to the point in discussion, namely, that of

danger of infection of zymotic disease by surface irrigation with crude sewage, it is well also to show that not only is there a danger by actual contact of sewage with cattle food, but that the food itself becomes unhealthy. The late Dr. Tidy, an eminent chemist, and also a medical man, puts these points before us with unswerving force. I will give a few quotations which will show that I have neither over-rated nor over-stated the dangers I have adverted to.

In his paper No. 1,768, vol. xxxiv., Society of Arts Journal, he states that he admits to the full the power of soil to purify sewage by oxidation, if there is intermittency of application; he doubts the efficacy of short rests or intervals, or of the soil remaining in contact with the glutinous contents of the sewage whereby aeration is prevented. He also quotes from what he describes to be an excellent paper by Dr. John M. H. Munro, read before the Society of Chemical Industry in 1885. Speaking of and against broad irrigation, he says: "There are several causes for the disappointing want of success of broad irrigation, but the chief among them is the inability of arable soil to deal with a continuous supply of sewage without great deterioration in its purifying and aerating properties, so that at least it becomes sewage-sick and inefficient as a purifying agent. This is not completely remedied by intermittent filtration or irrigation, for the land which has been often covered with crude sewage never recovers, even by rest, its original efficacy. The commonly accepted, and I believe the true, explanation of this is that the slimy suspended matter of the sewage gradually chokes the pores of the soil, forming a deposit impervious to air, and thus preventing the aeration essential to nitrification, and, on the other hand, encouraging putrefactive fermentation. Difficulties of a practical nature crowd upon us in considering this method of treatment. Three, at any rate, may be noted:—

1. The cost of preparing the land.
2. The difficulty of securing proper land, or of ensuring its

effective working at all times, in all weathers, with all kinds of sewage, and under all circumstances.

3. The fact that much of the solid filth of the sewage will, unless previously removed, accumulate on the surface, where it undergoes decomposition, and becomes, especially in hot weather, a formidable nuisance.

We now turn to the hygienic aspect of sewage irrigation. I shall speak of three classes of effects rendering sewage irrigation dangerous to the public health:—

1. Offensive and injurious emanations.
2. Pollution of subsoil water.
3. Distribution of undefecated sewage, containing the ova of entozoa" (and he might, and would now have added, bacterial pollution).

Of offensive and injurious emanations, Dr. Tidy states: "Of such emanations the evidence is ample." (Flower on Sewage Disposal, p. 13, *re* Aldershot Farm.) The River Pollution Commissioners (first report, p. 87) admit that odours do arise with land irrigated with sewage, day after day, for years. The Crengentinny meadows, near Edinburgh, can only be described as filthy, emitting a stink hardly endurable. The surgeon to the barrack adjoining the meadows described the stench (1868) as "sickening." Of the Croydon sewage-farm at Beddington, Dr. Creasy, Surgeon to the Orphan Asylum at Beddington, stated that "typhoid fever had been in every cottage on the estate," every disease, in fact, assuming a particular type, accompanied by what is called "a sewage tongue." In fact, the stink of sewage-irrigated ground and the malarious effects of the sewer gases evolved are matters of frequent complaint and litigation. Dr. Clouston traced an outbreak of dysentery in the Cumberland and Westmoreland Asylum (where there were 200 patients) to the effluvia from a sewage-farm (*Medical Times and Gazette*, June, 1865). The dysentery appeared soon after the sewage was used, 31 being attacked, of which 20 cases proved fatal. The farm was 300 yards from the female ward,

23

where the greater number of cases occurred. The sewage-farm was removed, and the dysentery disappeared. The following year some land near the asylum was again irrigated, when another outbreak of this disease occurred. The facts show incontrovertibly that many emanations may be and are the cause of dysentery, diarrhœa, and typhoid fever, and again the sanitarian recognises the importance of defecating the excreta of the typhoid patient as soon as evacuated, and of removing it from the sick room without delay. And why? To prevent the morbid materials being carried into the air during the evaporation of the liquor portion. It must, therefore, be an unscientific method to spread the sewage of a mixed population over the land, thereby increasing the area of evaporation.

The reasonableness of this will commend itself to any reader, and ought, also, to weigh seriously with those persons who are often too ready to make excuses for or to defend the *status quo,* which generally means unsanitariness more or less. If, then, zymotic germs in sewage spread over land, constitute a source of danger and infection to man, may we not safely infer, whilst waiting for more light and knowledge, that they also constitute a danger to cattle, and that both milk and animal tissue must be vitiated by such pernicious influences, and by being used as food, convey the virus and bacterial germs into the human economy, so causing zymotic and other forms of disease and a general lowering of health?

CHAPTER XXV.

ON THE REMOVAL OF STORM-WATER, THE FLUSHING AND VENTILATION OF SEWERS, AND ON RAILWAYS AS DATUM LEVELS IN SEWAGE-DISPOSAL.

I THINK the first thing should be to determine on the adoption of some system of sewage treatment and disposal, and then see if it is necessary or not to remove the storm-water. It is not only unnecessary in some circumstances, but better that it should not be removed; that can easily be demonstrated both in practice and theory; therefore it would be better to consider the question of storm-water removal, either concurrently with the adoption of a system of sewage disposal, or after some definite system has been decided upon. Such removal may or may not be necessary or even advisable according to the system adopted, and particularly with regard to the nature and extent of land in the neighbourhood to which sewage or effluents could be applied.

There are several objections to the removal of storm-water, the neglect of which incurs grave responsibility. First, the raising of the death-rate, and the lowering of the tone of urban vitality by slowly flowing sewage and unflushed sewers; such conditions greatly aiding the formation of putrefaction and poisonous gases by the rapid growth of bacteria in a too highly concentrated sewage, especially in hot weather.

Second, that in urban districts, and especially where there is much traffic, the first storm-washings contain large quantities of putrescible organic matter. Mr. Santo Crimp states that they are very foul and often contain as much as the sewage itself.*

Third, the removal of the storm water should be accompanied by the adoption of adequate flushing-tanks, which, in fact, are

* See Appendix C.

needed in hot and dry seasons, whether the storm-water mixes with the sewage or not.

Fourth, that where there is plenty of land needing irrigation upon which sewage or effluents can be applied, either by gravitation or by pumping, or by both, it is waste to run the organic matter of storm-water into a river even if it should be present only in a dilute state.

During the dry weather of the fortnight ending November 8th, 1891, I had occasion to examine the sewage of a flat-lying street-main sewerage at Leek, which represents the drainage of a district containing 7,000 persons. I received two separate gallons, one taken shortly after the other.

They were unusually viscous and charged with organic matter in solution. There was nothing abnormal as to the suspended matters, which were only sparingly present.

The sewage of No. 1 was fairly clear for so thick a liquid, but it had a very offensive odour.

The No. 2 gallon was of a purplish colour, no doubt from the presence of the dyewash from dyers' waste wash.

My analyses gave the following results:

Crude Sewage.		Free Ammonia. Parts per million.		Albuminoid Ammonia. Parts per million.
No. 1 bottle	......	120·00	......	12·00
No. 2 bottle	......	160·00	......	20·00

The No. 2 sample, after treatment with my ozonine process, had its organic matter removed from the effluent to the considerable extent indicated by the following amounts of free and albuminoid ammonia, giving a percentage of purification of 87·5.

After treatment.		Free Ammonia. Parts per million.		Albuminoid Ammonia. Parts per million.
No. 2 bottle	......	11·00	......	2·50

Now let us compare the state of this crude sewage with an average sample of Salford sewage:

Crude Sewage.		Free Ammonia. Parts per million.		Albuminoid Ammonia. Parts per million.
Leek No. 2	.. ...	160·00	......	20·00
Salford	......	7·20	......	2·70

On mentioning this unusual degree of impurity in the sewage to the Chairman of the Commissioners, he told me the sewer at the part from which the two gallons had been taken had been laid of unnecessarily large diameter, and it had very little dip. I think, then, we may safely infer from these facts the necessity of constantly looking to the flushing of the sewers both by the admission of stored flushing-water in dry weeks and storm-water of rainy seasons. Very much, of course, depends on the velocity of sewage-flow.

I think this exceptionally bad sewage occurring in rainless weather, points out very forcibly two things: first, the great risk to general health by removing the whole or any of the so-called storm-water; second, the great necessity which exists of regularly flushing the sewers in dry weather. Let us now see what danger co-existed with the foulness of this crude sewage. There were 20 cases of typhoid fever and one of diphtheria localised there, for none of which a cause could be given by the sanitary inspector. I ask, is it at all surprising that typhoid fever, or other zymotic diseases, should be found accompanying an abnormally impure state of sewage, and need there be any difficulty in assigning a cause for the outbreak? The locality has several times been the first to be visited with zymotic trouble.

Should it be deemed advisable to remove what is called the storm-water from the sewers, by which is meant both storm-water and that from ordinary rainfall, I am convinced that, generally speaking, only actual storm-overflows need be diverted.

At Kingston the excess over 45 gallons per hour per day is diverted, and this provision is instructive. I think storm-water falling upon back-areas should always flow into the sewers, for no part of a town requires such constant flushing and attention as back-yards, and the smaller house-drains which carry off cooking and washing drainage. I add a few considerations of the Government Referees on Metropolitan Sewers, which give interesting information. A rainfall of ·9 inch gave to one sewer 3·05 cubic feet per minute per acre; another as much as 20·50 cubic feet per minute. The referees suggested an interception of ·05 of rain per hour.

Mr. Santo Crimp states that of a quarter of an inch of urban rainfall per day only five-eighths reaches the sewers, "the remainder being absorbed, evaporated, or utilised." In the metropolis, it was found that there are only from 14 to 21 days per annum when this rainfall is exceeded.

The Flushing and Ventilation of Sewers.

It is hardly necessary to point out the necessity of flushing town sewers, a precaution of the first importance to the health of urban populations. It was found by the Westminster Commissioners that water passing along the sewers at a velocity of 6 inches a second, scours away all the sediment, and a velocity of 1 foot per second sweeps away fine gravel. They found that a head of water with a flood-gate was sufficient. At Leicester "the sewers are flushed at their two highest points from a thousand gallon tank filled by a hydrant with the water-company's water." At Salisbury "there are flushing wells at the corners of the streets, and the sewers are well cleaned by fortnightly flushings."*

The necessary corollary to the flushing of sewers is their ventilation, because of the rapid accumulation of foul air, charged at times with pathogenic germs. Sewers are best ventilated by shafts, and not by street sewer-grids. Absence of ventilation causes the accumulated gas to force its way backwards through the house-traps and drains, and there is frequently sufficient pressure to do this. The danger attending such escape is too evident not to force its seriousness upon all sanitarians and guardians of public health. When one considers the numerous outlets in a house through which sewer-gas pressure can intrude itself into upstairs and downstairs rooms, such as by soil-pipes, defective joints, rain-water pipes, the

* The Treatment and Utilisation of Sewage, Corfield, p. 205.

ordinary drains, overflow-pipes, and even into the water cistern, it seems wonderful there has been no compulsory legislative enactment for towns to add ventilating shafts to their sewerage systems.

In reference to this point, it is interesting to know that in America, such is the importance attached to having house-pipes hermetically closed,—bad plumbing is a penal offence.

The safest and surest way of all is for the sewer-air to be drawn from the sewers by fires at the highest parts of the sewers, and by passing through them to have its deadly properties decomposed; so purified, it would pass into and out of the shaft in an innocuous state. For this purpose a small and not a large fire is best, because too great a draught should not be created.

The only objection which I have seen made to ventilation through shafts with fires at their base is that air is sometimes sucked inward through the house and street traps into the sewers. To this I would reply, that the upward shaft-current might be so regulated as to moderate the up-cast; this may be done by automatic "dampers," or valves, and by keeping the fires in a moderate state, but it is surely better for air to get into the sewers than for sewer gases to be forced by internal pressure through the house-drains and traps into the interior of houses.

In wet weather, when the sewers are carrying off the rainfall, there is great danger of sewage gases being forced up the house-drains by the presence and flow of abnormal quantities of rain-water in the sewers.

Without good flushing and ventilation of sewers, who can wonder at the sudden outbreaks of zymotic diseases, the origin of which local authorities and their medical officers of health and nuisance inspectors are so often unable to specify or localise. The only way to minimise, if not altogether to prevent these outbreaks, is to have a good water-supply, good sewerage, which must be incomplete and dangerous if the sewers are not perfectly flushed and ventilated.

The reports of medical officers of health have proved that of all zymotic diseases typhoid fever is the one which has been the most kept in check by good sewerage and good water.

Outbreaks of typhoid fever have been abundantly proved to be owing to sewage polluted air, or bad drinking water; there is no doubt the germs of the typhoid bacillus are constantly present with other pathogenic germs, and they only require such cultivation as stagnant sewage for their rapid development. It has been shown that in the worst parts of Glasgow and other places fever was always present, and that not until better ventilation and drainage be adopted, could it be stopped. In Bedford, where cesspools abound, the Medical Officer of the Privy Council found there had been annually 30 deaths from typhoid fever and diarrhœal diseases, and they were attributed by him to decaying cesspool matters.

The infant mortality in Manchester is higher than in any other place, and is no doubt chiefly owing to pathogenic organisms being bred and distributed by their unhealthy pail-closet and ashpit system.

The average death-rate of Manchester for the 15 years prior to 1885 was 32 per 1,000, and in one of the years, when zymotic diseases held mastery, it reached 39 per 1,000, and this with a fairly good water-supply of excellent purity.

Typhoid fever visited Sherborne in consequence of sewer-gas entering the water pipes when the water-supply was cut off for a time.

In 1604 the plague at York killed one out of every three of the population, and in 1832 the cholera killed one out of every 142 people. Both broke out in a filthy part of the city called Hagworm. In 1604 there was no drainage, but instead open channels, and wide, stagnant moats; but in 1832 there had been an improved drainage.

Railway Levels and Sewage-disposal.

More than two years ago I suggested to the Leek Sanitary Committee the advisability as well as the possibility of conveying the sewage or effluent from the town in pipes along the sides of the railway, and distributing it at various points, until it was disposed of upon the farms through which the railway passed, thus making the railway a datum level. I showed that this method would secure much more disposal acreage than would suffice for a very much larger population than that of Leek. Since this suggestion, a paper has been read by Dr. B. W. Richardson, M.A., LL.D., F.R.S., &c., President of the Association of Public Sanitary Inspectors of Great Britain, on May 2nd, 1891, in which I am glad to see a much larger practical suggestion of this idea than I had formed. He proposes a national main-drainage scheme, by which sewage should be conducted along the railways in hermetically sealed pipes, and be distributed upon such lands as would be best suited to receive it, at no matter what distance. He argues with great force and clearness that the levels have been already laid down for the whole country by the railways. Pumping would have to be resorted to for low-level sewage work, and gravitation would take the sewage and effluent from high-levels to low ones. For further details, I refer my readers to his address, published by the Association.

At Bedford the sewage is pumped to a height of 13 feet and disposed of upon land. 950,000 gallons per day are raised to this height, and there is no pumping in the night. In many other places also the sewage is pumped to a greater or less height, as will be seen in the answers of many towns to the question in Chapter XIV.

CHAPTER XXVI.

ON THE AUTHOR'S NEW METHOD OF SEWAGE PRECIPITATION, BASIC PERSULPHATE OF IRON, FERRICUM, OR OZONINE.

THREE years ago I resumed my investigations on Sewage Treatment and Disposal, on being appointed a member of a sewage observation committee by the Sanitary Committee of the Leek Improvement Commissioners.

The object of this committee was to examine into the whole question by visiting sewage-works in other places, and reporting upon them, with a view of recommending a scheme for the disposal of the Leek sewage, which should satisfy the County Council, who had insisted on the removal of the sewage from the river Churnet.

After some interesting visiting work with my colleagues, I found it necessary to commence an independent inquiry in order to make my studies more complete and more thoroughly satisfactory. The result of such work has its record in the foregoing chapters. When the investigations were first undertaken, I had no thought or intention of occupying myself with original research into sewage treatment, or of adding a new method to the many already existing ones, although I had given the subject considerable attention for several years, even as far back as 1872, as recorded in the preface of my chemical examination of the effluents of several systems for the Tunstall authorities. I wished rather to confine myself to the examination of the several processes adopted in various parts of the country, and in making myself acquainted with the more advanced ideas of the present day. But a closer acquaintance with, and an increasing interest in the subject naturally brought forth suggestive reflections, resulting in an experimental effort in the direction of research.

I have shown the Buxton process to be the addition of a natural mineral water, highly charged with carbonate of iron (a protosalt) in solution, to lime, and then both conjointly run into the sewage. On coming into contact with the sewage this protosalt of iron is oxidised at the expense of the sewage: in its new nascent state the oxygen immediately acts upon the organic sewage matter, which falls down in a voluminous flocculent precipitate. I have fully described the process in Chapter XI., in which I have stated that on examination of the effluent resulting from this treatment I found by analysis that it contained 2·4 parts of free ammonia and ·5 parts of albuminoid ammonia, a state of purity sufficient to be allowed to pass into any river containing a volume of water about eight or ten times the volume of the sewage effluent.

It was from this decided and most interesting reaction at Buxton that I was led to examine the effects of the more basic forms of iron upon organic residual matter. Basic iron-salts certainly possess remarkable antiseptic and disinfectant powers, especially in the form in which I am about to describe, and can be employed at a more reasonable cost than any other precipitant, both upon heavy sewages and those not too highly charged with organic matter.

I have had much experience during the past 40 years as a Silk Dyer and Printer in the construction of the various salts of iron, both ferrous and ferric, and in their application to the organic fibres of wool, cotton, and especially of silk, in order to form the bases and mordanting of the best kinds of black dyes. Silk is an organic substance named fibroine, and has a composition indicated by the formula, $C_{15}H_{23}N_5O_6$. Some of the processes of dyeing silk black are very complicated, requiring the application of iron-salts many times repeated, and in various stages of the processes, at one time in a ferric state, and at another as ferrous salts, which act as mordants to further treatments and additions of colouring or vegetable matter. Many years ago I succeeded in constructing a salt of iron sufficiently delicate to obtain the required results of lustre and shade of certain special and complicated black dyes upon silk, and have since continued to manufacture and use this ferric salt in considerable quantity. It occurred to me that it might be possible to apply this idea of the joining chemically together of the *solid* organic matter of silk fibre with a suitable iron-salt, to the production of a similar reaction on organic matter in *solution*, and nearly three years ago I engaged myself in laboratory experiments with this end in view.

I had for some time known that several of the salts of iron possess the property of precipitating the organic matter of sewage to a greater or less degree, and I thought it might be worth while to follow up the idea and endeavour to discover such an iron-salt as would attain the maximum power of the precipitation of such organic matter, with or without additions, which would answer all the requirements desired in the treatment of sewage of various kinds. My experimental contributions to the subject in this direction resulted in the discovery of a method which has more than realised my anticipations, which I have since patented, and in which a specially prepared salt of iron is the chief ingredient, and forms, in combination with other substances, a satisfactory precipitant of organic and refuse-matter.

I found that the most effective precipitant was a specially constructed basic ferric-salt. This salt is so delicately constituted and unstable in the presence of organic matter in solution as to instantly combine with it, and in the case of sewage a dense precipitate immediately forms, leaving in a very short time a perfectly clear effluent. On some sewages this action is accelerated and the organic matter more completely precipitated by the addition of a small quantity of milk of lime or lime-water.

The reaction is precisely what occurs in the case of mordanting animal or vegetable fibres before dyeing; that is, a chemical combination of the mordant with the fibre, whereby a new substance is formed which more readily receives the after-addition of colouring matter, because of the affinity set up by the chemical mordant between the organic fibre and the tinctorial matter.

But there is perhaps rather more analogy in the formation of coloured lakes, in which processes the organic colouring-matter combines with the chemical mordant, and falls as a lake or precipitate, as in the case of carmine, which is a lake or precipitate formed by the action of an aluminous mordant, acting on the colouring matter of cochineal and separating it from its solution.

Alumina and its salts possess analogous properties in their effect upon sewage, but not to the same extent or completeness, as has been well demonstrated by the chemists and biologists of the Massachusetts State Board of Health, whose most important conclusions I have recorded in Appendix A; in fact, they act differently, and the results are not the same. Alum salts greatly assist in decolourisation, but as disinfectants and precipitants they are less operative as compared with one or two basic forms of iron-salts, especially the one which forms the basis and principle of my patent process for sewage treatment, and which is an excellent decolouriser, as has been amply demonstrated at Salford and Pendleton, where, for upwards of four months of the past year, 100,000 gallons of the sewage have been daily treated with this precipitant at the desire of the Rivers Conservancy Board of the Corporation of Salford, and where, after a commensurate trial upon the whole volume of the sewage, amounting to an average of 12,000,000 gallons per 24 hours, its adoption is contemplated. This sewage is largely composed of manufacturing refuse-wash, a very mixed product from bleachworks, dyeworks, paper-works, printworks, indiarubber-works, gasworks, &c., together with a quantity, perhaps half, of domestic sewage. It is generally highly coloured, often black, red, blue, or opaque dark-grey, by turns, and is hourly changing in consistency and colour. Whatever its state or colour has been, this iron-salt precipitant has never failed to decolourise and to deodorise it, removing from it at the same time a larger percentage of soluble organic matter than by any other process known to me. However foul-smelling the sewage was, whether from domestic or manufacturing sources, the smell was entirely removed if its chemical equivalent proportion of the precipitant had been used. With less than its equivalent a corresponding amount of organic matter remained in solution, but with the proper combining proportion, precipitation and clarification of the effluent begin immediately. This liquid precipitant, after being considerably diluted with water, is made to fall into and mix with the sewage in the most simple manner from the cistern containing it as the sewage passes from the sewers through the mixing channel or shoot, on its way to the settling tanks. In fact, it is as beautiful and simple an example of precipitation and subsidence as any made for a lecture illustration.

The action of the precipitant on dye-wash, however darkly coloured, is as remarkable as it is complete. The precipitant attacks it and decomposes it; what was colouring matter falls down as a lake, leaving a decolourised effluent.

The organic matter in chemical union with the iron-salt, together with colouring matter and matter in suspension, separate in a dense flaky precipitate, the greater part of which falls to the bottom of the settling-tanks in less than an hour, whilst the liquid effluent remains colourless and clear, and, after filtration through sand, as limpid as well-water.

On many days at Salford the effluent required no filtration at all. Filtration, however, certainly helps to reduce the remaining organic matter to a much lower extent by oxidation. This is especially the case when the filter contains a stratum of iron, such as polarite, carbide of iron, magnetic oxide of iron, or the ozonite filter of the Standard Sewage and Water Purification Company.

Where the unfiltered or tank-effluent cannot be discharged

upon land for further oxidation, it is always better to filter it it is then fit to be turned into any river or stream.

At Salford the condition is particularly severe, for the effluent must be such as can be safely admitted directly into the Ship Canal, a requirement effectually fulfilled by this basic salt of iron process.

Where an effluent can be used for intermittent irrigation, I strongly recommend a less proportion of the precipitant than its chemical equivalent of organic matter to be used, because two important purposes can thereby be served, namely, a decreased cost by a decreased quantity of precipitant, and more organic matter left in the effluent, enabling it to retain more fertilising power than if it were mainly a purer effluent; in fact, it would be better to remove but a small quantity of organic matter, where a sufficient amount of land could be available for intermittent irrigation, and the effluent applied either by gravitation or by pumping.

The raising of an effluent only a few feet by pumping brings in generally a much larger irrigating area, and the cost of pumping is not generally great, especially where there is water-power. In this way, uplands, which most require fertilisation, would reap the benefit, instead of running it upon wet alluvial or clayey soils, which are generally wet enough without it.

For this purpose effluents so graduated would be clarified and decolourised at various degrees of cost. I have found that with most sewages, especially those uncontaminated by manufacturing refuse and wash, less than one ton of the precipitant per million gallons is a sufficient equivalent for the amount of organic matter to be removed to the highest degree.

Now, the cost of one ton of the precipitant at the present time is about 30s., so that this quantity would represent a cost of 30s. per million gallons of sewage, and is therefore easily reckoned; but if we take a lower quantity of precipitant, say half a ton per million gallons, not by any means the lowest to produce a clarified and decolourised effluent on some sewages, then the cost of treatment per million gallons would only be 15s., and so on with any intermediate quantity. £2 per million gallons is a moderate sum to expend on sewage treatment; but I have thus shown that the process is considerably more economical than this limit.

When more precipitant than the chemical equivalent is used, the greater part of the iron in excess falls with the precipitate or sludge, and thereby improves it both as a manure and as a disinfectant. Even when properly balanced a little remains in the effluent, but subsides as a flocculent protoxide, in a few days gradually absorbing oxygen from organic matter in the water into which the effluent flows, and becomes an insoluble peroxide, helping, if there is much organic matter present, to further purify it. As in the case of the Manchester Ship Canal, where the water used for it would never be pure and its sedimentary matter frequently stirred up, it would be an advantage to it if Salford and other riparian towns used an excess of the iron precipitant, and it would pay the directors of the Ship Canal to have such excess of iron-salt to help to preserve its comparatively stagnant and slowly-flowing waters from putrefaction. One of the properties of basic persulphate of iron process is, that it is further-reaching than mere antiseptics, which only retard the growth of bacteria; it is thoroughly a disinfectant and a germicide, and completely destroys bacteria. After the precipitation, in my laboratory, of sewage swarming with bacteria of many species, I found the effluent wholly free from them, and in frequent examinations of the resulting sludge I failed to find any signs of the presence of bacteria or of animal or vegetable life of any kind.

Ozonine and the Cholera Bacillus.

The sterilising powers of this basic persulphate of iron process have during the past year received most important confirmation by the experiments of Dr. Bolz, an eminent biologist of Hamburg, who was actively engaged there throughout the great cholera visitation. I sent him two bottles of the precipitant, marked

"A" and "B," but prepared in different ways, requesting him to try their effects on the cholera bacilli, upon which I ventured to think they might prove destructive. My anticipations have been fully confirmed, as the following quotation from his letter, which is dated February 1st of the present year, will show:—

"I venture to send you the results I obtained by a good many experiments made with your sewage precipitants. I mixed 50 cubic centimetres of a choleraic child's fluid dejection with 5 cubic centimetres of a cholera bouillon-culture, and 1 cubic centimetre of your disinfectant 'B.' In 24 hours all bacilli had been destroyed.

"To have a pure trial with comma bacilli, I mixed 100 cubic centimetres of sterilised river-water with 5 cubic centimetres cholera bouillon-culture. The control-plate contained a great number of colonies of the comma bacillus. To this I added 5 cubic centimetres, or 5 per cent of your disinfectant 'B.' I did not find any bacilli after 10 hours time.

"Corresponding with these experiments, I made similar ones with your solution 'A,' and I obtained the following results: Adding to 100 cubic centimetres of virulent cholera dejections, 2½ cubic centimetres or 1 per cent of 'A,' I found no cholera bacilli on the following day, but 48 hours later I found a good many cholera germs, proving that their growth had been for some time arrested. On adding 10 cubic centimetres of a 5 per cent solution of 'A' to 50 cubic centimetres of cholera dejection, all comma bacilli proved destroyed in 48 hours, but a number of other intestinal bacilli had grown on the plates.

"One cubic centimetre, or 5 per cent of 'A,' added to 5 cubic centimetres of cholera bouillon, sterilised the latter in 10 hours, so I came to the conclusion about your sewage disinfectants that 'A' is four times stronger than 'B,' and that both are able to destroy the comma bacilli after 10 to 16 hours influence."

The importance of these trials cannot, I think, be exaggerated, especially at this period of the opening summer, when we may be probably visited with the scourge of cholera, and if my researches have only tended to this result they will not have been in vain.

Flügge found that 5 per cent solutions of carbolic acid and permanganate of potash severally destroyed anthrax spores in two days, and that one part of bichloride of mercury to 20,000 parts of water destroyed them in a day. He found that chloride of iron (he does not say whether ferrous or ferric) destroyed anthrax spores in six days, whilst absolute alcohol had no effect upon them. My basic form of the persulphate of iron is much more active than the chloride.

Mr. C. Lowe informs me that 1 per cent solution of carbolic or cresylic acid is also an effective germicide.

A true disinfectant, then, being a destroyer of bacterial life and miasma, the value of this process in this respect claims for itself the possession of considerable hygienic importance, if not an unworthy offering to Hygeia herself.* It at any rate deserves the attention of all medical officers of health, whose chief concern must ever be the prevention or lessening of zymotic and septic diseases. Accumulations of imperfectly precipitated sewage-sludge, or stationary sewage, over which a thin layer of this disinfecting precipitate could be applied, would absolutely remain without putrefactive decomposition; because bacteria, which are the sole cause, could not live or grow in its presence, and the danger of infection from bacteria or their spores would be greatly diminished, if not entirely removed, as well as the nuisance from sewage stenches; for there can be, of course, no sulphuretted hydrogen formed so long as ferric-oxide is present or being deposited. This idea is also not without its importance in lessening other malodorous nuisances, the products of decomposition, because much of such stench is occasioned by putrefactive and gaseous sulphur compounds.

The following description, taken from the specification of my

---

* Hygeia, the Goddess of Health.

patent, will fully explain the nature of the means I employ to precipitate the organic and other refuse matters in sewage:—

"This invention relates to a new or improved process for use in precipitating organic matter and manufacturing refuse from sewage, and consists in using a certain proportion of persulphate of iron, preferably in a basic form, as a bisulphate or monosulphate, or the intermediate or other basic forms of the persulphates of ferric oxide which have been prepared in a suitable manner, to which are added—according to circumstances, and the various kinds of sewage and other refuse waters or matters dealt with—varying proportions of an extra basic sulphate of iron (which is insoluble in water, but dissolves in acid), nitric oxide, protosalts of iron, salts of alumina, cresylic acid, and lime, or other similar earthy matter or alkalies.

"By this improved process, organic matter in solution in sewage or other refuse water or matter is precipitated and deposited together with suspended matter, either organic or mineral, leaving a clear effluent, not requiring filtration, unless a higher degree of purity is wanted.

"To carry my said invention into practice, I use the above mentioned persulphate or persulphates of iron in one or other form, with or without the addition of lime, alumina, salts, protosulphate of iron, cresylic acid, and any of the alkalies, such as potash, soda, ammonia, or their alkalic salts, according to the requirements of various states of sewage to be found in manufacturing localities, or of private firms, houses, or institutions.

"I take a strong or weak solution of persulphate of iron, preferably a basic form, this form being the most effective, in a liquid or solid state, or in a liquid or solid form conjoined. For ordinary sewage I take about one ton of one of the said basic persulphates of iron in a liquid or solid state, or about half a ton in a liquid and half a ton in a solid state for each million gallons of sewage, or 16 grains of the solution of one of the said persulphates of iron, for each gallon of sewage which does not contain manufactory wash or refuse; less quantity than this will suffice for weak sewage, but for sewage containing manufactory wash or refuse a larger quantity is more efficacious.

"When it is found that the sewage is such that the basic persulphate of iron is insufficient to precipitate the organic and chemical matter in a thorough enough degree, I add varying proportions of lime, alumina salts, persulphate of iron, cresylic acid, or alkalies, or salts of the alkalies, according to the nature of the sewage to be treated.

"When it is found requisite to use lime, I take about a quarter to half a ton of newly-burnt lime for each million gallons of sewage, according to the nature of the sewage, and I convert such lime by the addition of water into milk of lime, and stirring the same together by any suitable means, so as to thoroughly convert the lime into milk of lime, and allow the said milk of lime to flow or run out of the tub or cistern in which it is dissolved, into the sewage contained in a cell lined with brick or stone of about three cubic feet in capacity. From this cell it passes underneath the cell-wall, and rises up into another cell of about the same capacity; the sewage as it rises up into the second cell flows over the walls of a similar cell, into which my special preparation of persulphate of iron, &c., is also allowed to flow from a tub or cistern in regulated quantities, and it will be found that precipitation immediately takes place. The flow of sewage after leaving the conducting-shoot of two or more cells passes into settling-tanks made of a convenient form and of proper construction, and in pairs for alternate precipitation and cleansing. The bulk of the precipitant falls into the first and second tanks, and in a lessening degree into the other tanks, the flow of the precipitated sewage being over the top of the tank-walls, each set of tanks being four, six, or more in number. The best effluents are obtained when the tanks can be left a short time at rest without any flow, as the precipitate, being light, is easily carried forward by a current. From the tanks the precipitated sewage flows into a filter, if it should be found necessary, and is distributed by gravitation, or by pumping, for irrigation purposes, or conducted directly into a river or watercourse; the precipitant or sludge being collected and dealt with by filter-presses, or by other suitable methods, with or without the addition of fertilising matter.

"The employment of one or more of the additions mentioned above to the iron-salt depends on the nature of the sewage, and also upon the disposal of the effluent—whether for agricultural purposes or for direct river or canal entry."

One of the advantages of the application of this form of persulphate of iron is that, being a strong solution, it can be diluted to any degree without impairing its disinfecting and precipitating powers, and it is alike applicable to mansions, public institutions in town and country, and to private houses, as it is to sewage-disposal works on a large scale. A little daily used in the drains would disinfect the whole sewage arrangements, and effectively prevent epidemics arising from decomposing fæcal matter, garbage refuse, and scullery wash.

The idea of the use of iron-salts to sewage is not new; so far back as 1845 a patent was obtained by a Mr. Baronnet for persulphate of iron, but this could not have been my basic salt, for it was not then discovered; it probably would be the normal or neutral persulphate of the books, having the formula of $Fe_2(3SO_4)$, which is not the chemical construction of my specially prepared basic compound, nor one of the several basic forms which sulphuric acid forms with ferric oxide.

Mine is much less stable and of greater delicacy than the normal persalt.

Although this process of basic persulphate of iron has never before been used, there have been and still are other preparations of iron-salts used for sewage precipitation, but they are less effective.

Some time ago a paper was read at the Manchester Literary and Philosophical Society on the use of persalt of iron in the treatment of sewage, by Mr. Harry Grimshaw. The persalt he employed was the perchloride of iron (Fe_2Cl_6). It was termed Clarine, and was practically tried at Salford, but was not successful. My own experiments confirm the inefficacy of perchloride of iron as compared with the basic persulphate. Iron and hydrochloric acid in combination will not do the work of sewage precipitation of iron and sulphuric acid, nor can the perchloride be advantageously made to combine with the solid organic matter of silk fibre in dyeing.

The following list is a record of some patents taken out up to 1887, in which iron in some form or other is a component part of the process. The list, with the iron processes described in previous chapters, is as complete as I have been able to collate :—

Name of Substance.	Inventor.	Date.
Protosulphate of iron	Deboissien	1762
Ditto	Briant	1824
Sulphate of iron and zinc, with tan and water	Siret	1837
Persulphate of iron	Baronnet	1845
Waste salts of iron, lead, zinc, with pyroligneate matters, ashes, &c.	Brown	1847
Pyroligneate and perchloride of iron	Ellerman	1847
Fresh bark, sulphate of iron, and peat charcoal	Angely	1850
Lime, magnesian earth, sulphate of zinc or iron, and vegetable charcoal	Gilbee	1852
Sulphate of iron with lime in solution (recommended only), &c.	Letheby	1858
Sponge iron	Bischoff	1865
Iron finely divided in revolving cylinders	Anderson	1884

Name of Substance.	Inventor.	Date.
Iron sulphate, calcium sulphate	named Ferozone	1887
Perchloride of iron, &c.	Barry Patent Manure Co. ...	1887
Perchloride of iron, &c.	Bradbury	1887

It will be seen from the foregoing list that the antiseptic and purifying power of iron has been recognised for a long time, but the real crux is to determine in which form or admixture it is at once the most potent, the least costly, and the least detrimental, particularly in sludge-fertilising power. It is in the solution of this desiderata that my aims have been directed. With regard to its potency there is now no question, and I think it will be impossible to find any form of iron-salt so cheap, if its greater combining and precipitating powers are considered.

It is better to use a little milk of lime to the sewage before the addition of my iron-salt, because its full power is not exerted unless the sewage is in a neutral state, and still better if slightly alkaline; but the chief reason why the use of lime is of advantage is, that it reacts on the iron-salt and gives it a little more basicity and delicacy of action. It is then also much quicker and more complete in combining with the organic matter of the sewage. The lime should be added to the sewage shortly before the iron-salt, and not with it.

Where the sewage is acid from manufacturing refuse, as sometimes at Salford and other places, the necessity for the use of lime becomes emphasised, but in no case need the quantity exceed 8 grains per gallon or half a ton per million gallons. Generally, 4 grains per gallon or 5 cwt. per million gallons is quite enough, and much less than this would suffice for purely domestic sewage; for slightly alkaline sewage no lime is needed.

The following analyses of Mr. Carter-Bell, Analyst to the Salford Corporation, of the Salford and Pendleton sewage and effluent of November last, will show the action of my iron-salt and lime. The sewage was collected hourly for the 24 hours and mixed, thus giving an average day and night sewage.

"The Cliff, Higher Broughton,
Manchester, 15th Nov., 1892.

ANALYSES OF THE SALFORD SEWAGE AND EFFLUENTS, FILTERED AND UNFILTERED. PRECIPITANT, WARDLE'S OZONINE.

	Parts in 100,000.		
	Sewage.	Effluent.	Filtered effluent.
Total solid matter at 212 F.......	119	105	103
Suspended matter	7	nil	nil
Suspended mineral matter	5	nil	nil
Suspended organic ,, ...	2	nil	nil
Solids in solution	112	105	103
Mineral ,, ...	78	80	82
Organic ,, ...	34	25	21
Chlorine	26	21·4	19·2
Oxygen required for 15 minutes	3·4	·48	·24
,, 3 ,,	4·6	·76	·32
Free ammonia	1·8	1·80	1·10
Albuminoid ammonia	·48	·21	·10
Percentage of purification	...	56	79

The two effluents were clear and bright, and there was very little difference in appearance between the unfiltered and the filtered. The percentage of purification has been calculated upon the settled sewage, which is rather a severe test. If the percentage of purification had been calculated upon the shaken sewage, the numbers for the purification would have been much higher, making the unfiltered 69 per cent and the filtered 92 per cent.

These effluents, on being heated to various summer temperatures, showed no signs of secondary decomposition, and therefore are fit to turn into any river or canal."

The process is being used at Nuneaton, where the sewage amounts to 400,000 gallons per day. This sewage is a remarkably difficult one to treat, being composed of domestic sewage, and a very large quantity of refuse-wash from fell-mongers and wool-scouring industries, giving to the sewers peculiar organic and fatty matters very offensive and extremely difficult to deal with.

On some days the best results are obtained there with ozonine alone, at others with ozonine and one or more of the ingredients mentioned in my patent. The chief difficulty in treating this sewage is to prevent secondary decomposition in the river Anker, which is here small in volume, and which, up to the time I commenced, had, after receiving the effluent, been most offensive for several miles down the river.

By the kindness of the Corporation of Salford, I am able to forestall their report as far as it relates to the trials of my method at Weaste, during several months of last year, upon 100,000 gallons of the Salford and Pendleton sewage daily.

The full analysis of Mr. Carter-Bell, given on the previous page, will suffice to show the average composition of this sewage, and I need here only give the free and albuminoid ammonia and the percentages of purification of a number of Mr. Carter-Bell's official analyses.

It will be interesting to show also the variations of rainfall, which Mr. Carter-Bell has carefully taken during the continuance of these trials.

The rainfall having an influence on the sewage by diluting it, makes it more difficult to treat.

If sewage becomes very much diluted, or, say, if the albuminoid ammonia gets down to ·10 in 100,000 parts, precipitants cease to act upon it, and thus in the table of the percentages of purification it will be seen that less purification is effected, although it may be that the effluent may be in first-class condition, because containing so little albuminoid ammonia. Therefore we must not lay too great a stress on the percentage of purification, for this percentage must depend upon the amount of albuminoid ammonia originally in the sewage. The greater the amount of albuminoid ammonia in the sewage, the greater will be the percentage of purification, and this is really what is wanted—that is, to take bad sewages, and by means of chemical precipitation to raise them to 70 or 80 per cent of purification; whereas in many cases the weak sewages, containing so little as ·10 and ·20 per 100,000, would not require chemical precipitation, but might be simply allowed to settle in tanks, and then passed on to filter beds. If what I have here stated is kept in view, it will be found that a uniform percentage of purification is always obtained by my process.

The following table gives the rainfall, the albuminoid ammonia in sewage and effluents, and the percentages of purification of experimental trials with 100,000 gallons per day of the Salford and Pendleton sewage during the months of August, September, October, and November, 1892. The figures are from the official analyses of Mr. Carter-Bell. The quantity of my precipitant used was 10 to 15 cwt. of Ozonine per million gallons of Sewage in the earlier trials, and 20 cwt. per million gallons in the later ones.

Good results were obtained without the use of lime, but on the whole, when 5 cwt. of lime per million gallons were used, the percentage of purification was higher. The sewage on some days being so very acid from manufacturing waste-wash, it was necessary to add lime, but lime, as I have stated, may be dispensed with where the sewage is neutral, or slightly alkaline, and without much waste-wash.

PURIFICATION RESULTS ON THE SALFORD AND PENDLETON SEWAGE WITH THE WARDLE PROCESS, OR OZONINE.

Rainfall.	Parts per 100,000. Sewage. Alb. Amm.	Parts per 100,000. Tank Effluent. Alb. Amm.	Percentage of Purification.	Parts per 100,000. Filtered Effluent Alb. Amm.	Percentage of Purification.
·0	·62	·34	45	·17	73
·82	·44	·34	23	·13	70
·01	·52	·30	42	·19	63
·11	·76	·44	42	·18	76
·87	·24	·16	33	·14	41

Rainfall.	Sewage. Alb. Amm. (Parts per 100,000.)	Tank Effluent. Alb. Amm. (Parts per 100,000.)	Percentage of Purification.	Filtered. Effluent. Alb. Am. (Parts per 100,000.)	Percentage. of Purification.
·11	·44	·30	31	·14	68
·0	·52	·25	52	·14	73
·0	·46	·32	30	·16	65
·085	·50	·37	26	·13	74
·0	·52	·24	54	·14	73
·0	·32	·28	12	·08	75
·0	·28	·12	57	·08	71
·81	·40	·22	45	·20	50
·25	·60	·30	50	·12	80
·15	·28	·16	42	·08	71
·06	·44	·23	48	·10	77
·085	·38	·26	31	·11	71
·140	1·4	·10	28	·03	78
·025	·28	·14	50	·06	78
·02	·34	·26	23	·11	67
·31	·36	·24	33	·07	80
·75	·48	·22	54	·11	77
·57	·40	·21	48	·14	65
·15	·44	·30	32	·14	68
·0	·56	·30	46	·12	78
·0	·64	—	—	·09	85
·08	·28	·19	32	·18	71
·65	·48	·32	33	·14	70
·0	·56	·29	48	·14	75
·0	·48	·21	56	·10	79

APPENDICES.

A. Abstract from the Experimental Investigations by the State Board of Health of Massachusetts, on the Precipitation and Filtration of Sewage, and on the Differentiation of the Bacillus of Typhoid Fever at Lawrence, Massachusetts.

B. Bibliography of Sewage.

C. Mr. Santo Crimp on the Construction of Precipitation and Settling-Tanks, and on the Composition of Liquid flowing from London Street Surfaces.

D. Leek Sewage Disposal: Heights and Contours of the Neighbourhood; Geology of the Locality.

E. On the Estimation of Free and Albuminoid Ammonia.

F. Extracts from the "Hendon Times" and "Falkirk Herald."

G. A New Form of Filter by Mr. J. Corbett, Borough Engineer, Salford.

APPENDIX A.

Abstract from Experimental Investigations by the State Board of Health of Massachusetts, upon the Purification of Sewage by Filtration and by Chemical Precipitation, made at Lawrence, Massachusetts, by Hiram F. Mills, A.M., C.E., Member of the Board; Thomas M. Brown, M.D.; Allen Hazen; W. T. Sedgwick, Ph.D.; also on the Differentiation of the Typhoid Fever Bacillus, its Removal from Water by Filtration, and on its diagnosis in the Water and Sewage at Lawrence, Massachusetts, by Mr. G. W. Fuller. (*22nd and 23rd Annual Reports of the State Board of Massachusetts, for the years* 1890 *and* 1891. Published at Boston, by the Wright and Potter Printing Company, State Printers, 18, Post Office Square.)

The most recent, and in many respects the most scientific and detailed, study of the Purification of Sewage has been carried out by the Massachusetts State Board of Health during the years 1888 to 1892, and the conclusions arrived at with regard to the comparative results given by the different chemical precipitants are so important, that they are deserving of every consideration at the hands of all sanitary authorities whose duty it is to deal with sewage disposal. This is the more so as, for want of acquaintance with the results arrived at by this impartial and highly-equipped government body, needless expense may be incurred, and valuable time thrown away, by a repetition of an inquiry into the respective merits of chemical precipitants, which question appears to have been effectively solved by the Massachusetts Board.

It is a significant fact that the general conclusions of these American investigations, made independently of mine, exactly confirm my own observations and experimental conclusions, both in the laboratory and in the large-scale treatment of sewage.

The substances used for chemical precipitation were lime and salts of aluminium and iron, either alone or in combination; and the experiments were so conducted as to determine the quantity of each which, when used alone, would give the best result, and the best proportions in which any of them could be used together, and the resulting cost.

The most satisfactory results were those obtained with persulphate of iron; the second in order were with copperas and lime; and the least satisfactory were with alum.

Lime containing 70 per cent of available calcium oxide can be bought there for 37s. 6d. per ton. Ferrous sulphate, or copperas, containing 26 per cent of ferrous oxide, costs 41s. 8d. a ton. Aluminium sulphate, or crude alum, containing 14 per cent of alumina, costs £5 4s. 2d. a ton. The ferric-salt can be made by oxidising copperas with chlorine, or with sulphuric acid and nitrate of soda.

The approximate cost of these oxides in solution is as follows:—

Aluminium oxide (AlO)	4½d.	per pound.
Ferric oxide (Fe_2O_3)...........................	1½	„
Ferrous oxide (FeO)	1	„
Calcium oxide (CaO)	⅓	„

Using these figures, the cost of chemicals has been calculated for the different experiments. One hundred gallons of sewage daily for each inhabitant is assumed in calculating the annual cost.

Their experiments most carefully and thoroughly indicate that a certain *definite* amount of lime gives as good or a better result than either more or less; and that in general the more copperas, ferric sulphate, or sulphate of alumina used, the better the result, and that ferric sulphate and (to a certain extent) sulphate of alumina usually require no lime for precipitation, while with copperas a definite amount of lime must be used.

It then remained to compare the results obtained with the best amount of lime, and with equal values of the other chemicals, upon the same sample of sewage when used under favourable circumstances,

Two series of trials were made in this way, with the following results:—

1st Series.—*Results of Experiments with various Precipitants.*

Parts per 1,000,000.

Sewages and Effluents. Precipitants used per million gallons.	Free Ammonia.	Albuminoid Ammonia.	Percentage of Purification.	Settled.	Bacteria per cubic centimetre.
				Hours.	
No. 1 Sewage	3·15	·63	—	—	7,854
Effluent with 2,000 lbs. of Lime	1·84	·30	52·30	22	8,720
No. 2 Sewage	2·37	·59	—	—	—
Effluent with 2,000 lbs. of Lime	2·02	·34	42·37	22	—
No. 3 Sewage	2·36	·60	—	—	—
Effluent with 500 lbs. of Alum	2·00	·39	15·25	—	—
No. 4 Sewage	1·72	·55	—	—	—
Effluent with 500 lbs. of Alum and 800 lbs. of Lime	1·60	·34	38·00	—	—
No. 5 Sewage	2·00	·40	—	—	—
Effluent with 500 lbs. of Copperas and 600 lbs. of Lime	2·52	·31	22·5	—	—
No. 6 Sewage	2·28	·51	—	—	—
Effluent with 500 lbs. of Alum and 700 lbs. of Lime	2·24	·38	25·49	—	—
No. 7 Sewage	1·90	·67	—	—	—
Effluent with 700 lbs. of Copperas and 720 lbs. of Lime	1·90	·35	47·76	—	—
No. 8 Sewage	1·75	·57	—	—	218,960
Effluent with 400 lbs. of Sulphate of Ferric-oxide, *i.e.*, $Fe_2(SO_4)_3$, the Non-basic or Neutral Per-sulphate	1·50	·19	66·66	—	6,080
No. 9 Sewage	1·75	·57	—	—	218,960
Effluent with 400 lbs. of Sulphate of Ferric-oxide and 1,000 lbs. of Lime	1·65	·17	70·17	—	8,940

2ND SERIES.—*Percentage of Soluble Organic Matter Removed by Chemicals of Equal Value.*

	No. of Series.	0·3 Dollars or 30 Cents.				40 Cents.	
		Lime.	Copperas and Lime.	Ferric Sulphate.	Sulph'te of Alumina.	Ferric Sulphate.	Sulph'te of Alumina.
Soluble Albuminoid Ammonia removed.	1st.	27	35	31	27	42	27
Do.	2nd.	17	24	34	14	41	31
Do.	Av'ge.	22	29	32	20	41	29
Soluble loss on Ignition removed	1st.	9	18	29	29	48	14
Do.	2nd.	0	24	26	30	41	26
Do.	Av'age	4	21	28	30	45	20
Turbidity removed...	1st	80	70	80	75	83	78
Do.	2nd	74	70	82	78	82	76
Do.	3rd.	77	70	81	77	83	77
Bacteria removed ...	1st.	93	38	92	91	93	93
Yeast Cells removed..	1st	92	98	95	82	95	90

If we take the percentage of albuminoid ammonia removed to represent organic matter, we find that, in addition to the suspended matter, the following amounts of soluble organic matter have been removed :—

Soluble Organic Matter removed by Chemicals at Equal Costs, per head annually.

Lime..................................	Costing	30	cents,	removed	22	per cent.
Copperas and Lime................	„	30	„	„	29	„
Aluminium Sulphate................	„	30	„	„	20	„
Do. do.	„	40	„	„	29	„
Ferric Sulphate	„	30	„	„	32	„
Do.	„	40	„	„	41	„

In gauging the cost in England of commercial copperas, aluminium sulphate, and ferric sulphate, it may be noted that the first contains about 50 per cent of real ferrous sulphate, the second about 50 per cent of real aluminium sulphate, and the third about 80 per cent of real ferric sulphate. The price will, of course, vary with the facilities for delivery.

The Massachusetts Report comes to the following general conclusions :—

"The lime process has little to recommend it, owing to the large amount of lime required, and the difficulty of accurately adjusting the lime to the sewage. Very close supervision is required to obtain a good result, and even then the result is inferior to that obtained in other ways.

"Precipitation by copperas is also somewhat complicated, owing to the necessity of getting the right amount of lime mixed with the sewage before adding the copperas. When this is done, a good result is obtained. The amount of iron, however, left in the effluent is much greater than with ferric sulphate, owing to the greater solubility of the ferrous hydroxide.

"Ferric sulphate and alumina sulphate have the advantage over both lime and copperas, that their addition in concentrated solution can be accurately controlled, and the success of the operation does not depend on the accurate adjustment of lime or any chemical to the sewage.'

Using equal values of the different precipitants, applied under the most favourable conditions for each, upon the same sewage, the best results were obtained with ferric sulphate.

"The results with ferric sulphate have been on the whole more satisfactory than those with alumina. This seems to be due in part to the greater rapidity with which precipitation takes place, and in part to the greater weight of the precipitate. From the greater ease with which ferric sulphate is precipitated, it would give a good result with a sewage that was not sufficiently alkaline to precipitate alum at once."

In the latter case lime would be necessary. It should be noted that the ferric sulphate would throw down all the noxious sulphur compounds, and would, of course, absolutely prevent the formation of sulphuretted hydrogen, while, on the other hand, alumina has no power to combine with sulphur.

My basic ferric persulphate (*ozonine*) was not known to these American chemists, but by virtue of the fact that it is a basic salt, and not the ordinary neutral persulphate, it possesses the properties alluded to in the Massachusetts Report in an enhanced degree, having greater precipitating power, and can be produced at an equally cheap rate.

Typhoid Fever Bacillus—Differentiation and Diagnosis.

Mr. G. W. Fuller, State Board Biologist in Charge, has conducted for several months past a series of investigations relating to the life-testing of the typhoid fever bacillus, its deportment in sewage or water, and the possibility of its removal from water by filtration, as well as devising means for its diagnosis in the Merrimac River at Lawrence. He has succeeded in differentiating this bacillus from the other species of bacilli found in this water, notwithstanding that no fewer than five species of bacilli under the usual tests present the same characteristics. His results will be found at page 635 of the 23rd Annual Report of the State Board of Health of Massachusetts, published last year at Boston.

He found that when the water of this river became contaminated with these germs they were able to survive for three weeks, but in diminishing numbers, long enough for them to pass from the Lowell sewers to the service pipes of the Lawrence water supply.

These researches are most interesting, and are most important. It was found that the prevailing microscopic flora of the Merrimac River do not number more than about thirty species.

In the experiments to prove whether any bacteria are found in the river-water effluent passed through the filter, it was found that in 27 determinations there were removed by filtration 97·3 per cent of bacteria.

In a series of quantitative determinations during May and August, it was found that the number of bacteria shown in the following table were present in the city water; these were compared with the number found in the media indicated in the following table :—

City water	115
Effluent, filtered or city	5·25
In outlet-pipe and under-drains	320
In the air	100

The species found were the following: Bacillus cereus (Frankland); Bacillus subtilis (Cohn); Bacillus cloacæ (Jordan); Bacillus fluorescens liquefaciens (Flugge); Bacillus candicans (Frankland); Bacillus coli communis (Escherich); Bacillus ubiquitus (Jordan); Bacillus aurantiacus (Frankland).

Of these, Bacillus candicans and coli communis formed more than 80 per cent in the water and 40 per cent of those in the effluent, but these species were very prominent in the under-drains. It was shown that even the small number in the effluent owed their origin to the under-drains and outlet-pipe, and that they do not pass through the filter, which was composed of sand, with a layer of loam eight inches thick, and has a filtering power of from one and a half to two million gallons per acre daily. The conclusions arrived at of the property of this filter are that it is capable of removing all bacteria from the Merrimac River-water by intermittent filtration.

A few elucidations on the morphology of the typhoid bacillus, made by Mr. Hiram F. Mills as part of these valuable investigations, possess much interest. When grown in agar tubes at 68 degrees F. for one week, it appears as a plump bacillus, with rounded ends about 1 μ* long, and 0·6—0·8 μ in diameter.

When grown in bouillon at 98 degrees F. for two days, it is a bacillus 1·5—2·5 μ long, and 0·5—0·6 μ in diameter, but without either spore formation or involution forms. These bacilli from bouillon cultures possess very lively movements, both of rotation and translation ; in agar cultures at 98 degrees F., the movements are less marked, whilst at a temperature of 68 degrees F. movement is not uniformly present. The limits of temperature in which it can be grown are from 50 degrees to 114 degrees F.

One of the most reliable tests in determining this species of bacillus is that, in peptone with potassium nitrate, it reduces the nitrate into nitrite. Another conclusive diagnosis is by Dr. Smith's test, by which he has discovered that the presence of glucose enables facultative anaërobes to grow luxuriantly in the absence of oxygen. The *modus operandi* is described in the Report to which I am referring, at page 642. The three tests which have been found to be highly characteristic of the bacillus of typhoid fever are (after non-liquefaction): (1) Non-coagulation of milk. (2) Non-formation, or formation of a very slight amount of acid in milk. (3) Production of a turbidity without gas, in the Smith test.

* The mark μ signifies a micromillimetre—that is, 0·001 of a millimetre.

APPENDIX B.

Bibliography of Sewage.

The literature of this subject is too voluminous for full tabulation here. In vol. xii. of the Index Catalogue of the Library of the Surgeon-General's Office of the United States Army, it occupies as much as 36 pages up to the year 1891. This work was published in Washington, and for want of space I refer my readers to it, giving here only the outlines of its contents. Since that time no important work on Sewage Treatment has appeared, except those alluded to in the text of this book, but many papers have been read or published.

The books and papers are principally English, American, German, and French, and are recorded under the following heads :—

Sewage, 47 works and communications, enumerated at p. 918.

Chemical Treatment and Disinfection of Sewage, 74 works and communications, enumerated at pp. 918—920.

Disposal of Sewage, 1,500 works and communications, pp. 920—949

Methods of Sewage-Disposal in Special Localities, English and Foreign, pp. 927—944.

Sewage-Farming and Broad Irrigation, p. 944.

Sewage-Filtration, p. 945.

Sewer-Gases, p. 946.

Sewers, Ventilation and Disinfection of, p. 947.

Sewers and Sewerage, p. 949.

Sewers and Sewerage in Special Localities, p. 959.

Also many other communications under the heading of Disinfectants, Nightmen, Offal, Rivers Pollution, &c.

Amongst the more important books relating to the subject must be mentioned :—

Micro-organisms and Disease. By E. Klein, M.D., F.R.S.

Practical Sanitation. By George Reid, M.D., Medical Officer, County Council, Staffordshire.

Sewage Treatment, Purification, and Utilisation. By J. W. Slater, F.E.S. Containing a list of patents for Sewage Treatment.

The Report of the General Work of the State Board of Health of Massachusetts, for the years 1888—1890 and 1891. Published, in two volumes, by the Wright and Potter Printing Company, Boston, in 1890 and 1892.

Sewage Treatment and Sludge Disposal. A pamphlet by Santo Crimp. 1893.

APPENDIX C.

MR. SANTO CRIMP ON TANK CONSTRUCTION.*

IT is an undoubted fact that great improvements have been made in the construction of settling tanks. The earlier ones were, as a rule, constructed with flat bottoms, and the arrangements for drawing off the clarified water were of a very defective character, the outlets being frequently placed near the bottom of the tanks. Settling tanks should not only serve the purpose of clarifying the liquid, but they should be so constructed that the settled sludge may be removed with a minimum of labour. This object may be accomplished by constructing the tanks with segmental bottoms in cross section, and with a longitudinal fall towards the inlet of about 1 in 100. At the outlet end the effluent water should be allowed to pass over a weir, whilst for cleansing purposes a floating arm-valve should be provided for the purpose of drawing off the clarified water down to the level of the sludge, without disturbing the latter. The sludge should, of course, be swept into a reservoir near the sewage inlet end, and dealt with in any suitable manner. If it is intended to filter-press the sludge, large gratings with a mesh of about three-quarters of an inch should be provided in the tanks, in order to intercept all large substances, otherwise much difficulty will be experienced in pumping the sludge.

The following table from p. 7 of the same pamphlet by Mr. Santo Crimp of the present year shows how exceedingly polluted are street rain-washings.

* Sewage Treatment and Sludge Disposal, p. 14.

Composition of Liquid flowing from London Street Surfaces, 1892.

Composition.	Wood Pavement.		Macadam	
Appearance	Dark colour		Slate colour.	
Odour	Strong urine		Urine.	
Chlorine	37·8 grains per gall.		17·08 grains per gall.	
Free ammonia	4·82	„	2·48	„
Albuminised ammonia.	2·974	„	1·740	„
Oxygen absorbed by matters in solution in 15 minutes	0·476	„	0·268	„
Ditto, in 4 hours	3·464	„	1·968	„
Suspended matter—				
Mineral loss	686·40	„	1414·40	„
On ingition	58·40	„	54·4	„
Dissolved solids—				
Mineral loss	323·50	„	125·00	„
On ignition	82·00	„	27·00	„

Such putrescible matter is not therefore fit to be turned unpurified into any river or stream used for drinking purposes either by man or beast. He further observes "the liquid escaping from the storm-overflows on such occasions will contain an undue amount of polluting matters in suspension to the possible detriment of the stream into which it is discharged."

APPENDIX D.

Leek Sewage Disposal. Heights and Contour Lines of the Neighbourhood. Geology of the Locality.

I have taken out the heights and contour lines of a number of low and high-level parts of the town of Leek and the neighbourhood in order to assist in the consideration of the practicability of distributing the sewage by broad irrigation, or purified effluent, over a wider area of land in the vicinity of the town than is at present irrigated, which, in order to be reached, would require the raising of the sewage by mechanical or water-power, should such a distribution at any time commend itself to the Commissioners.

The remarks on this subject in the following pages apply with still greater force to the disposal of sewage effluents over an extended area of land, at only a few feet of elevation, should a precipitation process be adopted; it is within the reach of fertilisation at no more cost, if as much, than would be needed if the construction of filter-beds should be considered desirable, a step which I in no way advocate, because the land above the gravitation alluvial-levels is remarkably fit to receive benefit from whatever amount of organic matter is left in the effluent after precipitation.

The following observations on levels are purely local, but they introduce a discussion on pumping sewage and effluents, by various methods, to high levels, which is intended not only for local but for general consideration; and as the latter subject is occupied with an examination of the cost of pumping for a population of 14,000 people, it may be of more general service to the smaller towns of the country. For much larger populations the more expensive appliances requisite constitute problems of engineering only to be dealt with, according to varying conditions and localities, by competent engineers.

Feet above sea.

455 Road under railway bridge at Churnet-Grange Farm.
466 Spicer's-Stone Farmhouse.
463 Cheddleton, at the Churnet Bridge.
466 6in. Level of Churnet at Wall-Bridge, lower side.
475 Leek-Brook.
484 Wall-Bridge, height of road at the bridge.
486 Canal at Leek Wharf and Barnfield Bridge.
487 Canal Bridge at Wall-Grange and Wall-Bridge Farm.
496 5in. Level of railroad at Leek Railway Station.
500 Turnpike road opposite Leek-Wharf.
500 Considerable part of Bank Farm.
500 Entrance to Gaunt's Plantation at Spicer's-Stone, and greater part of Spicer's-Stone Farm.
500 Lower half of Birchall-Dale.
501 Little-Birchall House.
504 White Lion Inn.
505 Macclesfield-Road, between White-Lion and Mill-Street.
506 Churnet Dyeworks.
506 North end of Little Birchall Turnpike Embankment.
512 Milestone at Sheep-House.
514 Leek Railway Bridge—that is, the turnpike level.
515 White's Bridge, at Bridge-End.
518 Junction-Road, at Broad Street.
523 Wall-Lane Farm and banks.
538 Rifle-range Flagstaff at Wall-Grange Canal Tunnel.
540 The Hollies, near Cheddleton.
546 Higher half of Birchall-Dale, at the turnpike-gate level.
549 Broad Street entrance to Mellor's garden.
550 Greater part of Bank-Farm.
555 Abbey Bowling-green.
569 Lower end of the row of weaving cottages, Broad Street.

570 Alsop Street, lower end.
574 Wall-Grange Turnpike, at upper gateway by Bratt's house, near Ladderedge.
577 Brook at North end of Sheep-House Farm.
583 Highfield Lodge.
584 Lower Folker, a little lower down than the farmhouse.
594 Wellington Silk-Mill.
600 Sheep-House Barn.
600 The Leek Cemetery Mound.
600 Cheddleton-Heath.
600 Three-fourths of Wall-Grange Farm.
600 Westwood Hall, a little below on the east side.
600 Harracles, top line of lower woods skirting the Churnet.
600 Lower part of Pickwood.
600 Top of Birchall Hills, at Leek-Brook.
601 Ashenhurst Mill.
608 Compton.
627 Ball-Haye Hall.
641 Ladderedge Brick-kilns.
644 Leek Church.
648 Fowl-Church, a little south, near the Ropewalk.
660 Portland Street Mill.
663 Ashcombe Park.
667 Haregate-Terrace, Ball-Haye Green.
673 Upper Folker Farmhouse.
700 Top of Ballington Wood.
700 Ashenhurst Hall.
700 Top of Cheddleton Park.
700 The high table land of Harracles Farm.
700 Table land above Pickwood.
720 Crown-Point, near Cheddleton.
731 Ladderedge Reservoir.
732 Top of Ladderedge.
734 Leek Union.
748 The Hillswood Farm.

752 Leek-Moor Claypits and Brick-kilns.
771 Lowe-Hill Bridge.
771 The Waste, Buxton Road.
780 Little Longsdon.
800 Lowe-Hill House.
863 Top of Hillswood.
879 Knivedon.
1000 Gun Hill.
1190 Cloud Hill.
1500 Morridge at Blakemere.
1686 Shuttlingslow.
1670 The Roches.
1750 Axe Edge.

Some of the heights given above are to show the general lie and altitudes of the neighbourhood in comparison with others, and not, of course, for sewage-disposal purposes.

From the levels I have just given it will be seen that Macclesfield-Road at Leek is 21ft. higher than Wall-Bridge, and 19ft. higher than the railway-bridge at Barnfield, 18ft. higher than the canal-bridge at Wall-Grange, so there will be no difficulty in bringing the north and west sewage down the valley if such a course proves desirable.

The question of altering the levels of the sewage-outfalls may also have to be considered, and it may be desirable to have to discuss whether we may not with advantage have high-level, mid-level, and low-level outfalls, or two of these. For example, instead of the south sewage being all sent down to the railway station, the upper part of it may be diverted at the lower end of Alsop Street, and made to follow the contour of the land of that level, or somewhat lower. The lower end of Alsop-Street is 56ft. higher than the Leek-Station railway bridge. This would give a large irrigating area to the south; or the sewage might be diverted at Mellor's garden-gate in Broad-Street, which is 35ft. higher than the railway. This would very materially increase the irrigation area; in fact, it would irrigate the higher half of Birchall-Dale, being only 3ft. higher, whilst the lower half of the dale is 49ft. below this point. I am not including the high mounds in the dale, but the average higher and lower contours; by similar arrangements many acres of additional land, much more suited to broad or effluent irrigation, and in much greater need of it than the valley alluvia, would become available—so much so, that in my opinion, if the sewage could be elevated and delivered upon the many massive sandy banks and mounds, say for one to two miles down the Churnet valley, broad irrigation would dispose of the sewage of at least a population twice as large as Leek, and that in being so utilised upon such porous soil, subsoil, and rock as are to be found on the Churnet valley sides and plateaux, not only there would be no effluent to send into the river, but that farmers would willingly pay for the use of sewage or effluent so applied. Its value in this respect is nowhere more strikingly shown than it is upon the land in the big-Birchall Farm, known as the ox-pasture—a meadow of 35 acres of grazing land. Below the present level of the sewage delivery, on the lower portion of this field—in extent about eight acres—there are two kinds of land; the lower part by the railway is an alluvial soil not greatly needing irrigation, but between this and the sewage-carrier is a long strip of land, with a sandy soil, lying directly upon the Triassic or New-red sandstone rock. The most casual observation will show how excellent is the herbage of this part, which receives a portion of the town sewage, as compared with the same soil and herbage above the level of the carrier, or even upon the alluvium near the railway, and what splendid pasturage it affords to cattle, who much prefer it to the upper or lower parts of the field, and they are seldom seen away from it. The grass in both the places away from the sewage is coarse and irregular, and the cattle will not eat it if they can get any in the irrigated parts; and so it would be with the higher lands I have referred to if the sewage or effluent could be raised upon them.

Visits to towns which have pumping stations have quite

convinced me that this mode of sewage application is not only worthy of serious consideration, but that it need not be at all as costly, as is erroneously supposed.

Geology of the Locality.

Both the Geology and Physiography of the neighbourhood of Leek are intimately connected with the subject of drainage and sewage disposal. The River Churnet, with its numerous brooks and smaller tributaries, all resulting from a high watershed, with many valleys, large and small, serve as the water-drainage arteries and veins of the district. The very varying altitudes of the neighbourhood point to great and varied bygone geological changes, chiefly glacial, resulting in acute surface configurations and varying soils, which have to be taken into consideration in dealing with sewage disposal.

Briefly, the geology is this: our oldest rocks are those of the lower carboniferous system underlying the coal-measure horizon, yielding Yoredale sandstone and shales, as at Knivedon, the Waste, &c.; Millstone-grit, as at the Roaches, Ladderedge, Mossley, Horton, &c., all more or less capped with brick boulder clay, of the glacial period, containing water-worn pebbles and boulders of older rocks. The soils resting on these clays are more or less clayey, and are unlike the soils which lie upon the Triassic or New-red sandstone beds, which occupy and rest unconformably in a palæozoic trough in the Churnet-valley at Leek, and a few miles lower down the valley. It is singular to note the almost absence of the glacial brick-clays on this New-red sandstone, but it would be outside the scope of this paper to discuss the causes of so interesting a fact. Between these two great formations in this locality—*i.e.*, the lower Carboniferous and the Triassic—a vast thickness of rock has been removed by denudation, including the whole of the Millstone-grit formation of about 2,000ft. thick, except portions which have been left, as at the Roaches, Ladder-edge, Rudyard, Wetley-Rocks, &c., also the whole of the coal measures, about 6,000ft. in thickness, except those portions which have remained in detached basins, as at Shafferlong, Goldsitch, and Cheadle, as well as the whole or part of the Permian formation.

It is owing to these geological facts that we find the soils and subsoils of the locality so varying in their composition and structure—clays of several degrees of consistency in the alluvium of the Churnet valley, to the tougher clay soils of the Boulder-clay, and light, sandy, and porous soils which rest on the still more porous Bunter-beds of the New-red sandstone formation; and it is, I repeat, upon the latter soils and rocks that we could with the most benefit and safety distribute the sewage effluent, and there is a large area of land of this kind to the south of Leek which is available by pumping for this mode of fertilisation and sewage disposal.

APPENDIX E.

On the Estimation of Free and Albuminoid Ammonia.

In order to correctly estimate the amount of organic matter contained in crude sewage, sewage-effluent, or potable water, a process is employed which has for its end the ascertaining of the amount of albuminoid ammonia—or, as some chemists prefer to call it, organic ammonia—which can be evolved by the action of alkaline permanganate on the organic matter present.

The process adopted is Professor Wanklyn's, and is that known as the Nesslerising test. Its principle, to quote Professor Wanklyn's own words, is "the measurement of the nitrogenous organic matter in waters, by the quantities of ammonia yielded by the destruction of the organic matter." To do this, the sewage or organic matter has to be treated at boiling point with permanganate of potash and excess of caustic potash.

The organic matter is destroyed by this combustion process, or rather is converted into albuminoid ammonia, the quantities of which have to be estimated by a most delicate process, known as the Nessler test, which is a solution of iodide of potassium and per-iodide of mercury, with potash to render the solution alkaline.

The crude sewage, effluent, potable water, or other liquid to be examined for ammonia is first subjected to distillation in a retort until a portion of the distillate has come over; portions of this distillate are then Nesslerised, and the amount of free ammonia is thereby shown; the permanganate of potash and the other reducing chemicals are then added to the retort, the contents boiled and distilled; the distillate is Nesslerised, the result showing the quantity of albuminoid ammonia. The *modus operandi* will be found at pages 32 to 41 of Prof. Wanklyn's eighth edition of his "Water Analysis." So delicate is this test that the thousandth part of a milligramme may be measured.

Dr. Burghardt succinctly explains the meaning and effect of this method of estimating the amount of organic matter in the following words :—

"*Free Ammonia.*—This is the ammonia formed to a great extent, but not entirely, from the nitrogenous matters in the sewage by oxidation and reduction. It is quite harmless, as it exists mostly as ammonia salts.

"*Albuminoid Ammonia.*—This is ammonia obtained in the process of analysis by a special splitting up of the so-called albuminous nitrogenous matters in solution and suspension, the nitrogen of the organic matter being caused to combine with hydrogen, thus producing ammonia, which is distilled off as gas, and its amount carefully estimated. Albuminoid ammonia, therefore, mostly represents any sewage contamination in a water; consequently the estimation of its amount in an effluent is an important consideration."

APPENDIX F.

EXTRACTS FROM THE "HENDON TIMES" AND "FALKIRK HERALD," ON THE INTERNATIONAL WATER AND SEWAGE PURIFICATION COMPANY'S SYSTEM.

The *Hendon Times* of February 28th, 1891, contains a copy of a letter from the Hendon Local Board to the Local Government Board, from which the following is an extract:—

"With reference to that portion of your letter of the 19th ult. which refers to my Board having entered into an arrangement with the International Water and Sewage Purification Company for the treatment of the sewage at the outfall works, I am desired by my Board to state that they found that the scheme which was sanctioned by the Local Government Board for purifying the sewage, by a combined process of chemical treatment of lime and sulphate of alumina, by land filtration, did not produce a satisfactory effluent—and many complaints were made, by the owner of the fishery at the Welsh Harp of the destruction of the fish by reason of the use of chemicals, and they therefore, after careful consideration, entered into a contract with the before-mentioned Company for the treatment of sewage, which, they are happy to say, is carried out much more economically and with a purer effluent produced by the process adopted by the Company."

The *Falkirk Herald* of 4th April, 1891, contains an account of the system adopted by the Stirling District Lunacy Board for purification of the Asylum sewage, from which the following is an extract:—

"The Stirling District Lunacy Board have recently adopted the system of the International Water and Sewage Purification Company Limited, of 7, Victoria Street, Westminster, for disposal of the Asylum sewage at Larbert, to supersede the system of land irrigation formerly in use there, which polluted the local streamlets. The International system has now been long enough in operation at Larbert to warrant an opinion being publicly expressed about the results it has produced there.

Samples of the crude sewage and of the purified effluent from the filters, taken from the works in the presence of the Board, have been submitted to Professor Ivison Macadam, F.R.S.E., for analysis, and the results show that the Board have every reason to be satisfied with the adoption of the new system. The sample of sewage submitted for analysis, before passing into the tank, was very impure, and contained an enormous quantity of noxious matter; the test was therefore a very severe one. The analyses show that a reduction of more than 98½ per cent of the putrescible matter in the sewage was brought about by the International process of purification. Professor Macadam, in his report, states that the excellent results produced are much better than any he has previously obtained, and are better than any published results that he is aware of. He says: 'The Larbert effluent is not, of course, pure water, but I am satisfied that the system of purification now adopted is the best available. It converts the liquid from a grossly polluted and extremely offensive body into a clear water, containing some salts, but devoid of substances which can enter into a state of active putrescence.' The Lunacy Board were compelled to adopt some other system of sewage disposal to supersede the plan of applying it to the Asylum lands, as, in addition to other disadvantages of that process, actions for damages were threatened on account of the pollution of the streams into which the irrigated land drained. Now, the purified effluent tends rather to purify the stream adjoining than otherwise, as it is much better in quality than the water flowing past in the burn. The cost of the ferozone for precipitation, and labour in attending to the works, will be much less than the cost of keeping the irrigation pumping engine going, while the capital expenditure on the tanks and filters is insignificant. A comparatively small quantity of sludge remains to be disposed of, and it is applied to the land as manure when and where required."

APPENDIX G.

A NEW FORM OF AERATING-FILTER FOR SEWAGE, &c.

By J. Corbett, Borough Engineer, Salford.

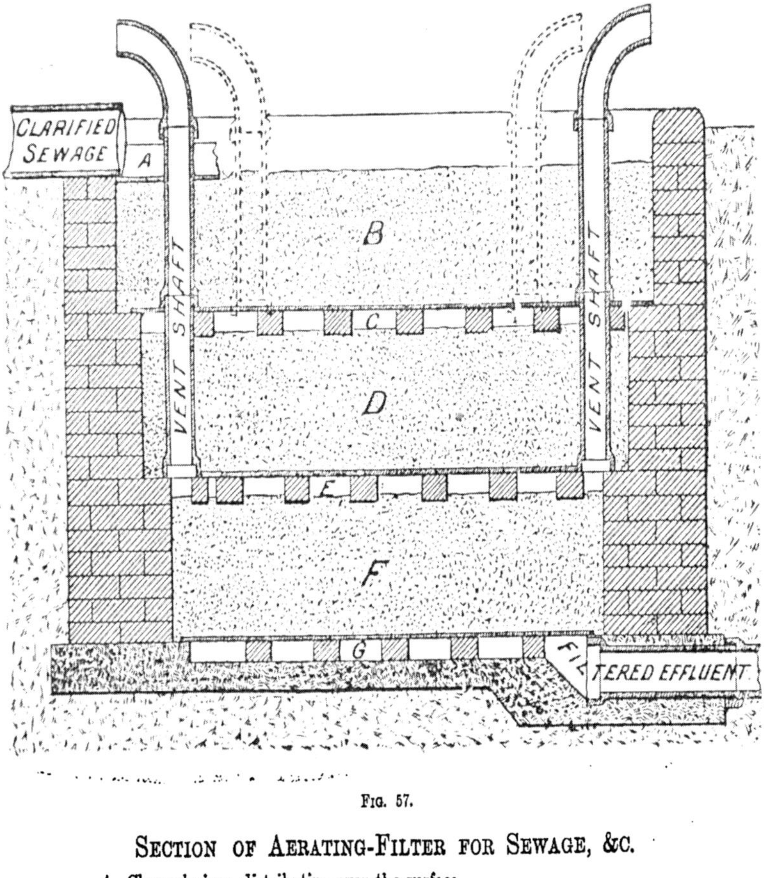

Fig. 57.

Section of Aerating-Filter for Sewage, &c.

A. Channel-pipes, distributing over the surface.

B. Filtering material, fine cinders or sand.

C. Air-space, formed by slate or tile floor on bricks spaced apart, with vent shafts.

D. Second filter-bed, similar to B.

E. Second air-space, similar to C.

F. Third filter-bed, similar to B and D.

G. Third air-space, vented by effluent pipe.

Four filters of this form have been on trial at the Salford Sewage Works for some months past. Two are filled with coarse sand, fine pea-gravel, and a little coarse gravel over the tiles. The other two are filled chiefly with fine sifted cinders, with coarse cinders over the tiles, and a little sand over the top to prevent the cinders from floating about when freshly flooded.

The results have, so far, been eminently satisfactory, effluents from various processes having been materially improved. The effluent from sewage-treatment by precipitation showed a purification of 68 per cent in the sand filters, and 82 per cent in the cinder filters, in the item of albuminoid ammonia.

APPENDIX H.

London Sewage Treatment and Disposal.

The system which has been in practice for some time is that of partial precipitation and disinfection.

The sewage amounts to 150 million gallons daily. There are two sewage outfalls; one at Barking, which brings the sewage from the north side of London, the sewage from the south side being received at Crossness.

The treatment at both outfalls consists of using four grains of lime per gallon and one grain of protosulphate of iron, to obtain some degree of clarification. After a subsidence of one to two hours the effluent is treated with a disinfecting fluid resembling Condy's, prepared from manganate of soda and sulphuric acid. Great attention is paid to both the ventilation and flushing of the London sewers. The separate system is condemned; the storm water unites with the sewage.

There are 13 settling tanks, each 30ft. wide, and averaging 1,000ft. long; they hold 20,000,000 gallons, or about 2¾ hours sewage flow. The sewage is dealt with on the continuous-flow system.

The sludge produced yearly is about 2,000,000 tons, or 2,500,000 cubic yards. It is carried to the sea 50 miles below the works to the Barrow Deep by five steamers, at a cost of 3d. per cubic yard.

The total cost of both works at Barking and Crossness will be about £1,000,000; the yearly cost is estimated at £110,000, equal to a rate of ·78 of a penny in the pound.

Lieut.-Col. Jones remarked, in a discussion at the Society of Arts on Dr. Tidy's paper on the Treatment of Sewage, that the Metropolitan Board, in minimising the precipitants to this attenuated degree, were only starving down the chemicals, avowedly with the object of reducing the amount of troublesome sludge, rather than that of the saving of cost of chemicals.

The amount of purification must, of course, be very small with such an insufficient quantity and defective nature of the chemicals used.

APPENDIX I.

ON THE ACTION OF INFUSORIA IN PURIFYING AND DEODORISING SEWAGE.

DURING the past few months I have allowed a number of half Winchester quart bottles of sewage, chiefly from Salford and Nuneaton, to accumulate on my laboratory table; some of the bottles were stoppered, others were unstoppered. I have observed in all of them a gradual decrease and finally an extinction of bacteria, and, up to a certain time, an increase of infusoria, particularly the genus kerona, chilodon, paramæcium, and monads, also a somewhat abundant growth of minute species of green algæ and their spores, and a corresponding purification of the sewage.

Without having had leisure enough for a systematic examination of these interesting changes, I have come to the conclusion that during the time the bacteria were absorbing the organic impurity of the sewage, infusoria rapidly increased and fed upon the bacteria, and that they in their turn for want of food died; for only a very few minute monads now remain in the liquid, which is perfectly odourless and clear, and almost free from organic matter.

The absence of bacteria, ærobic and anærobic, and in fact almost so of infusoria, is remarkable, and suggests that this branch of the subject is fully deserving of further research.

Purification took place in every bottle set aside, there being no exception to the rule.

INDEX.

A

C

G

H

I

J

K

L

M

N

O

P

Q

R

S

T

V

JOHN HEYWOOD, Excelsior Printing and Bookbinding Works, Manchester.

Sanitary Plumbing.

IMPORTANT ADVANCE
IN
SOIL PIPES
AND
CONNECTIONS.

FIRST CLASS AWARD

INTERNATIONAL SANITARY EXHIBITION, LONDON.

SOIL PIPES OF EVERY APPROVED DESIGN.

ELECTRIC LIGHT PILLARS AND BRACKETS.
ELECTRIC WIRE WAYS.

STABLE FITTINGS—A SPECIALITY.

Illustrated Catalogue, Price List, and Estimates on application.

WM. BIRCH'S AUTOMATIC SEWAGE CLEANSING, FILTERING, AND SLUDGE COMPRESSING PLANT,

FOR ALL CLASSES OF POLLUTION WHEN NOT TOO ALKALIED.

Treating 7,000 Gallons per hour and upwards to 20,000 according to class of pollution.

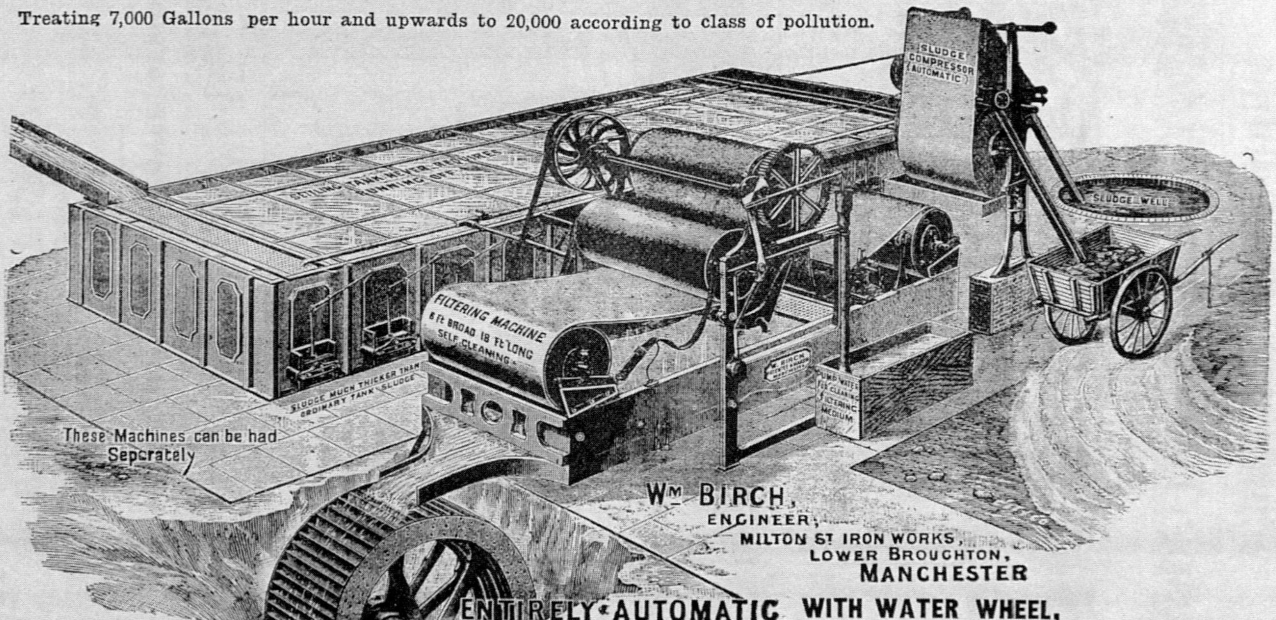

Wm BIRCH,
ENGINEER,
MILTON St IRON WORKS,
LOWER BROUGHTON,
MANCHESTER

ENTIRELY AUTOMATIC WITH WATER WHEEL,

OR ONE QUARTER HORSE POWER WILL DRIVE THIS PLANT WITHOUT THE WATER WHEEL.

PUBLIC HEALTH.

MAIN SEWERAGE & HOUSE DRAINAGE.

ALBION CLAY CO., LIMITED,

MANUFACTURERS OF

STONEWARE SANITARY PIPES,

PATENT PARAGON PIPES,

SOUND GAS AND WATERTIGHT JOINTS.

TRUE ALIGNMENT OF INVERT.

FIRM BEARING ON BED.

The Sockets of the PATENT PARAGON PIPES are made eccentric to the pipe, by which the spigots have a solid bearing, and a true invert is maintained at the joints. The sockets are thickened at the lower part and grooved for the packing material, and made flat at the base to give the pipes a firm bearing, and maintain them in true line on their bed.

DEFECTS OF ORDINARY SOCKET PIPES.

However carefully Ordinary Socket Pipes are laid, it is practically impossible to maintain them in true line. The sockets are concentric with the pipe, and the large annular space allows the spigots to drop, which lowers the invert at the joints of the pipes, and the ledges thus formed check the flow of the sewage and hold the solid matter, where it gradually accumulates and eventually stops the drain, which soon becomes a mere Underground Cesspool, exceedingly offensive and dangerous to health.

Mr. John Phillips, C.E., in his "DRAINAGE AND SEWERAGE OF TOWNS" (in reference to Ordinary Socket Pipes), says:—"Owing to the short lengths of the pipes, irregularity of shape, and multiplication of joints, they are often laid in undulations, instead of to regular inclination; and owing to the large size and irregular form of the sockets, the diameters of which are much greater than those of the butt ends, there are protuberances which check the sewage current, and cause hair, paper, rags, and other things to catch the joints, fæcal matter to deposit, and the drain or sewer to stop up. Hence through the joints both sewage and sewage gases constantly escape, so that while the sewage sinks into and saturates the ground beneath, the gases emanating from it rise through the permeable ground above, and issue through the chinks of the flooring, into the houses, the heat of which promotes the draught, and this is what really takes place, more or less, from every line of pipes as now put down."

These serious evils arising from defective construction are avoided by the use of the Patent Paragon Pipes. Efficient Sewerage and Drainage is secured in the simplest and best manner at the lowest cost. A true invert with firm bed is maintained in all cases, and sound gas and watertight joints are easily made by ordinary workmen, who need little supervision, as the Trade Mark stamped on the upper part of the pipes guides the men in laying them, and is a clear indicator that they are laid in true line on their bed. And there being no ledges at the invert, the solids are carried freely to the outfall, without the necessity for excessive flushing, which wastes much water, and adds greatly to the expense and difficulty of dealing with the extra volume of sewage at the outfall. Nuisance and danger to health from stagnant sewage and foul gases is minimised, as, the drains being free from obstruction, the solids are carried to the outfall before decomposition, and the ventilation of the sewers is thus more easily effected.

SOLE MANUFACTURERS FOR THE UNITED KINGDOM OF

SYKES'S PATENT JOINT PIPES,

FOR MAKING WATERTIGHT JOINTS IN WATERLOGGED GROUND.

With these, watertight sewers can be made in waterlogged ground at less cost than by any other method, combined with safety and permanency.

ALBION WORKS, WOODVILLE, BURTON-ON-TRENT.

Works' Telegrams: "Albion, Woodville."

LONDON DEPOT: MIDLAND RAILWAY, EUSTON ROAD, ST. PANCRAS.

CHIEF LONDON OFFICE: 18, NEW BRIDGE STREET, BLACKFRIARS, E.C.

Telegrams, "SEWERAGE, LONDON." Telephone, 2958.

ON SILK:

ITS ENTOMOLOGY, HISTORY, & MANUFACTURE,

AS EXEMPLIFIED AT THE

ROYAL JUBILEE EXHIBITION, MANCHESTER, 1887.

BY

THOMAS WARDLE, F.C.S., F.G.S., &c.

PUBLISHED BY

MR. EDWARD BUMPUS, BOOKSELLER,

5 & 6, HOLBORN BARS, LONDON, E.C.

NEW AND IMPORTANT WORK

ON

RECENT DEVELOPMENTS AND IMPROVEMENTS

IN

FRENCH SILK POWER LOOM WEAVING,

BY

THOMAS WARDLE, F.C.S., F.G.S., &c., &c.,

President of the Silk Association of Great Britain and Ireland.

In this work will be found some important information relating to SILK POWER LOOM WEAVING, acquired during a recent visit of the author to Lyons. Description of various kinds of Power Looms will be given, with improvements in Jacquard arrangements. The book will be illustrated by a series of drawings of plain and figured stuffs which the author saw woven in Lyons.

The price of the Book is FIVE SHILLINGS net.

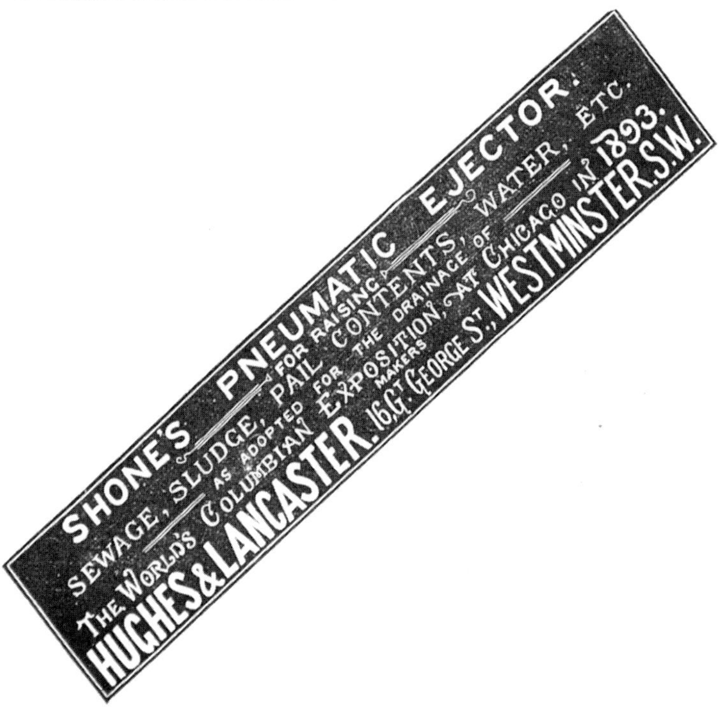

IMPROVED PLANT and PROCESS
PATENTED
FOR THE

PURIFICATION
of
Effluent Waters
from
Dye works. Paper works &c

AND FOR THE

PURIFICATION AND SOFTENING

of Hard Water, rendering it suitable for use in Steam Boilers, and for Manufacturing purposes generally.

ARTESIAN WELLS, from 9 to 45 inches diameter, bored by our Improved Steam Earth Boring Machine.

PUMPING MACHINERY, STEAM ENGINES, BOILERS, &c.

MATHER & PLATT L^D.

HYDRAULIC ELECTRICAL & GENERAL ENGINEERS

SALFORD IRON WORKS

MANCHESTER.

ROBERT HALL & SONS,

Attendance at Manchester Royal Exchange,
No. 8 Pillar, every Tuesday and Friday.

Telegraphic Address: "HALL, BURY."

BURY,

Nr. MANCHESTER.

Office:
3, HOPWOOD AVENUE, Manchester.
Every Tuesday & Friday.

Telegraphic Address: "HALL, BURY."

POWER LOOMS

FOR

SILK WEAVING.

ENGLISH OR FRENCH SYSTEM.

We are also prepared to make Special Looms or Machines for Manufacturers.

MAKERS OF ALL KINDS OF

PREPARING & FINISHING MACHINERY,

FOR WEAVING PURPOSES.

LETTERS TO BE ADDRESSED:

HOPE FOUNDRY, BURY,

LANCASHIRE.

FRENCH AND GERMAN CORRESPONDENCE.